Sustainable Materials and Technology

Sustainable Materials and Technology (SMT) book series publishes research monographs (both edited and authored volumes) showcasing the latest developments in the field and comprehensively covering topics such as:

- Recycling of waste into useful material and their energy applications
- Catalytic action of Nano oxides for efficient carbon reforming process
- Sustainable technologies for plastic transformation
- Bifunctional nanoparticles for sustainable water splitting applications
- Sustainable dying and printing
- New materials from waste
- Sustainable Manure Management and Technology: Potentials, Uses and limitations
- Sustainable Mechanical Engineering Approach
- Sustainable biochemistry for the improvement of health
- Sustainable development of Mechanical recycling of automotive components
- Sustainable-waste recycling and conversion in useful materials for different applications
- Sustainable development of inexpensive Nano-photo catalysts
- Sustainable development of recycling of discarded lithium ion batteries
- Modern sustainable cement and concrete
- Sustainable adsorbent for hazardous removal
- Sustainable superior electromagnetic shielding materials
- Excellent sustainable nanostructured materials for energy storage device
- Sustainable development of heavy metal detoxification from water
- Carbon dioxide utilization for sustainable energy
- Sustainable development in green syntheses of materials
- Environment friendly and sustainable cloth for garments application
- Sustainable design and application of eco-materials
- Nanoparticles for sustainable environment applications
- Sustainable remediation of industrial contaminated water towards potential industrial applications
- Biomaterials for sustainable bioremediations

Mohammad Jawaid · Anish Khan
Editors

Conversion of Electronic Waste in to Sustainable Products

Springer

Editors
Mohammad Jawaid
Laboratory of Biocomposite Technology
Institute of Tropical Forestry and Forest
Products Universiti Putra Malaysia
Serdang, Selangor, Malaysia

Anish Khan
Center of Excellence for Advanced
Materials Research
King Abdulaziz University
Jeddah, Saudi Arabia

ISSN 2731-0426 ISSN 2731-0434 (electronic)
Sustainable Materials and Technology
ISBN 978-981-19-6540-1 ISBN 978-981-19-6541-8 (eBook)
https://doi.org/10.1007/978-981-19-6541-8

This Springer imprint is published by the registered company Springer Nature Singapore Pte Ltd.
The registered company address is: 152 Beach Road, #21-01/04 Gateway East, Singapore 189721, Singapore

Contents

Design of a Proper Recycling Process for Small-Sized E-Waste

Emanuele Caroline Araujo dos Santos, Carline Fabiane Stalter, Luciana Kaercher, Daiane Calheiro, Feliciane Andrade Brehm, and Carlos Alberto Mendes Moraes

Abstract Small electronic waste has been addressed in this chapter. With this, issues such as consumption and generation, composition and recycling techniques were raised. The equipment/waste addressed were cell phones and smartphones, LED lamps, computers, and electrical wires and cables, which were chosen due to their great generation, for being more current technologies, their great applicability and quantity, and variety of valuable and critical materials in their compositions. All this waste shows a notable quantity and variety of precious and technological metals, and of rare-earth elements, all metals of great interest and research today. All except electrical wires and cables show considerable portions of gold, for example, which is a precious metal of great applicability, and which has been achieving high yield values, being, with this, one of the metals most studied by researchers. On the other hand, electrical wires and cables are waste, which is present in almost every WEEE, and are rich in copper and PVC, which are materials of great use in the most diverse areas. In any case, there is a need for the development of viable techniques of recovery of these metals, as well as the development of viable industrial processes. Important steps, such as disassembly and mechanical processing, should be developed, as they enable better revenues, in addition to the development of more sustainable and productive recovery procedures. As well as the development of designs that aim to facilitate the end of life of these products, seeking to meet the precepts of the tripod of sustainability, industrial ecology, and, more recently, the circular economy strategy, where all materials, not only metals, of this waste are recovered, valued, and recycled.

Keywords Small-sized WEEE · Recovering · Valuation · Recycling · Industrial ecology

E. C. A. dos Santos · C. F. Stalter · L. Kaercher · D. Calheiro · C. A. M. Moraes (✉)
Graduate Program of Mechanical Engineering, University of Vale do Rio dos Sinos - UNISINOS, São Leopoldo, Rio Grande do Sul, Brazil
e-mail: cmoraes@unisinos.br

F. A. Brehm · C. A. M. Moraes
Graduate Program of Civil Engineering, University of Vale do Rio dos Sinos - UNISINOS, São Leopoldo, Rio Grande do Sul, Brazil

M. Jawaid and A. Khan (eds.), *Conversion of Electronic Waste in to Sustainable Products*, Sustainable Materials and Technology,
https://doi.org/10.1007/978-981-19-6541-8_1

1 Introduction

According to the European Union Directive 2012/19/EU, electronic equipment is any device that depends on electric current or electromagnetic fields for its correct operation, including any apparatus used for the generation, transfer, and measurement of electric currents and fields. The waste electrical and electronic equipment (WEEE) refers to the post-consumer devices that have been discarded at the end of their life, as well as their components, subsets, and consumable materials (European Union 2012; Işildar et al. 2019; Forti et al. 2020).

According to Xavier et al. (2020a), the adoption of categorization of these post-consumer products is important, because it allows to segregate of them, according to their specificities, and enables an optimization in disassembly processes, since each category requires specific methods and tools.

Forti et al. (2020) indicate that a classification of electronic equipment should contemplate the function of the products, as well as the composition (considering hazardous and valuable materials) and issues related to their end of life. Added to this, between each category the average size and life of the products should be taken into consideration (Forti et al. 2020).

The European Union Directive 2012/19/EU has divided electrical and electronic equipment into 10 categories, considering from household equipment to medical equipment, as well as of measurement and control. Table 1 presents the 10 categories created by the directive, as well as examples of equipment.

However, in the second edition of E-waste Statistics, Forti et al. (2018) divided the EEE into 54 categories, in which they focused on the characteristics of the products; however, Forti et al. (2020) grouped them into 6 general categories, focusing on their post-consumer management. This division, for example, brings advantages in terms

Table 1 Classification of electronic equipment—European Directive 2012/19/EU

Category	Examples
Large household appliances	Refrigerators, washing machines, stoves, etc.
Small household appliances	Toaster, vacuum cleaner, fryer, hair dryer, etc.
Computer and telecommunications equipment	Computer, laptop, printer, mobile phone, etc.
Consumption equipment	Radio, television, musical instruments, etc.
Lighting equipment	Fluorescent lamps, LED lamps, etc.
Electric and electronic tools	Saws, sewing machines, welding machines, tools, etc.
Toys and sports and leisure equipment	Video games, etc.
Medical equipment	Radiotherapy, cardiology, dialysis equipment, pulmonary ventilators, etc.
Monitoring and control instruments	Thermostat, heating regulators, etc.
Vending machines	Distributors of hot drinks, bottles, cans, etc.

Source Adapted from European Union (2012)

Table 2 EEE classification presented at GEM 2020

Category	Examples
Temperature exchange equipment	They are freezing and cooling equipment. Some examples are: refrigerators, freezers, air conditioners, and heat pumps
Screens and monitors	Examples are: televisions, monitors, laptops, and tablets
Lamps	Include fluorescent lamps, high-intensity discharge lamps, and LED lamps
Large equipment	Some examples are: washing machines, dryers, dishwashers, electric stoves, large printers, copiers, and photovoltaic panels
Small equipment	They range from household appliances such as vacuum cleaners, microwaves, ventilation equipment, toasters, small electrical and electronic tools to small medical and monitoring devices, and control instruments
Small IT and telecommunication equipment	Included: mobile phones, Global Positioning System (GPS) devices, pocket calculators, routers, personal computers, printers, and phones

Source Adapted from Forti et al. (2020)

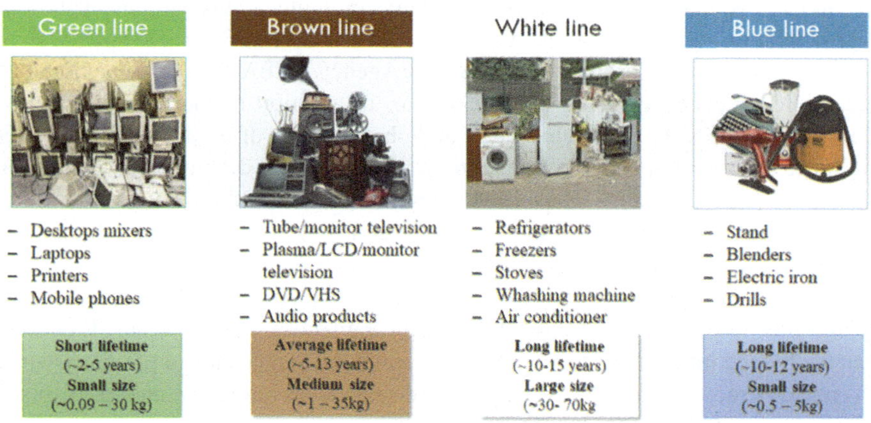

Fig. 1 EEE classification according to ABDI (*Source* Adapted from ABDI 2013)

of logistics, physical space, and disassembly methods in cooperatives, associations, or recycling companies. Table 2 shows the classification made by Forti et al. (2020).

The Brazilian Agency for Industrial Development[1] ABDI (2013), on the other hand, divides these equipment into broader categories, considering lifetime, size, and diversity of materials. Figure 1 shows this classification.

[1] Agência Brasileira de Desenvolvimento Industrial.

Xavier et al. (2020a, b) presented a new proposal (adapted from Xavier et al. 2017) classification of this equipment. They created eight categories, in which they contemplated: household appliances, electronics/electroportables, monitors, computers and telecommunications, wires and cables, batteries, lighting equipment, and photovoltaic panels. In addition to the authors considering batteries as a category, there was the insertion of the category "wires and cables," which, according to the authors, are significant in value, since they have copper in their composition. Apart from that, photovoltaic panels were included, since, according to the authors, there will be an increase in their disposal already in this third decade of the twenty-first century (Xavier et al. 2020a, b).

According to the 2020 Global E-waste Monitor (GEM 2020) report, around 53.6 Mt of WEEE were generated in the world in 2019, meaning a generation per capita of 7.3 kg/hab. In addition, it is estimated that this amount will reach 74.7 Mt of WEEE generated in the year 2030 (Forti et al. 2020).

GEM 2020 also reports that, of the total WEEEs generated worldwide in 2019, about 32.4% (17.4 Mt) were small equipment, and about 24.4% (13.1 Mt) were large equipment. And the categories of temperature exchange equipment, screens and monitors, small IT and telecommunications equipment, and lamps collaborated with 10.8 Mt, 6.7 Mt, 4.7 Mt, and 0.9 Mt, respectively (Forti et al. 2020).

According to Forti et al. (2020), between 2014 and 2019 there was a growth (of mass generation) of 4% in the generation of waste of small equipment and lamps and 5% in large equipment, while small IT and telecommunications equipment showed a growth of only 2% (Forti et al. 2020).

According to the European Environment Agency—EEA (2021), the amount of electronic products increases rapidly due to, in addition to the normal trend of consumption growth, shorter lifetimes, and obsolescence of products. Both shorter lifetimes and obsolescence occur due to fast technological development, marketing incentive, product quality drop, need for greater energy efficiency, among other reasons (EEA 2021). With this increase in the consumption of electronic equipment and generation of WEEEs, recycling techniques must be developed.

According to Chancerel and Rotter (2009), the recycling process can be conceptualized as a series of procedures that aim for the recovery of materials. Processes can be divided into (1) collection, (2) pre-screening, (3) mechanical pre-processing, (4) reuse or recovery of parts, materials, and substances or energy, and (5) disposal (Chancerel and Rotter 2009).

The WEEE recycling cycle generally begins by collecting and/or delivering end-of-life equipment. From this, a screening step takes place, in which whether the equipment is suitable to be reused or not is evaluated, and then a categorization of the types of equipment available is carried out (Gerbase and Oliveira 2012; Moraes et al. 2014).

The techniques used for recycling an electronic equipment can be mechanical, chemical, or thermal (Veit et al. 2008; Gerbase and Oliveira 2012; Oliveira 2012). The main ones are mechanical, pyrometallurgical, hydrometallurgical, electrometallurgical, and biometallurgical processes (Gerbase and Oliveira 2012; Kunrath 2015).

In general, the WEEEs recycling flow can be divided into three steps (Cui and Forssberg 2003; Oliveira 2012; Moraes et al. 2014; Giese et al. 2018):

- **Pre-treatment or disassembly**: the process of separating parts, usually manual;
- **Processing (separation or concentration)**: in this step, in order to increase the concentration of the materials, physical processes and unit operations, such as comminution, particle size separation, magnetic and electrostatic separation, flotation, etc., occur. Moraes et al. (2014) include metallurgical processes, such as pyrometallurgy and hydrometallurgy, in this step. There is also the process of biohydrometallurgy (Giese et al. 2018).
- **Refining**: the step at which the materials are recovered. This step can include metallurgical processes, followed by refining processes such as solvent extraction, precipitation, and electrolysis. The biosorption technique is currently used here (Giese et al. 2018).

Pyrometallurgy is the process in which high temperatures are used in order to concentrate materials; with this, it is possible to separate target materials, such as gold and copper, from organic materials and other contaminants in the form of slag and emissions. This process can generate pure metals, alloys, or secondary compounds; in addition, after the refining steps, secondary phases of other metals can be generated (Oliveira 2012; Gerbase and Oliveira 2012).

Hydrometallurgy is the process in which a leaching of metals through a solution is made. In this process, the metals are dissolved in a leaching solution, acidic or alkaline, becoming available to be recovered through some refining process, such as electrolysis, for example (Oliveira 2012; Gerbase and Oliveira 2012).

Biohydrometallurgy or biolyxiviation, on the other hand, is the process by which metals of interest are separated by means of solutions containing microorganisms. It is a biotechnological process that uses naturally occurring microorganisms or their metabolites to solubilize and recover metals from mineral sources and waste through chemical or bacterial oxidation (Gerbase and Oliveira 2012; Oliveira 2012; Pradiian 2013; Giese et al. 2018).

The flow chart in Fig. 2 illustrates and summarizes the entire WEEEs management and recycling flow.

2 Mobile Phones

Mobile phones and smartphones as we know them today, with connectivity and advanced computing resources, began to arrive on the market in the mid-90s, but only became popular from 2007 on (O'dea 2021a).

Li et al. (2010) define a smartphone as a mobile phone that offers more advanced computing capability and connectivity compared to other mobile phones. Usually, it is equipped with a camera, has Wi-Fi connectivity, an interface for the installation of new applications, touch screens, and other functions equivalent to computing

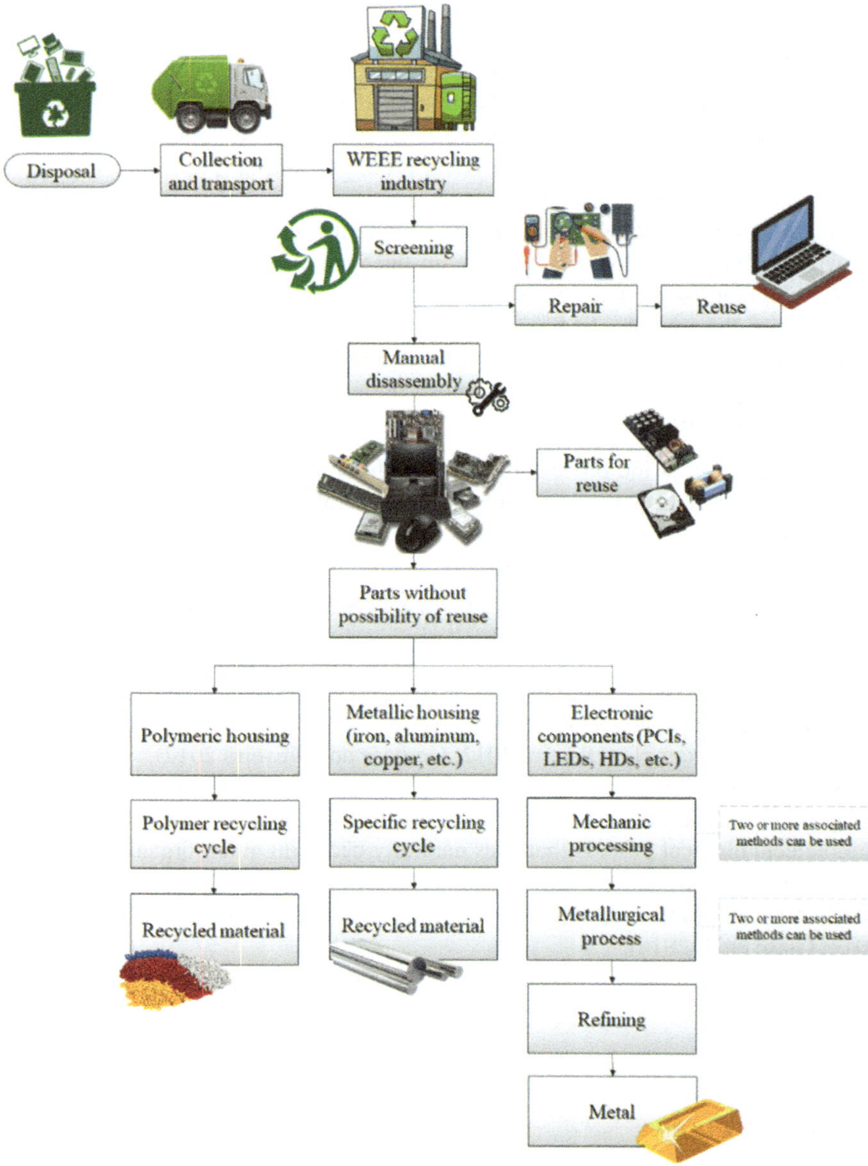

Fig. 2 Flow chart of the management and recycling process of WEEE (*Source* Authors)

capabilities found in personal computers. As smartphones have several telecommunications and computer functions, they have replaced feature phones (conventional phones) at a very accelerated growth rate (Mejame et al. 2016).

With this, the smartphone industry has been growing constantly, including the consumer market, development of new models, and suppliers (O'dea 2021a).

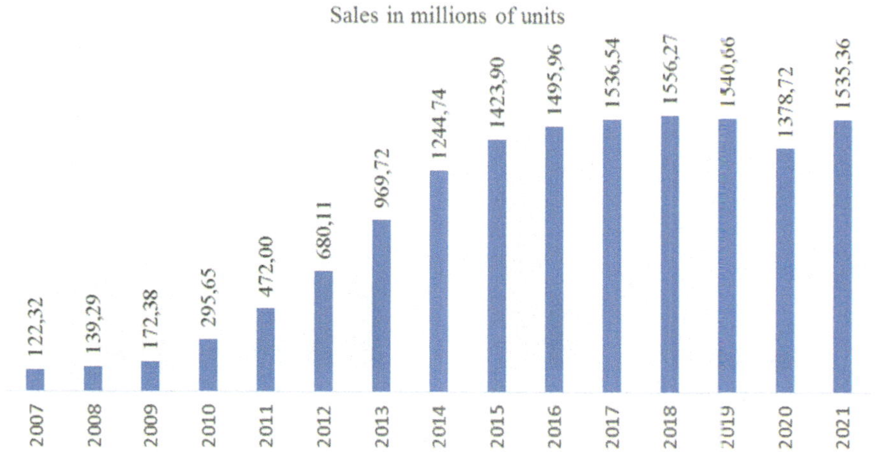

Fig. 3 Number of smartphones sold to end users worldwide from 2007 to 2020 (in millions of units) (*Source* Adapted from O'dea 2021b)

According to Araújo (2013), mobile telephony is considered an immature market, that is, it is not stable and very fast growing.

In 2020, about 1.38 billion smartphones were sold worldwide and this number is expected to reach 1.48 billion units sold by 2023 (O'DEA 2021a, b). Figure 3 shows the growth of smart device sales from 2007 to 2020.

Through the graph in Fig. 3, a steady growth in sales of these devices can be observed. In 2020, however, there was a drop due to the coronavirus pandemic (COVID-19) that began this year; but an increase in global sales is expected for 2021 (O'dea 2021b).

In 2018, about 38% of the world's population owned a smartphone; in 2020, the penetration rate rose to 46.5% (O'dea 2021a, b). According to O'dea (2021a), many users own more than one smartphone, that is, there is a much larger number of subscriptions than would be expected. With this, there are about 6.4 billion smartphone subscriptions in 2021, and this number is expected to reach 7.5 billion in the year 2026 (O'dea 2021a).

The greater access of the population, combined with constant technological advances and the search for better quality products and services, leads consumers to quickly change devices, such as, in the case of mobile phones, for lighter, more functional, and modern ones (Kasper 2011).

Sarath et al. (2015) state that the average phase or time of use of a mobile phone is less than 3 years in developing countries and less than 2 years in developed countries, which leads to a very frequent disposal of these devices. In Brazil, the exchange of mobile phone devices occurs, in general, every 2 years (Borges et al. 2016). Figure 4 shows the actual, projected, and desired lifetime of smartphones and other products.

As shown in Fig. 4, smartphones have the shortest real life, which can be explained by the fast technological development of these products in recent years. According

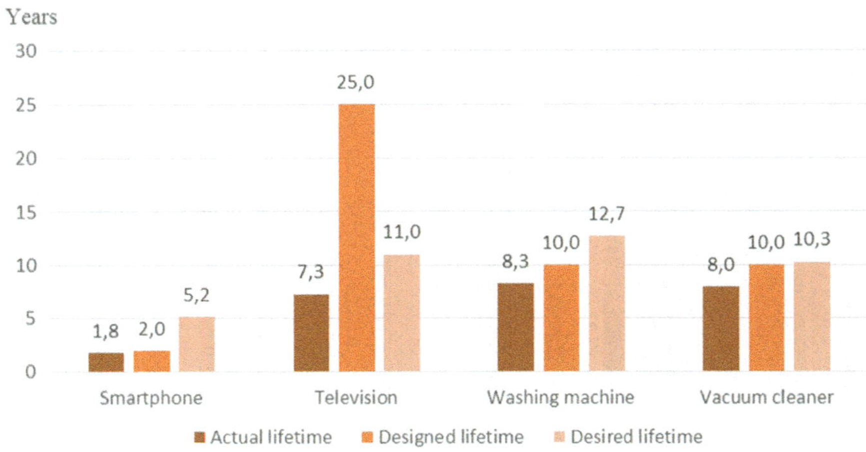

Fig. 4 Real, projected, and desired lifetime of smartphones, televisions, vacuum cleaners, and washing machines (*Source* Adapted from EEA 2021)

to Araújo (2013), this characteristic of fast technological change in this equipment indicates that, even when these markets become saturated, sales growth will still become possible, since consumers will tend to buy new and more technologically advanced products. That is, it is a market that is always renewing itself.

The availability of new smartphone models on the market can significantly add to the portion of electronic waste, which may also generate environmental impacts. Just like other WEEEs, smartphones can contain toxic, rare, and precious materials. With this, these post-consumer appliances can cause impact to resource availability, human health, and the environment. Therefore, adverse environmental impacts caused by this waste must be avoided through adequate, effective, and preventive management (Mejame et al. 2016).

According to UNEP (2009), various components can be found in electronic equipment and may be manufactured in metals, polymers, or other substances. A cell phone can contain about 23% of its mass in metals, the rest being polymers and ceramic material (UNEP 2009). According to the European Commission (2017), a smartphone demands 50 different types of metals to meet requirements for size, lightness, and functionality.

These metals range from base metals—such as copper and tin—to precious metals—such as silver, gold, and palladium (UNEP 2009). Many of these metals are considered critical (they have industrial and technological importance and present a supply risk) (Cui and Forssberg 2003; Peiró et al. 2011; Nancharaiah et al. 2016; Ayres and Peiró 2017; European Commission 2017; Zhuang et al. 2015; Işildar et al. 2019).

In addition, the concentration of some of these metals in WEEEs may be higher than in natural ores. Umicore (2016) compares how, in primary mining, there is about 1–5 g/t of gold, while 150–200 g/t can be found in printed circuit boards and 250–300 g/t in mobile phones.

In mobile phones and smartphones, these metals can be scattered in the various parts that make them up. Normally, a cell phone device can be divided into larger parts, being: printed circuit board (PCB), a display unit (screen), a battery, and a box/cover; in addition, there are also antenna and accessories such as headphones, for example (IPMI 2003; Gu et al. 2019).

As reported in the literature, the polymer fraction represents the largest portion by mass of cellphone devices (UNEP 2009; Bollinger 2010; GSMA 2015; Holgersson et al. 2018). According to the GSMA (2015), this fraction accounts for about 45% of the mass of the device.

According to Santos (2016), the main polymers used in mobile phones are thermoplastics, among which the main one is the copolymer of acrylonitrile butadiene styrene (ABS); the blend of polycarbonate (PC) and ABS is the most used, especially in the housing, because it has properties such as being a flame retardant. However, although these thermoplastics are recyclable, the individual recycling process of each one of them is difficult, as they are mixed or connected inside the device, being recycled together due to that (Santos 2016).

Bookhagen et al. (2020) characterized three different models of smartphones (without battery), with release years between 2011 and 2012, and analyzed their composition. The authors found that, on average, the three smartphones investigated contained 45% of metals, 32% of glass, and 17% of plastics by weight, as well as a portion of 6% of mixed materials that were named from others (Bookhagen et al. 2020). Furthermore, the authors investigated 53 metal elements contained in the three smartphone models and quantified them according to Fig. 5.

It is noted that, in order to achieve certain technological improvements, the massive use of metals is required. The technological advancement of this equipment comes with the increase in the use of metals, some of which are precious, which increases the economic potential of recycling this equipment (Gu et al. 2019).

In addition to the commonly known metals, such as aluminum and copper, some metals—called high-tech metals due to their chemical and physical properties—are indispensable in modern equipment (such as mobile phones), since they can enable some functionalities. Some of these metals are antimony, cobalt, lithium, tantalum, tungsten, and molybdenum. Tungsten, for example, due to its high density, is used as a counterweight at the end of the axis of the small motor that acts as the vibrator in mobile phones (GSMA 2015; Cordella et al. 2020).

Figure 6 shows the main parts of a smartphone, as well as the critical metals used in each of them.

According to Fig. 6, we can visualize the diversity of critical metals that are used in a single smartphone, as well as how each metal is important due to its properties, and, consequently, assigns important functions to the equipment. In addition to these metals, mobile phones and smartphones also contain other metals, such as aluminum, copper, gold, silver, and among others. Next, each part will be better described.

Electronics: in addition to gallium, nickel, and tantalum, which are used in connections and various devices, silicon dioxide doped with phosphorus, antimony, arsenic, boron, indium, or gallium is used for the production of the chips. Copper, gold, and silver make up the microelectronics and wiring of phones. For welds that

> 1g	> 0.1 g (< 1g)	> 0.01g (< 0.1g)	<0.01g		Contained, not investigated
Mg	Sr	Zr	Li	Ga	H
Cr	Ba	V	Be	In	Na
Fe	Ti	Ta	Rb	Ge	K
Ni	W	Mo	Sc	Pb	Ca
Cu	Mn	Co	Hf	Bi	B
Al	Zn	Ag	Y	Sb	C
Si	Sn	Au	Nb	As	P
	Nd	Pr	Pd	Te	S
		Dy	Pt	La	Cl
			Cd	Ce	Br
			Hg	Sm	N
			Eu	Gd	O
			Tb	Ho	F
			Er	Yb	
			Lu		

Fig. 5 Elements contained in the three smartphone models (*Source* Adapted from Bookhagen et al. 2020)

previously used tin and lead, from the 2003 RoHs, these were replaced by mixtures of tin, silver land gold (Compound Interest 2014; Cordella et al. 2020; Venditti 2021).

Microphones and speakers: in addition to the nickel that is used in the diaphragms of the microphones, in these components there are usually magnets produced from neodymium-iron-boron alloys, and, in some cases, dysprosium, praseodymium, or gadolinium can be used in these alloys. These metals, in addition to terbium, can also be used in vibration units (Compound Interest 2014; Cordella et al. 2020; Venditti 2021).

Cover: covers can be produced from metal or polymers or a mixture of the two. When made of metal, magnesium alloys are usually used, but aluminum, copper, and iron/steel alloys can also be used, since polymers can be ABS or PC/ABS. Apart from that, brominated flame retardants can also be found, although their use is being minimized (Kasper et al. 2010, 2011; Compound Interest 2014; Santos 2016; Cordella et al. 2020; Venditti 2021).

Batteries: batteries today are made mostly of lithium ions; they consist of an aluminum casing, inside which there is a positive electrode (usually a lithium-cobalt oxide), a negative electrode (which is usually graphite), and an organic solvent that acts as an electrolytic fluid (Compound Interest 2014; Cordella et al. 2020; Venditti 2021).

Screens: they can be divided into three parts, which are: a screen, a capacitive layer (touch screen), and a glass cover. The glass is basically composed of aluminosilicate and may contain potassium ions to strengthen it. To produce the touch sensitivity

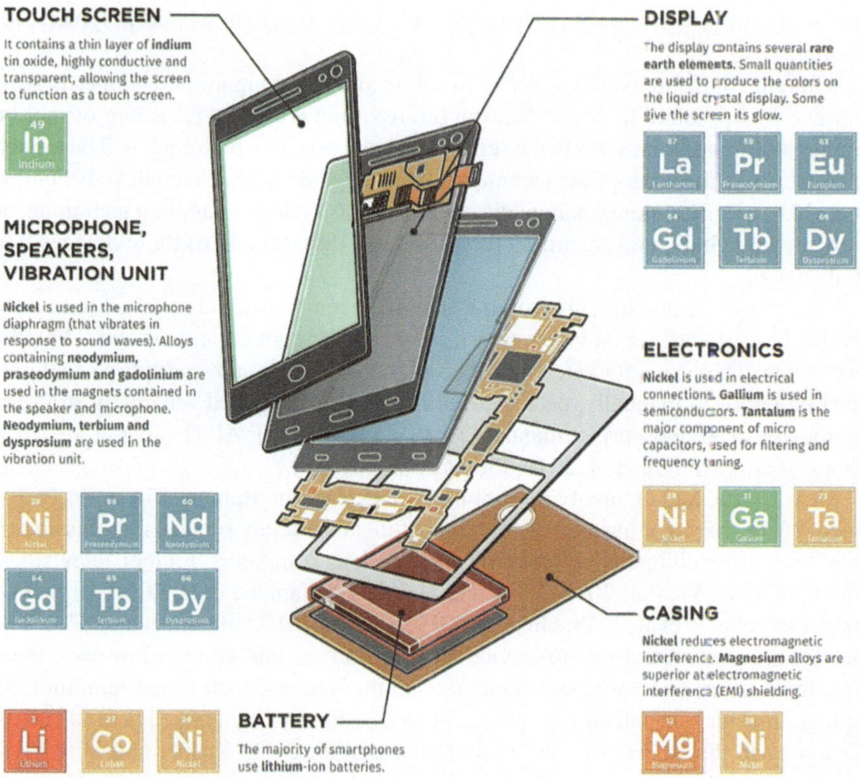

Fig. 6 Critical metals used in every part of a smartphone (Venditti 2021)

property, a thin, transparent and conductive layer of mixed indium and tin oxides is used. And, in order to produce the colors of the canvas, a varied composition of rare-earth elements is used, such as yttrium, lanthanum, terbium, praseodymium, europium, dysprosium, and gadolinium. In addition, one can also find silver, gold, and palladium on touch screens (Compound Interest 2014; Cordella et al. 2020; Venditti 2021; Konyalioğlu and Bereketli 2021). Currently, there are several image transmission technologies being used in smartphones, which are aimed not only at image and technological improvements, but also energy savings, such as liquid crystal screens (LCD), light-emitting diodes (LEDs), organic light-emitting diode (OLED), and active matrix OLED (AMOLED) (Bizzarri et al. 2018; PODE 2020; Cordella et al. 2020; Golzar-Ahmadi and Mousavi 2021).

In view of the above, the importance of the disassembly step for this type of waste can be seen, as each part has very important metals in its composition, which, with the disassembly and proper treatment of each of them, makes it easier to recover these metals. With disassembly, it is possible to better concentrate the metals of interest, in which the techniques and/or recovery parameters can often be different for each one; in addition, it is possible to better recycle other materials such as polymers, for example.

Although there are works aimed at grinding and recycling mobile phones in their entirety, it is possible to notice that the future of the complete recycling of mobile phones and smartphones lies in the separation into parts, in which each will be treated in isolation by the appropriate techniques (Bachér et al. 2015; Gu et al. 2019; Flerus et al. 2019). Furthermore, manual disassembly is still the most applied technique, as there are still challenges regarding the automated disassembly of these devices (Gu et al. 2019).

In any case, a pretreatment involving at least the removal of the battery is necessary for this type of waste, as it involves risk of fire and/or contaminations caused by the electrolyte. Therefore, for safety reasons, all mobile phone and smartphone batteries must be removed manually, packaged, transported, and treated separately from the equipment in an appropriate manner (Jang and Kim 2010; Al-Thyabat et al. 2013; Zhang et al. 2018; Gu et al. 2019; Flerus et al. 2019).

Most of the studies involving mobile phones and smartphones focus on PCBs, since it is the part that has higher fractions of metals, mainly precious, in this equipment, as well as compared to other WEEEs, such as computers, routers, televisions, among others (Ha et al. 2014; Priya and Hait 2018; Yamane et al. 2011; Hira et al. 2018; Holgersson et al. 2018; Gu et al. 2019; Silveira 2019, Silveira et al. 2020). The most interesting metals of this device are gold, silver, and copper; however, there are some studies that have been focusing on other metals, such as gallium, indium, germanium, nickel, palladium, among others (Quinet et al. 2005; Ha et al. 2010; Kasper et al. 2011; Ha et al. 2014; Stuhlpfarre et al. 2016; Kasper and Veit 2018; Kasper et al. 2018; Xiu et al. 2015; Priya and Hait 2018; Silveira 2019; Gu et al. 2019; Flerus et al. 2019; Panda et al. 2020; Flerus and Friedrich 2010; Silveira et al. 2020).

With regard to batteries, the focus of the studies has been to find safe and environmentally viable techniques for the recovery of lithium and cobalt mainly (Al-Thyabat et al. 2013; Horeh et al. 2016; Zhang et al. 2018; Heydarian et al. 2018; Gu et al. 2019; Dos Santos et al. 2019; Chandran et al. 2021; Costa et al. 2021). Another part that has aroused interest in researchers are the screens, both those of LCD and others (LED, OLED, AMOLED), in which the indium, mainly, but also other metals such as silver, strontium, copper, and molybdenum, has been targeted (Dodbiba et al. 2012; Hashimoto 2015; Fontana et al. 2015; Silveira et al. 2015; Zhuang et al. 2015; Jimenez et al. 2016; Golzar-Ahmadi and Mousavi 2021; Pourhossein et al. 2021).

On the other hand, the polymer fraction—as well as some accessories such as microphones, keyboards, contacts, and chargers, for example—appears in some studies, but in a more isolated way (Ha et al. 2010; Monteiro et al. 2007; Kasper et al. 2010; Kasper 2011; Santos 2016; Zazycki et al. 2017; Gu et al. 2019).

According to Gu et al. (2019), the recovery processes of materials contained in cell phone equipment generally involve at least two of the following steps:

(1) Pre-treatment: involves a disassembly step and/or comminution.

(2) Separation: step in which fractions are separated, either by particle size, density, magnetism, and/or electrostatic.
(3) Extraction: step in which the separation of the material of interest occurs, which may involve two processes: (i) leaching (hydrometallurgy), (ii) purification/concentration of the metals of interest (Gu et al. 2019).

In fact, there are several works that involve at least two of the mentioned techniques. As mentioned before, a disassembly stage is commonly used; Gu et al. (2019) carried out a review of works involving recycling of the various parts of mobile phones (between 2005 and 2019), in which they reviewed 68 studies and found that half of the works involved a disassembly stage, while other works used parts already previously separated.

According to Gu et al. (2019), a comminution step is very important, because it facilitates the steps of mechanical separation and material extraction, as it increases the surface area, thus facilitating the separation of metals by chemical leaching, for example. With this, many works opt for a crushing and/or grinding step (Kasper 2011; Kasper et al. 2011; Xiu et al. 2015; Silveira et al. 2015; Horeh et al. 2016; Flerus et al. 2019; Panda et al. 2020; Pourhossein et al. 2021; Golzar-Ahmadi and Mousavi 2021). In addition, some studies show satisfactory results regarding the separation of fractions that are rich in metals of interest only with mechanical treatments (Kasper 2011; Silveira et al. 2012; Li et al. 2017a, b; Silveira 2019; Silveira et al. 2020).

Li et al. (2017a, b) performed Eddy current separator tests for the separation of polymeric fractions from metals in PCBs, from cellphones, that were grinded. They obtained a separation efficiency of about 95.54%. Finally, the authors recommended the use of electrostatic separation to give continuity to the separation of metals (Li et al. 2017a, b).

Kasper et al. (2011) used grinding, granulometric, magnetic, and electrostatic separation techniques in the PCBs of mobile phones, and obtained average concentrations of 60% copper. Silveira (2019), on the other hand, performed grinding, granulometric, magnetic, and electrostatic separation in the PCBs from smartphones, and obtained fractions with concentrations of 40–52% copper. However, to concentrate a metal of interest from the rest, metallurgy techniques become necessary (Kasper et al. 2011; Silveira 2019).

In relation to metallurgical techniques of metal extraction, there is a variety of studies involving pyro, hydro, biohydro, and electrometallurgical routes, in which the combination of two of them is often used (Ha et al. 2010, 2014; Kasper 2011; Kasper et al. 2011, 2018; Silveira et al. 2012; Xiu et al. 2015; Silveira et al. 2015; Horeh et al. 2016; Kasper and Veit 2018; Flerus et al. 2019; Gu et al. 2019; Panda et al. 2020; Flerus and Friedrich 2010; Golzar-Ahmadi and Mousavi 2021; Pourhossein et al. 2021; Andrade et al. 2022).

Kasper (2011) used different techniques (crushing, grinding, mechanical processing, electro-obtaining, and leaching with alternative solvent) to separate copper, gold, and silver, and obtained recovery percentages of about 95%, 30% and 17%, respectively. In addition, the author also carried out recycling tests of the polymers from the housing and PCBs.

According to Andrade et al. (2022), although conventional pyro and hydrometallurgy techniques are still the center of research, alternative techniques such as biohydrometallurgy, use of alternative solvents that are less aggressive to the environment, as well as techniques involving ultrasound and mechanochemistry, among others, have been studied, aiming at a more environmentally friendly procedure (Andrade et al. 2022). According to Gu et al. (2019), it is essential to develop automated disassembly techniques, since this type of waste is increasing, as well as develop a recycling route analyzing profitability, evaluate the life cycle, and explore mechanical/physical techniques.

3 LED Lamps

LED lamps are classified as WEEE after their end of life, as they fall under electronic equipment, due to being dependent on a printed circuit board for their operation (European Union 2012; ABNT 2013; Xavier and Carvalho 2014). According to Pourhohssein and Mousavi (2018), LED lamps will be directly inserted into the reverse logistics flow of WEEEs and should be classified as light waste.

The recycling of this waste is important from the point of view of environmental effects and for the recovery of materials, as they have materials of great economic value in their composition, such as gold, silver, copper, and platinum, for example, in addition to metals in greater volume and value such as iron and aluminum; with this, they can be considered as a secondary ore, and, in some cases, have higher precious metal contents than the ore of origin itself (Pradiian 2013; Nicolai 2016; Baldé et al. 2017; Pourhossein and Mousavi 2018; Xavier et al. 2020a, b; Santos 2021).

LEDs may have higher concentrations of some metals than those found in natural ores, such as gold, silver, copper, aluminum, tin, and gallium (Cenci et al. 2020). According to Cenci et al. (2020), when recycling tubular and bulb-type LED lamps, the economic return could be of US$2,405.99 and US$2,595.02 per ton, respectively.

Regarding the gravimetric composition of LED lamps, it can be observed that there is an evolution and change in relation to composition and design as time passes; however, it can be noted that LED lamps are also composed mostly of polymers, which make up their housing, and metals such as aluminum, which are mainly used for cooling.

Hendrickson et al. (2010), when performing the disassembly and mass balance of three LED lamps manufactured in 2008, observed that they had varied materials in their composition; however, a significant amount of aluminum was highlighted, followed by plastic and glass. In relation to aluminum, the whole material represented about 45–60% of the total mass of the lamp. On the other hand, the content of plastic and glass accounted for 13–27% and 9–13%, respectively. In addition, other materials such as copper, porcelain, steel, and gold were found (Hendrickson et al. 2010).

6 years later, a 9.5 W, 85.5 g, bulb-type LED lamp was disassembled by Gassmann et al. (2016), and the mass distribution was verified in the seven parts found: heat sink

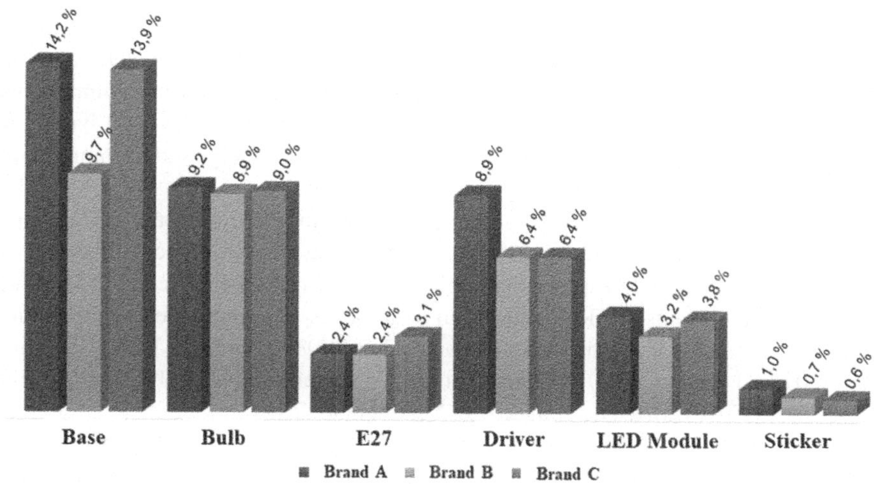

Fig. 7 Mass distribution (%) of bulb-type LED lamps (*Source* Adapted from Santos et al. 2019)

(42.3%), plastic body (21.3%), PCB (16%), glass bulb (15%), LED module (3.5%), socket (1.9%), and 10 LEDs (0.32%).

Santos et al. (2019), on the other hand, carried out the disassembly of seven bulb-type LED lamps of three different brands in the marketing year 2017. First, the lamps were divided into five large parts: bulb, base, LED module, driver (PCB), and Edison screw (E27). The seven lamps that were analyzed in this study had no heat sink (Santos et al. 2019).

Figure 7 shows the percentage of mass distribution of the lamp parts that were disassembled by Santos et al. (2019).

Cenci et al. (2019) disassembled and characterized six different brands of LED lamps, of which three brands were of tubular type lamps and three of bulb-type. In all, 60 lamps were disassembled and, finally, they could be separated into five parts (polymer carcass, metal carcass, PCB, PCB components, and LED module).

The tubular lamps showed the very similar design and composition, with polycarbonate polymer housing and aluminum metal housing. On the other hand, the bulb-type lamps showed some variations both in design as well as in the materials used in some parts, such as the use of polyester or polyamide for polymer housing, for example, and the contact pin being aluminum or nickel. All presented a polycarbonate bulb (Cenci et al. 2019).

Finally, it was found that polymers and aluminum are the most commonly found materials in lamps. Polycarbonate, polyester, polyamide, and aluminum respectively accounted for about 48.13, 5.08, 1.77, and 23.04% of the total mass of the lamps (Cenci et al. 2019).

Rebello et al. (2018, 2020) performed a gravimetric composition of LED lamps, and divided them into metal, glass, polymer, printed circuit boards, and electronic components; although it is a mixed sample, in which 16 different models of lamps were obtained through donations. The results found are consistent with those obtained by Santos et al. (2019) and Cenci et al. (2019), in which the most prevalent material was polymer (37%), followed by metals (18%) and PCBs and electronic components (13%). In this work, the authors also reported the presence of glass and batteries, because many of the donated lamps were emergency ones, widely used in commercial establishments (Rebello et al. 2018, 2020).

Another important point to be addressed is that, in addition to the LED lamps showing high added value materials inside, reinforcing the importance of their recycling, they also present hazardous materials, such as lead, copper, arsenic, and brominated flame retardants (BFRs), which can be toxic, bioaccumulative, carcinogenic, and can cause serious environmental impacts (Gois 2008; Fraunhofer IZM 2012; Lim et al. 2013; Gosavi et al. 2013; Tuenge et al. 2013; Pieroni et al. 2017; Pourhossein and Mousavi 2018; Bossche et al. 2019; Annoni et al. 2019).

Gassmann et al. (2016) presented a proposal for a recycling flow for LED lamps, in which they first propose a coarse fragmentation of the entire lamp, citing the electrohydraulic fragmentation method (EHF) as efficient for fragmentation of WEEEs; after that, each material can be separated by specific processes. And finally, the LEDs would be stored, waiting for when a recovery technology of their critical materials will be available and consolidated (Gassmann et al. 2016).

Rahman et al. (2017) note that some studies focus on recovering metals contained in LEDs, but not for complete recycling of the lamp. With this, the authors suggested a process for recycling this waste, in which, first, the disassembly of the product in parts or by types of materials is carried out.

Some techniques of separation of materials are based on size or type (ferrous metals, non-ferrous metals, and non-metals such as plastic), in which one can divide the lamp into the printed circuit board, LED chip, and materials in bulk. After that, each part goes through distinct processes, such as comminution, mechanical processing, and pyrolysis (Rahman et al. 2017).

One point to be considered is the possibility of contamination of LED lamps with mercury from fluorescent lamps, if they are discarded in the same place (Gassmann et al. 2016; Rahman et al. 2017). According to Gassmann et al. (2016), it may not be clear to the consumer the difference between the technologies of the lamps, as they have very similar formats, so the joint disposal of them often occurs.

Furthermore, according to Bossche et al. (2019), the recycling of LEDs is a challenge, since they are connected to larger devices and are difficult to disassemble; in addition, the semiconductor chips are embedded in polymer boxes. With this, Sanchez Junior (2018) defends the need to invest in research to explore the recovery techniques of precious metals such as gold and silver, metals whose recovery rewards the cost, such as gallium and indium, as well as rare-earth elements.

Disassembly is a very important step in the process of recycling and reverse logistics of any electronic equipment waste, because it is in this step that the first screening of materials takes place, in many cases manually, in which toxic components, those of easy separation and those of greater added value, can be separated (Gouveia et al. 2014; Moraes et al. 2014; Silveira et al. 2019). With this, the equipment can be separated into several parts. Many of these are composed of a single material and can go directly to their recycling flow or, in some cases, be reused (Silveira et al. 2019; Santos et al. 2021a, b).

Some elements, especially critical ones, are present in low concentrations (in the range of μg) in finished products, which makes it difficult to detect and quantify them, making classification and separation of these materials challenging at the end of product life (Buchert et al. 2009; Ayres and Pieró 2017; Santos et al. 2019). Santos (2021) and Santos et al. (2021b) state that the step of disassembly and separation of the LEDs can be crucial for the recovery of the critical materials contained in them, as, in this way, they can be better concentrated, thus making them another part that can be commercialized in the recycling market.

However, some difficulties may be encountered, such as the heterogeneity of this equipment, according to brand, for example, as well as the design, which is not thought of considering the end of its life; this makes the disassembly stage of each part a difficult task and, consequently, a challenge for the recycling of this type of waste (Silveira et al. 2019; Tansel 2017; Santos 2021; Santos et al. 2021a).

A simplification of the design of electronic products in general, combined with an automation of the disassembly process, would mean benefits in relation to the time and efficiency of disassembly, and, therefore, the recovery of materials (Silveira et al. 2019; Knoth et al. 2002), which can result in social, economic, and environmental gains.

Gassmann et al. (2016) argue that the diversity and complexity of the geometry of these lamps should be foreseen during reverse logistics, and that classification modules should be created in order to direct the flow of the recycling process to be used.

That said, Santos et al. (2021a) found design differences when disassembling LED lamps of different brands, which led to differences and difficulties in the way of disassembling them. Finally, the authors suggested adopting a standard for the assembly of this equipment (Santos et al. 2021a).

According to Hendrickson et al. (2010), the standardization in the connection of the parts would be a benefit for the disassembly and remanufacturing of its components, in addition to a decrease in the variety of materials in structural parts in order to obtain a better homogenization, thus facilitating recycling.

Santos et al. (2021a, b) and Santos (2021) divided the LED lamps of the bulb-type into up to eight parts, being five basic parts: bulb, polymer base, led board, PCB, and E27 screw, plus extra parts such as aluminum alloy base, aluminum alloy plate, and screws, in addition to somewhere an adhesive was found. Figure 8 shows a disassembled bulb-type LED lamp.

With this, Santos et al. (2021b) have presented destination suggestions for almost all lamp parts/materials, except LEDs, which do not yet have a consolidated recycling technology. Figure 9 shows the suggested flows for each part of the lamp after manual disassembly.

Regarding the recovery of metals from LEDs, there are several studies being carried out involving techniques, with or without prior comminution, of mechanical processing, pyro, hydro, or biohydrometallurgy, mainly aimed at gallium, indium, copper, and gold (Zhan et al. 2015; Maneesuwannarat et al. 2016a, b; Nagy et al.

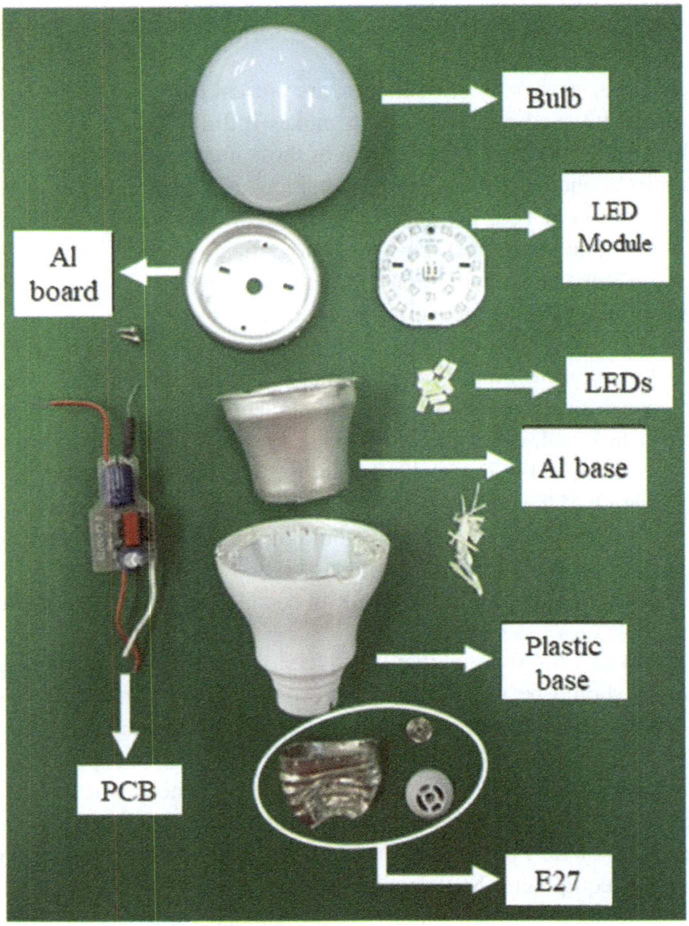

Fig. 8 Disassembled bulb-type LED lamp (*Source* Santos et al. 2021a)

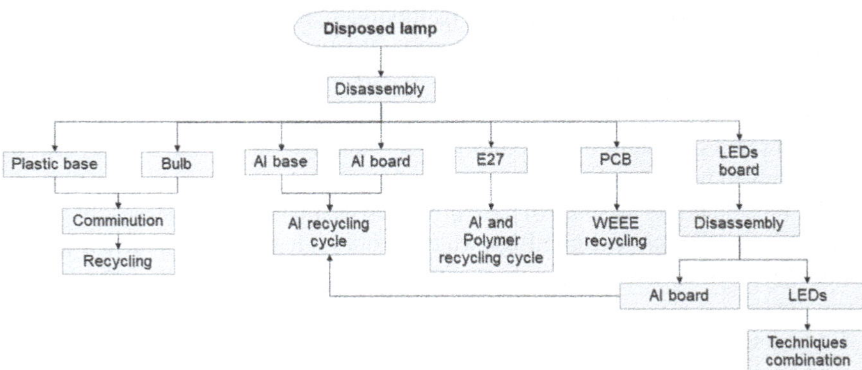

Fig. 9 Recycling destination of each part of the lamp (*Source* Santos 2021b)

2017; Pourhossein and Mousavi 2018, 2019; Maneesuwannarat et al. 2019; Bossche et al. 2019; Zhan et al. 2020; Oliveira et al. 2021; Cenci et al. 2021).

However, as mentioned previously, there is not a consolidated, viable, and industrial technology yet for the recovery of metals contained in LEDs, but, with the increase in consumption and disposal of these lamps, allied to the concentration of metals found in these devices, the development of an economically and environmentally viable technique becomes necessary (Santos 2021; Santos et al. 2021a, b).

4 Computers

Computers and laptops are two of the most consumed electronic equipment, and were part of the more than 53 million tons of waste of this type that was generated in the year 2019 worldwide (Forti et al. 2020).

It is estimated that worldwide computer shipments totaled 79.4 million units in the fourth term of 2020, which is an increase of 10.7% compared to the fourth term of 2019. The research also states that this increase was potentialized by the Coronavirus pandemic, which had already spread around the world in the last half of 2020 (Gartner 2021).

As technology advances and makes innovative products available to the market, tons of waste from obsolete equipment are also generated (Al Razi 2016). When disposed of incorrectly, this waste can affect human health, as well as contaminate water, soil, and air through the release of hazardous compounds (Cayumil et al. 2016).

Computers are composed mostly of metal and plastic, as shown in the graph in Fig. 10. However, in printed circuit boards, for example, more than 50 chemical elements may be present, among them precious metals such as silver and gold, and heavy metals such as mercury and lead (Lu and Xu 2016).

Despite the heterogeneity of the components found in a computer, much of it can be recycled and reintroduced into the production chain. In addition to the economic benefits, it is still possible to reduce the negative environmental impacts of both the inadequate disposal as well as the extraction of virgin raw materials (Van Eygen et al. 2016; Zhang and Xu 2016; Tabelin et al. 2021; Al Razi 2016; Rene et al. 2021).

Next, the main materials found in computers will be better addressed:

Polymers

Polymeric components are widely used in computers with the aim of making them lighter. They are mostly located on the monitor housing, keyboard, mouse housing, and CPU housing (Suresh et al. 2018). The main plastics used are: ABS, HIPS, PPO, PE, and PMMA, as well as some blends such as ABS/PC and ABS/OS (Freegard et al. 2006; Hadi et al. 2015). However, the recycling of these materials faces a serious challenge: flame retardants, which are added in order to meet fire safety standards (Forti et al. 2020; Suresh et al. 2018). There is also the high cost associated with the

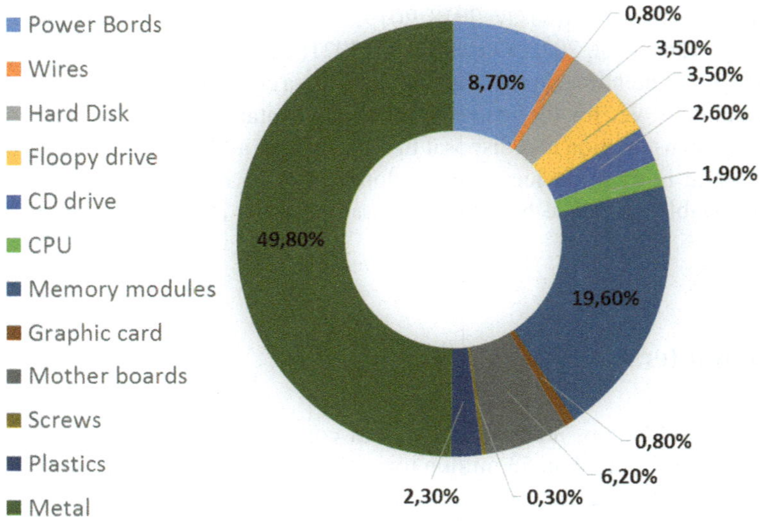

Fig. 10 Components present in computers (*Source* Adapted from Suresh et al. 2018)

segregation of plastics containing retardants compared to those that do not contain them, since percentages above 0.1% do not allow it to be used for the manufacture of any other product (Forti et al. 2020).

The recycling of plastic materials can be divided into four categories: primary, secondary, tertiary, and quaternary. In both primary and secondary recycling, mechanical processes are applied, in which the polymers are sorted, ground, and processed. However, primary recycling is done in pre-consumer waste, that is, those generated within industries, while secondary addresses post-consumer waste (Rahimi and Garcia 2017; Singh et al. 2016).

Tertiary recycling uses chemical processes to recover petrochemical components present in plastic waste. The quaternary method, on the other hand, approaches incineration in order to generate energy (Rahimi and Garcia 2017; Kumar et al. 2011). Moreover, energy recovery is problematic from an environmental and human health point of view, since this process produces toxic substances, such as light hydrocarbons NOx, sulfur oxides, and dioxins (Dogu et al. 2021; Al-Salem et al. (2009) 2010).

The alternative to the use of these polymers is controlled incineration. In this way, it is possible to generate energy without releasing dioxins and furans into the atmosphere, thus avoidin3g damage to the environment and human health (Forti et al. 2020).

Precious Metals and Heavy Metals

The recycling of precious metals, such as gold, silver, and platinum, as well as any other material, has great environmental importance. However, in the case of precious metals, there is also a strong economic interest, due to their high added value (Roslan et al.2017). The recovery of these metals can be carried out through hydrometallurgical processes, through which leaching agents are used to dissolve the metal of interest (Rene et al. 2021). For Chancerel et al. (2009), it is possible to find higher concentrations of precious metals in printed circuit boards than in mining mines. A great example of the recovery potential of these metals is the innovative project created by Tokyo, called the "Tokyo 2020 Medal Project", which was responsible for the production of about 5000 medals from electronic scrap. These silver, gold, and bronze medals were used in the awards of athletes at the Tokyo Olympics.

However, heavy metals such as lead, mercury, and chromium, which are extremely harmful to health and the environment, are also present in computers (Suresh et al. 2018). The liberation of these elements can occur during the recycling steps (Han et al. 2019). Therefore, it is very important to have effective regulation to prevent contamination, as well as having remediation methodologies if necessary. In this way, it is possible to control the widespread contamination of people and the environment (Soetrisno and Delgado-Saborit 2020).

Rare-Earths

The rare-earths are formed by a group of 17 chemical elements, of which 15 belong to the lanthanide series, and the remaining two are the transition metals scandium and yttrium. These elements are present in various components of a computer, such as the battery, HD, screen, boards, and among others (Goodenough et al. 2018).

Several studies have been developed to recover these rare-earths and reinsert them in the production of these components (Munchen 2018; Kumari et al. 2018, Behera and Parhi 2016, Reisdörfer et al. 2019). The element that has been the focus of many studies is neodymium, present in the magnets that make up the HDs. There are approved methods capable of separating it from the other components of magnets, such as iron and boron. Such processes can be pyrometallurgical, hydrometallurgical, pyrometallurgical, hydrogen decrepitation, and gas phase extraction (Binnemans et al. 2013). However, the most studied method is the hydrometallurgical one, in which chemicals are used to leach the previously ground magnets. Table 3 shows some of the chemicals used as leaching agents.

Some of the rare-earth elements show a perspective of an increase in demand, as is the case of neodymium and dysprosium. Therefore, the recovery of these materials from discarded equipment presents itself as a good opportunity to meet the market and, at the same time, reduce the environmental impacts of the extraction and processing processes of the main ore (Filippas et al. 2021). Bastnasite, monazite, and xenotime are the minerals responsible for containing about 95% of all rare-earth

Table 3 Leaching agents used in the recovery of rare-earths

Leaching agents	Author
Sulfuric acid	Munchen (2018)
Hydrochloric acid	Kumari et al. (2018)
Acetic acid	Behera and Parhi (2016)
Citric acid and maleic acid	Reisdörfer et al. (2019)

element resources in the world (Gupta and Krishnamurthy 2005). For Binnemans et al. (2013), the recycling of rare-earths has the potential to reduce a portion of the primary extraction in the future.

As technologies advance, bringing faster, lighter, and more utilitarian computers, so does obsolescence. Software and memory constraints, for example, result in the generation of tons of waste. This waste has a range of metallic and non-metallic materials in their composition, and elements that, if properly segregated, have great recycling potential.

Regardless of the material addressed, the need for investments for the development and improvement of disassembly methodologies, as well as recycling, is evident. In this sense, recycling cooperatives have a fundamental role in disassembly and segregation. For this reason, there is an urgent need for support and attention to this sector through well-defined public policies.

Regarding the recovery of materials, the effort of the researchers in trying to find alternatives that make it possible to reintroduce some materials into the production chain is observed. However, some processes are still very aggressive, which can cause negative environmental impacts greater than the benefit of the recovery. In this sense, evaluating the life cycle of processes helps in making decisions, allowing for definitions of more appropriate raw materials and processes.

Computers and laptops are composed of several critical chemical elements that are essential for their operation. If actions are not promoted at a global level to recycle these elements, the low supply will affect prices and slow down technological development. In addition, incorrect disposal and uncontrolled extraction of virgin raw materials bring serious negative environmental impacts.

5 Electrical Wires and Cables

Electrical wires and cables are indispensable parts of the electronic industry and can be considered as a type of WEEE, which receives great interest and studies regarding its recycling, because they have important metals such as aluminum and copper in their composition, as well as recyclable polymers, such as polyvinyl chloride (PVC) (Lambert et al. 2015; Anastassakis et al. 2015; Li et al. 2017a, b; Suresh et al. 2017; Xu et al. 2019; Çelik et al. 2019; Xavier et al. 2020a, b; Blinová and Godovčin 2021; Kaercher and Moraes 2021; Kumar et al. 2021).

According to Blinová and Godovčin (2021), most residential electronic products, such as washing machines, refrigerators, vacuum cleaners, mobile phones, computers, televisions, and among others, have electrical cables. In addition, there are discarded cables or shavings in telephone line repairs, as well as waste from the production of these components themselves (Çelik et al. 2019; Kaercher and Moraes 2021).

According to Christéen (2007), in Sweden, approximately 40,000 tons of wires and cables become waste every year. In Germany, on the other hand, about 150,000 tons of cable waste are generated annually (Çelik et al. 2019).

According to Matozinhos (2012), waste from wires and cables can be classified into two types:

(a) **Pre-consumer waste**: they are industrial scraps, that originated in the wire and cable industry;
(b) **Post-consumer waste**: they are generated from electronic equipment in repairs or at the end of a lifetime; scraps are generated from installations of electrical networks, telephony, or cable TV; in addition, there are also cables originating from renovations; in the electronic and automobile industry.

An electrical cable is formed by an electrical conductor that has the function of channeling the electrical flow, whose most commonly used material is copper; aluminum may also be used in some cases. And an insulating material, which serves to cover the conductive material and isolates it from external contact in order to contain the flow; the most commonly used thermoplastic material is polyvinyl chloride (PVC) and, as a thermoset, we use cross-linked polyethylene (XLPE) and Ethylene Propylene (EPR). In addition, some auxiliary elements, such as armor and outer covering layers, can be used in order to provide a longer lifetime to the cable (Pita and Castilho 2018; Top Cable 2020a, b). Figure 11 shows the main parts that an electrical cable may contain.

Furthermore, cables can be composed of only a single solid wire (non-flexible and without covering) or a set of thin, malleable wires (flexible and covered with insulating material); they can also be composed of a single conductor (single-core cable) or multiple conductors (multi-core cable) (Top Cable 2020a, b).

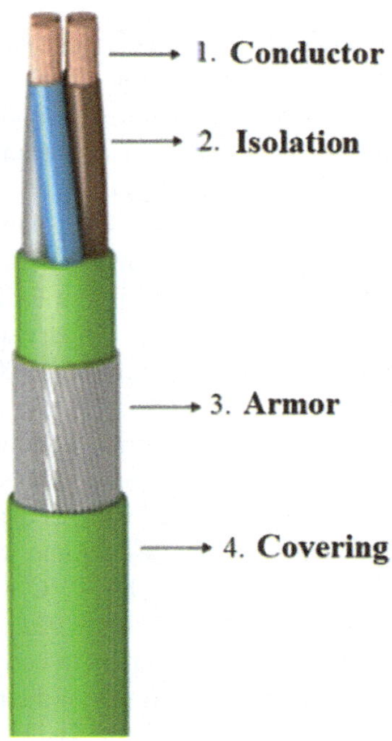

Fig. 11 Main parts of an electrical cable (*Source* Top Cable 2020a, b)

Until recently, the practice of melting the polymer cover that surrounds these wires was used in order to recover the inner copper or aluminum. However, this technique is highly harmful to the environment, as well as to the health of the worker, as it releases toxic gases such as hydrogen chloride, dioxins, furans, carbon monoxide (CO), sulfur dioxide (SO_2), polycyclic aromatic hydrocarbons (PAHs), as well as heavy metals and ash (Moksin et al. 2011; Abcobre 2016; Çelik et al. 2019; Kaercher and Moraes 2021; Gupta et al. 2021).

According to the Brazilian Copper Association (ABCOBRE), when these cables burn—which are below the combustion temperature, which occurs between 250 and 700 °C—especially in the open air, hydrocarbons and particulate materials are released, and may contain lead, since this heavy metal is used as a stabilizer in polymeric PVC covers (ABCOBRE 2016). In the case of the preponderant metal, it may have a low yield under uncontrolled conditions, due to its oxidation, and, in case of fusion, form slag of the main metal.

In addition, burned polymers, apart from forming harmful substances, are wasted materials that could be recycled. In an electrical cable, for example, metal corresponds to between 40 and 90% of the total mass, while the rest consists of polymers (Moksin et al. 2011).

Copper is the third most important industrial metal, following iron and aluminum, respectively (EURIC 2020). It is a very versatile base metal, used in the most varied applications, such as civil construction and architecture, manufacturing of pipes, electrical components and cables, industrial machines, connectors, brakes, and transport vehicles; in addition, it is fundamental to information and communication technologies worldwide (EURIC 2020; Garside 2021; Blinová and Godovčin 2021).

Garside (2021) indicates that, in 2020, the world consumption of refined copper was 23.4 million metric tons; however, only 20 million metric tons of refined copper were produced in mines around the world, but the global production was of 23.9 million metric tons, which means that 3.9 million metric tons come from secondary sources (Garside 2021).

Copper is extremely recyclable and virtually all products made of this metal can be recycled, and, when recycled, it does not lose its chemical and physical properties (ICSG 2014). According to EURIC (2020), 44% of the European Union's copper demand comes from secondary sources, in addition, they have invoiced around €1.91 billion by exporting around 986,000 tons of scrap copper to other countries (EURIC 2020).

Due to its characteristics and wide applicability, PVC is one of the most consumed thermoplastic polymers in the world (PVC4Cables 2017; Blinová and Godovčin 2021). In 2016, the world consumption of PVC was 42,931.1 thousand tons, and it is expected that this volume will reach 55,715 thousand tons in 2022 (Fernández 2021). In addition, only PVC wires and cables accounted for about 7% of global PVC consumption (Tiseo 2021).

Taking into account both economic and environmental issues, it can be seen that the recycling or recovery of the materials of these post-consumer cables is essential (Çelik et al. 2019). In the last two decades, the main focus of the recycling of wires and cables has been copper and aluminum, due to the high added value of these conductors, which makes their separation and recycling quite attractive; apart from that, it has also become important for the ecological issue now (Díaz et al. 2018). However, there are studies developing cable and wire recycling techniques aimed at recovering copper and PVC (Pita and Castilho 2018; Xu et al. 2019; Blinová and Godovčin 2021; Kumar et al. 2021).

Li et al. (2017a, b) and Blinová and Godovčin (2021) state that first, it is necessary to carry out a manual and rough separation between the different types of cables. After this, a separation technology is used, in which the copper core is separated from the plastic cover (Li et al. 2017a, b).

According to Blinová and Godovčin (2021), there are several simple techniques for thick cables that can be applied, as these can be easily disassembled or shredded into small pieces, whereas thinner cables require some combinations of techniques (steps).

Some of the most commonly used techniques for cable recycling involve mechanical processing, such as automated stripping, crushing, grinding, sieving, gravity separation, density separation, use of magnetism, electrostatic separation, flotation, use of high-pressure water jet, cryogenic treatment, ultrasound, Eddy current separation, use of a sensor, among others. In addition, the use of hydro and pyrometallurgical techniques may be involved (Xiao et al., 2015; Hlosta et al. 2017; Blinová and Godovčin 2021; Gupta et al. 2021).

In any way, Hlosta et al. (2017) emphasize that in any WEEEs recycling process it is necessary to perform the separation of polymeric and metallic fractions. With this, in addition to separating and recovering precious metals, polymers can be comminuted into suitable equipment without damage to them and, thus, fractions free of "contamination" are acquired (Hlosta et al. 2017).

Table 4 presents some works that focus on recovering copper and polymers from electrical cables and wires.

With Table 4, it can be observed that mechanical processing techniques are the ones that are mostly addressed, since they can be considered simpler and cheaper, and easily adaptable (Xiao et al. 2015; Li et al. 2017a, b; Blinová and Godovčin 2021). However, there is also the need to associate different techniques to obtain high recovery rates.

Kaercher and Moraes (2021) opted for only mechanical procedures, in which they approached crushing techniques, by knife mill, grinding, by ball mill, sieving, and second sieving, in which they obtained results of up to 99% copper separation. According to the authors, it is important to develop more accessible techniques, from the point of view of energy consumption, atmospheric or effluent generation, and emissions, since in many cases these materials are received by cooperatives (Kaercher and Moraes 2021).

Figure 12 shows the material obtained by Kaercher and Moraes (2021) after the sieving technique, in which 99% copper was obtained in the last sieve.

Li et al. (2017a, b) conducted a review and evaluation of some cable recycling technologies for copper recovery. They noted that the stripping technology is a well-accepted form of processing for uniform cables of larger diameters. This technology

Table 4 Works aimed at the recovery of materials from electrical wires and cables

Author	Method	Material of interest	Separation efficiency
Pita and Castilho (2018)	Mechanical techniques (grinding, jigging, vibrating table, and foam flotation)	Copper Polymer	97% 98%
Hlosta et al. (2017)	Mechanical technique (gravity separator)	Copper Polymer	53.7% 46.3%
Kumar et al. (2021)	Mechanical/chemical technique (wet ball-milling—using solvents—and jigging separation)	Copper Polymer (PVC)	90% 100%
Xu et al. (2019)	Chemical and mechanical techniques (solvent dilation and centrifugation)	Copper Polymer (PVC)	100% 100%
Çelik et al. (2019)	Mechanical and thermal technology (crushing, electrostatic and gravity separation, casting, and refining)	Copper	99.6%
Kaercher and Moraes (2021)	Mechanical techniques (crushing, grinding, sieving, and resieving)	Copper	99%
Lambert et al. (2015)	Mechanical techniques (crushing and separation by density) and Biolixiviation (using *Acidithiobacillus ferrooxidans*)	Copper	<90%
Bedeković and Trbović (2020)	Mechanical techniques (electrostatic separation)	Aluminum	91.20%

Fig. 12 Crushed, sieved, and resieved electrical cables (*Source* Kaercher and Moraes 2021)

involves a stepper motor, in which the wire attached to a wheel moves through a cutter (Xiao et al. 2015; Li et al. 2017a, b; Blinová and Godovčin 2021). Smaller diameter and mixed cables, on the other hand, required more complex technologies, often using thermal and chemical recovery techniques to obtain refined materials (Li et al. 2017a, b; Blinová and Godovčin 2021).

However, recovery technologies that use thermal and chemical techniques can generate severe environmental impacts, such as toxic gas and toxic effluents emissions, requiring treatment (Li et al. 2017a, b; Blinová and Godovčin 2021). According to Blinová and Godovčin (2021), leaching techniques are not advised as a first approach for electrical wires and cables, as large volumes of effluent would be generated, in addition to the consumption of chemicals. This technology should be used as a second step in order to obtain purer fractions of metals (Blinová and Godovčin 2021).

According to Blinová and Godovčin (2021), the heterogeneity of this type of waste, in which it is possible to find many types of polymers and metals, in addition to variations in diameters, should be predicted for an adequate recycling technology that adapts to the different types of cables. Furthermore, it is necessary to carry out a segregation in several steps (Blinová and Godovčin 2021). One step that must be included in the segregation process is the removal of the plugs (Kaercher and Moraes 2021).

Finally, the optimal technology must be economically and environmentally viable, as well as fully automated (Blinová and Godovčin 2021). In addition, according to Matozinhos (2012), both the properties of the recycled products as well as their origin and the state of conservation of the waste will define the commercial value of the product. The recycling methodology of electrical cables and wires should add economic value to the recycled products (Kaercher and Moraes 2021).

6 Final Remarks

Given the above and the examples reviewed in this chapter, it is possible to notice that small-sized electronic waste presents itself as a large source of valuable raw materials; although small in size, this waste often presents more precious and critical metals than larger ones, and even than what is found in natural ores.

It is evident the need to recover the materials from these sources, from the point of view of the scarcity of natural resources and pollution of the environment; given that many of the materials from this waste are critical, both for being in high demand as much as for the possible restrictions on their offer, and, on top of that, the improper disposal of the post-consumer equipment may cause damage to the environment and human health due to the toxic materials used in them.

However, although there are many studies regarding metal recovery, this practice can still be considered a challenge and for some of this waste, there is still no consolidated metal recovery technology. Some obstacles must be faced, such as heterogeneity and product design, hindering mechanized processes of disassembly and segregation; incentive by public authorities; technical, economic, and environmental feasibility, because in many techniques there has been high energy expenditure; consumption of chemicals; generation of toxic effluents and atmospheric emissions; or, on the other hand, high process time and difficulty of implementation on a large scale.

The disassembly and segregation step is of paramount importance for all this waste, because it values materials, concentrates metals, and avoids the loss of many of them. However, it is still almost exclusively done manually, which entails a process delay.

Due to the great design difference of each type of product, disassembly methodologies as efficient as possible should be developed, as well as disassembly benches and or disassembly lines for each type of waste; the ideal solution would be the development of automated systems. However, without a new mentality of product development, in which one begins to think about their end of life and to see them as materials that are sources of raw materials and not waste, this progress will be difficult to achieve.

In addition, techniques that are less aggressive to the environment and more economically viable should advance in studies, as is the case of the use of less aggressive organic solutions, closed cycle processes, the use of microorganisms, mechanical processing, and among others. Finally, it is necessary to seek to recycle all the materials from this waste, not only metals, thus meeting the guidelines of industrial ecology and circular economy.

References

Agência Brasileira de Desenvolvimento Industrial (ABDI) (2013) Logística Reversa de Equipamentos Eletroeletrônicos—Análise de Viabilidade Técnica e Econômica

Al Razi KMH (2016) Resourceful recycling process of waste desktop computers: are view study. Resour Conserv Recycl 110:30–47

Al-Salem SM, Lettieri P, Baeyens J (2009) Recycling and recovery routes of plastic solid waste (PSW): a review. Waste Manag 29(10):2625–2643 S0956053X09002190 https://doi.org/10.1016/j.wasman.2009.06.004

Al-Salem SM, Lettieri P, Baeyens J (2010) The valorization of plastic solid waste (PSW) by primary to quaternary routes: from re-use to energy and chemicals. Prog Energy Combust Sci 36:103–129. https://doi.org/10.1016/j.wasman.2009.06.004

Al-Thyabat S, Nakamura T, Shibata E, Lizuka A (2013) Adaptation of minerals processing operations for lithium-ion (LiBs) and nickel metal hydride (NiMH) batteries recycling: critical review. Miner Eng 45:4–17. https://doi.org/10.1016/j.mineng.2012.12.005

Anastassakis GN, Bevilacqua P, Lorenzi L (2015) Recovery of residual copper from low-content tailings derived from waste electrical cable treatment. Int J Miner Process 143:105–111. https://doi.org/10.1016/j.minpro.2015.09.011

Andrade DF, Castro JP, Garcia JA, Machado RC, Pereira-Filho ER, Amarasiriwardena D (2022) Analytical and reclamation technologies for identification and recycling of precious materials from waste computer and mobile phones. Chemosphere 286. https://doi.org/10.1016/j.chemosphere.2021.131739

Annoni R, Lange LC, Amaral MCS, Silva AM, Assunção MC, Franco MB, Souza W (2019) Light emitting diode waste: potential of metals concentration and acid reuse via the integration of leaching and membrane processes. J Cleaner Prod 246. https://doi.org/10.1016/j.jclepro.2019.119057

Araújo MG (2013) Modelo de avaliação do ciclo de vida para a gestão de residuos de equipamentos eletroeletrônicos no Brasil. Tese (doutorado). Rio de Janeiro, Universidade Federal do Rio de Janeiro

Associação Brasileira de Normas Técnicas (ABNT) (2013) NBR 16156: Resíduos de equipamentos eletroeletrônicos—Requisitos para atividade de manufatura reversa. Rio de Janeiro

Associação Brasileira do Cobre (ABCOBRE) (2016) Reciclagem consciente. São Paulo, nov. Disponível em http://www.cecil.com.br/cartilha_reciclagem_consciente.pdf. Acesso em 17 out 2021

Ayres RU, Pieró LT (2017) Material efficiency: rare and critical metals. The Royal Society Publishing, Fontainebleau, France, Janeiro

Bachér J, Mrotzek A, Wahlström M (2015) Mechanical pre-treatment of mobile phones and its effect on the Printed Circuit Assemblies (PCAs). Waste Manage 45:235–245. https://doi.org/10.1016/j.wasman.2015.06.009

Baldé CP, Forti V, Gray V, Kuehr R, Stegmann P (2017) The Global E-waste Monitor—2017, United Nations University (UNU), International Telecommunication Union (ITU) & International Solid Waste Association (ISWA), Bonn/Geneva/Vienna. Disponível em https://www.itu.int/en/ITU-D/Climate-Change/Documents/GEM%202017/Global-E-waste%20Monitor%202017%20.pdf. Acesso em 6 jun 2021.

Bedekovi G, Trbovi R (2020) Electrostatic separation of aluminium from residue of electric cables recycling process. Waste Manage 108:21–27

Behera SS, Parhi PK (2016) Leaching kinetics study of neodymium from the scrap magnet using acetic acid. Sep Purif Technol 160:59–66

Binnemans K, Jones PT, Blanpain B, Gerven TV, Yang Y, Walton A, Buchert M (2013) Recycling of rare earths: a critical review. J Clean Prod 51:1–22

Bizarri C, Spuling E, Knoll DM, Volz D, Bräse S (2018) Sustainable metal complexes for organic light-emitting diodes (OLEDs). Coord Chem Rev 373:49–82. https://doi.org/10.1016/j.ccr.2017.09.011

Blinová L, Godovčin P (2021) Importance of recycling the waste-cables containing copper and PVC. Research papers faculty of materials science and technology in Trnava—Slovak University of Technology in Bratislava 29(48)

Bollinger LA (2010) Growing cradle-to-cradle metal flow systems—an application of agent-based modeling ans system dynamics to the study of global flows of metals in mobile phones. Thesis. Leiden University

Bookhagen B, Bastian D, Buchholz P, Faulstich M, Opper C, Irrgeher J, Prohaska T, Koeberl C (2020) Metallic resources in smartphones. Resour Policy 68. https://doi.org/10.1016/j.resourpol.2020.101750

Borges G da R, Petry JF, Machado DN, Rasia ICRB (2016) Um estudo preliminar dos fatores perceptivos subjacentes da preferência dos brasileiros por aparelhos celulares. Perspectivas em Gestão & Conhecimento 6(1):255–269

Bossche AV, Vereycken W, Hoogerstraete TV, Dehaen W (2019) Recovery of gallium, indium, and arsenic from semiconductors using tribromide ionic liquids. ACS Sustain Chem Eng 7(17):14451–14459. https://doi.org/10.1021/acssuschemeng.9b01724

Buchert M, Schuler D, Bleher D (2009) Critical metals for future sustainable technologies and their recycling potential. UNEP. Disponivel em https://www.oeko.de/oekodoc/1070/2009-129-en.pdf. Acesso em 17 out 2021

Cayumil R, Khanna R, Rajarao R, Mukherjee PS, Sahajwalla V (2016) Concentration of precious metals during their recovery from electronic waste. Waste Manage 57:121–130. https://doi.org/10.1016/j.wasman.2015.12.004

Çelik C, Arslan C, Arslan F (2019) Recycling of waste electrical cables. Mater Sci Eng Int J 3(4):107–111. https://doi.org/10.15406/mseij.2019.03.00099

Cenci MP, Mello BCF, Schmidt LM, Veit HM (2019) Caracterização de Lâmpadas LED para a Reciclagem. In: Fórum Internacional de Resíduos Sólidos, 10., 2019, João Pessoa. Anais eletrônicos [...]. Instituto Venturi, João Pessoa. Disponível em http://www.institutoventuri.org.br/ojs/index.php/firs/article/view/1658/922. Acesso em 21 dez. 2019

Cenci MP, Berto FCD, Schneider EL, Veit HM (2020) Assessment of LED lamps components and materials for a recycling perspective. Waste Manage 107:285–293. https://doi.org/10.1016/j.wasman.2020.04.028

Cenci MP, Berto FD, Camargo PSS, Veit HM (2021) Separation and concentration of valuable and critical materials from wasted LEDs by physical processes. Waste Manag 120(2021):136–145. https://doi.org/10.1016/j.wasman.2020.11.023

Chancerel P, Meskers CEM, Hagelüken C, Rotter VS (2009) Assessment of precious metal flows during preprocessing of waste electrical and electronic equipment. J Ind Ecol 13:791–810

Chancerel P, Rotter S (2009) Recycling-oriented characterization of small waste electrical and electronic equipment. Waste Manage 29:2336–2352

Christeen J (2018) Swedish cable waste for recovery in China or Sweden. A study of the economic and environmental impact, Environ Technol Manag http://extra.ivf.se/Cable_Program/download/Report_Jonas.pdf

Chandran V, Glosh A, Patil CK, Mohanavel V, Priya AK, Rahim R, Madavan R, Muthuraman U, Karthick A (2021) Comprehensive review on recycling of spent lithium-ion batteries. Mater Today Proc 47:167–180. https://doi.org/10.1016/j.matpr.2021.03.744

Comissão Europeia (2017) Study on the review of the list of critical raw materials—critically assessments. European Commission, Luxembourg. Disponível em https://op.europa.eu/en/publication-detail/-/publication/08fdab5f-9766-11e7-b92d-01aa75ed71a1/language-en. Acesso em 15 Nov 2019

Compound Interest (2014) The chemical elements of a smartphone. Disponível em https://www.compoundchem.com/2014/02/19/the-chemical-elements-of-a-smartphone/. Acesso em 17 out 2021

Cordella M, Alfieri F, Sanfelix Forner J (2020) Guidance for the assessment of material efficiency: application to smartphones, EUR 30068 EN, Publications Office of the European Union, Luxembourg. ISBN 978-92-76-15411-2, https://doi.org/10.2760/037522, JRC116106

Costa CM, Barbosa JC, Gonçalves R, Castro H, Del Campo FJ, Lanceros-Méndes S (2021) Recycling and environmental issues of lithium-ion batteries: advances, challenges and opportunities. Energy Storage Mater 37:433–465. https://doi.org/10.1016/j.ensm.2021.02.032

Cui J, Forssberg E (2003) Mechanical recycling of waste electric and electronic equipment: a review. J Hazard Mater 99:243–263

Díaz S et al (2018) Recycling of polymeric fraction of cable waste by rotational moulding. Waste Manage [S.I.] 76:199–206. Disponível em https://www.sciencedirect.com/science/article/pii/S0956053X18301624. Acesso em 17 out 2021

Dogu O, Pelucchi M, van de Vijver R, Van Steenberge PHM, D'hooge DR, Cuoci A, … Van Geem KM (2021) The chemistry of chemical recycling of solid plastic waste via pyrolysis and gasification: state-of-the-art, challenges, and future directions. Progress Energy Combust Sci 84:100901. https://doi.org/10.1016/j.pecs.2020.100901

dos Santos CS, Alves JC, Silva SP, Sita LE, Silva PRC, Açmeida LC, Scarminio J (2019) A closed-loop process to recover Li and Co compounds and to resynthesize $LiCoO_2$ from spent mobile phone batteries. J Hazard Mater 362:458–466. https://doi.org/10.1016/j.jhazmat.2018.09.039

Dodbiba G, Nagai H, Wang LP, Okaya K, Fujita T (2012) Leaching of indium from obsolete liquid crystal displays: comparing grinding with electrical disintegration in context of LCA. Waste Manag 32(2012):1937–1944 https://doi.org/10.1016/j.wasman.2012.05.016

European Union (2012) Directive 2012/19/UE of the European Parliament and of the Council. Of 4 July 2012 on waste electrical and electronic equipment (WEEE). Official J L 197:59. Disponível em http://eur-lex.europa.eu/legal-content/PT/TXT/PDF/?uri=OJ:L:2012:197:FULL&from=PT. Acesso em 23 jul. 2019. https://doi.org/10.2478/rput-2021-0001

EURIC (2020) Metal recycling factsheet. Disponivel em https://circulareconomy.europa.eu/platform/en/knowledge/metal-recycling-factsheet-euric. Acesso em 17 out 2021
European Environment Agency—EEA (2021) Europe's consumption in a circular economy: the benefits of longer-lasting electronics. Disponível em https://www.eea.europa.eu/publications/europe2019s-consumption-in-a-circular. Acesso em 17 out 2021
Fernández L (2021) Consumption volume of polyvinyl chloride worldwide from 2016 to 2022 [Online]. Disponível em https://www.statista.com/statistics/887934/polyvinyl-chloride-consumption-volume-worldwide/#statisticContainer. Acesso em 17 out 2021
Flerus B, Friedrich B (2010) Recovery of gallium from smartphones—Part II oxidative alkaline pressure leaching of gallium from pyrolysis residue. Metals 10. https://doi.org/10.3390/met10121565
Flerus B, Billmann L, Bokelmann K, Stauber R, Friedrich B (2019) Recovery of gallium from smartphones—Part I: thermal and mechanical pretreatment. In: 10th European metallurgical conference, EMC, Düsseldorf, Germany. Disponível em https://publications.rwth-aachen.de/record/780775. Acesso em 17 out 2021
Fontana D, Forte F, Carolis R, Grosso M (2015) Materials recovery from waste liquid crystal displays: a focus on indium. Waste Manage 45:325–333. https://doi.org/10.1016/j.wasman.2015.07.043
Forti V, Baldé CP, Kuehr R, Bel G (2020) The global e-waste monitor 2020—quantities, flows, and the circular economy potential. United Nations University (UNU)/United Nations Institute for Training and Research (UNITAR)—co-hosted SCYCLE Programme, International Telecommunication Union (ITU) & International Solid Waste Association (ISWA), Bonn/Geneva/Rotterdam, 2020. Disponível em http://ewastemonitor.info/. Acesso em 15 ago 2021
Fraunhofer IZM (2012) Project: cycling resources embedded in systems containing light emitting diodes. Fraunhofer Insttitute, Germany. Disponível em http://www.cyc-led.eu/?LMCL=NcYj0n. Acesso em 15 set 2019
Freegard K, Morton R, Lund I, Huisman J, Studds P, Freer E (eds) (2006) Develop a process to separate brominated flame retardants from WEEE polymers, Interim Report 2. The Waste and Resources Action Programme (WRAP), Banbury, UK
Garside M (2021) Global refined copper consumption distribution by region 2019 [Online]. Disponível em https://www.statista.com/statistics/693466/distribution-ofglobal-refined-copper-consumption-by-region/. Acesso em 17 out 2021
Gassmann A, Zimmermann J, Gauß R, Stauber R, Gutfleisch O (2016) Led lamps recycling technology for a circular economy. Latest LPR Magazine. n. 25 de ago. Disponível em https://www.led-professional.com/resources-1/articles/led-lamps-recycling-technology-for-a-circular-economy. Acesso em 15 jul 2019
Gartner (2021) Gartner says worldwide PC shipments grew 10.7% in fourth quarter of 2020 and 4.8% for the year. Stamford, 2021. https://www.gartner.com/en/newsroom/press-releases/2021-01-11-gartner-says-worldwide-pc-shipments-grew-10-point-7-percent-in-the-fourth-quarter-of-2020-and-4-point-8-percent-for-the-year
Gerbase AE, Oliveira CR (2012) Reciclagem do Lixo De Informática: Uma Oportunidade para a Química. Quim Nova 35(7):1486–1492
Giese EC, Xavier LH, Lins FAF (2018) Biomineração Urbana: O futuro da reciclagem de resíduos eletroeletrônicos. Brasil Mineral 385:36–39
Gois A (2008) LEDs na Iluminação Arquitetural. Ed. 1. Lighting Now
Golzar-Ahmadi M, Mousavi SM (2021) Extraction of valuable metals from discarded AMOLED displays in smartphones using Bacillus foraminis as an alkali-tolerant strain. Waste Manage 131:226–236. https://doi.org/10.1016/j.wasman.2021.06.006
Gosavi RA, Knudsen GA, Birnbaum LS, Pedersen LC (2013) Mimicking of estradiol binding by flame retardants and their metabolites: a crystallographic analysis. Environ Health Perspect 121(10):1194–1199. https://doi.org/10.1289/ehp.1306902

Gouveia N, Ferron MM, Kuno R (2014) Os impactos dos resíduos de equipamentos eletroeletrônicos na saúde. In: Carvalho TCMB, Xavier LH (Org) Gestão de Resíduos eletroeletrônicos – uma abordagem prática para a sustentabilidade, 1 edn. Elsevier, Rio de Janeiro, pp 129–148

GSM Association (GSMA) (2015) eWaste in Latin America: statistical analysis and policy recommendations, November 21

Gu F, Summers PA, Hall P (2019) Recovering materials from waste mobile phones: recent technological developments. J Cleaner Prod 235. https://doi.org/10.1016/j.jclepro.2019.117657

Gustavo Reisdörfer D, Bertuol E, Tanabe E (2019) Recovery of neodymium from the magnets of hard disk drives using organic acids. Miner Eng 143105938-S0892687519303498 105938. https://doi.org/10.1016/j.mineng.2019.105938

Gupta CK, Krishnamurthy N (2005) Extractive metallurgy of rare earths. CRC Press, USA

Gupta AK, Kumar A, Singh VP (2021) Study on important techniques and processes for the management of waste electrical wires. In: Phanden RK, Mathiyazhagan K, Kumar R, Davim JP (eds) Advances in industrial and production engineering—select proceedings of FLAME 2020, 1 edn. Springer Nature Singapore Pte Ltd., pp 147–158

Filippas A, Sempros G, Sarafidis C (2021) Critical rare earths: the future of Nd & Dy and prospects of end-of-life product recycling. Materials today: Proceed 374058-4063 S2214785320369078 https://doi.org/10.1016/j.matpr.2020.09.210

Ha VH, Lee J, Huynh TH, Jeong J, Pandey BD (2014) Optimizing the thiosulfate leaching of gold from printed circuit boards of discarded mobile phone. Hydrometallurgy 149:118–126. https://doi.org/10.1016/j.hydromet.2014.07.007

Ha VH, Lee J, Jeong J, Hai HT, Jha MK (2010) Thiosulfate leaching of gold from waste mobile phones. J Hazard Mater 178:1115–1119. https://doi.org/10.1016/j.jhazmat.2010.01.099

Hadi P, Xu M, Lin CS, Hui CW, McKay G (2015) Waste printed circuit board recycling techniques and product utilization. J Hazard Mater 283234–243 S0304389414007705 https://doi.org/10.1016/j.jhazmat.2014.09.032

Han Y, Tang Z, Sun J, Xing X, Zhang M, Cheng J (2019) Heavy metals in soil contaminated through e-waste processing activities in a recycling area: implications for risk management. Process Saf Environ Prot

Hashimoto H (2015) Estudo da Extração de Índio a partir de Telas de Cristal Líquido (LCD). 2015. 106f. Dissertação (Mestrado em Ciências) - Universidade de São Paulo, São Paulo

Hendrickson CT, Matthews DH, Ashe M, Jaramillo P, Mcmichael FC (2010) Reducing environmental burdens of solid-state lighting through end-of-life design. Environ Res Lett 5. https://doi.org/10.1088/1748-9326/5/1/014016

Heydarian A, Mousavi SM, Vakilchap F, Baniasadi M (2018) Application of a mixed culture of adapted acidophilic bacteria in two-step bioleaching of spent lithium-ion laptop batteries. J Power Sources 378:19–30. https://doi.org/10.1016/j.jpowsour.2017.12.009

Hira M, Yadav S, Morthekai P, Linda A, Kumar S, Sharma A (2018) Mobile phones—an asset or a liability: a study based on characterization and assessment of metals in waste mobile phone components using leaching tests. J Hazard Mater 342:29–40. https://doi.org/10.1016/j.jhazmat.2017.08.008

Hlosta J, Zurovec D, Kratochvil M, Botula J, Zegzulka J (2017) WEEE sorting processes and separation of copper wires with support of DEM modeling. J Polish Miner Eng Soc 18(1):15–164

Holgersson S, Steenari B, Björkman M, Cullbrand K (2018) Analysis of the metal content of small-size Waste Electric and Electronic Equipment (WEEE) printed circuit boards—part 1: internet routers, mobile phones and smartphones. Resour Conserv Recycl 133:300–308. https://doi.org/10.1016/j.resconrec.2017.02.011

Horeh NB, Mousavi SM, Shojaosadati SA (2016) Bioleaching of valuable metals from spent lithium-ion mobile phone batteries using Aspergillus niger. J Power Sources 320:257–266. https://doi.org/10.1016/j.jpowsour.2016.04.104

International Copper Study Group—ICSG (2014) The World Copper Factbook 2014. Disponivel em https://copperalliance.org/wp-content/uploads/2012/01/ICSG-Factbook-2014.pdf. Acesso em 17 out 2021

IPMI (2003) Internacional Precious Metals Institute. Environmentally sound management used mobile phones. Disponível em http://ipmi.org/pdf/IPMI_Guidance_Used_Mobile_Phones.pdf Acesso em 09 set de 2017

Işildar A, Hullebusch EDV, Lenz M, Laing GD, Marra A, Cesaro A, Panda S, Akcil A, Kucuker MA, Kuchta K (2019) Biotechnological strategies for the recovery of valuable and critical raw materials from waste electrical and electronic equipment (WEEE)—a review J Hazard Mater 362:467–481

Jang Y, Kim M (2010) Management of used & end-of-life mobile phones in Korea: a review. Resour Conserv Recycl 55:11–19. https://doi.org/10.1016/j.resconrec.2010.07.003

Jimenez GL, Moraes VT, Soares JA, Espinosa DCR (2016) Extração de Índio de Telas de LCD 71(71):627–634. In: 71th ABM Annual Congress, Rio de Janeiro. Disponível em https://doi.org/10.5151/1516-392X-27731. Acesso em 17 out 2021

Kaercher L, Moraes CAM (2021) Separação do fio metálico do recobrimento de plástico em fios elétricos e de equipamentos eletroeletrônicos. Braz J Dev 7(3):30723–30735. https://doi.org/10.34117/bjdv7n3-682

Goodenough KM, Wall F, Merriman D (2018) The rare earth elements: demand global resources and challenges for resourcing future generations. Nat Resour Res 27(2):201–215. https://doi.org/10.1007/s11053-017-9336-5

Kasper AC (2011) Caracterização e Reciclagem de Materiais Presentes em Sucatas de Telefones Celulares. Dissertação (mestrado). Porto Alegre—Universidade Federal do Rio Grande do Sul

Kasper AC, Veit HM (2018) Gold recovery from printed circuit boards of mobile phones scraps using a leaching solution alternative to cyanide. Braz J Chem Eng 35(03):931–942. https://doi.org/10.1590/0104-6632.20180353s20170291

Kasper AC, Berselli GBT, Freitas BD, Bernardes AM, Veit HM (2010) Estudo da reciclagem das Blendas PC+ABS e PC+ABS+EPÓXI provenientes de sucatas eletrônicas. Braz J Environ Sci (Online) 16:26–35, 2010. Disponível em http://rbciamb.com.br/index.php/Publicacoes_RBCIAMB/article/view/386. Acesso em 17 out 2021

Kasper AC, Berselli GBT, Freitas BD, Tenório JAS, Bernardes AM, Veit HM (2011) Printed wiring boards for mobile phones: characterization and recycling of copper. Waste Manage 31:2536–2545. https://doi.org/10.1016/j.wasman.2011.08.013

Kasper AC, Veit HM, García-Gabaldón M, Herranz VP (2018) Electrochemical study of gold recovery from ammoniacal thiosulfate, simulating the PCBs leaching of mobile phones. Electrochim Acta 259:500–509. https://doi.org/10.1016/j.electacta.2017.10.161

Knoth R, Brandstotter M, Kopacek B, Kopacek P (2002) Automated disassembly of electr(on)ic equipment. In: Conference record 2002 IEEE international symposium on electronics and the environment, San Francisco, CA, USA. https://doi.org/10.1109/ISEE.2002.1003282

Konyalioğlu AK, Bereketli I (2021) Rare earth elements' waste management for smartphones' touch screens by system dynamics modelling. Res Square. https://doi.org/10.21203/rs.3.rs-458455/v1

Kumar S, Panda AK, Singh RK (2011) A review on tertiary recycling of high-density polyethylene to fuel. Resour Conserv Recycl 55:893–910. https://doi.org/10.1016/j.resconrec.2011.05.005

Kumar H, Kumagai S, Kameda T, Saito Y, Yoshioka T (2021) One-pot wet ball-milling for waste wire-harness recycling. J Mater Cycles Waste Manage 23:461–469. https://doi.org/10.1007/s10163-020-01163-7

Kumari A et al (2018) Recovery of rare earths from spent NdFeB magnets of wind turbine: leaching and kinetic aspects. Waste Manage

Kunrath JL (2015) Resíduos Eletroeletrônicos: Um diagnóstico da Cadeia de Processamento. 2015. 125f. Dissertação (Mestrado em Engenharia de Minas, Metalurgia e Materiais)—Universidade Federal do Rio Grande do Sul, Porto alegre

Lambert F, Gaydardzhiev S, Léonard G, Lewis G, Bareel P, Bastin D (2015) Copper leaching from waste electric cables by biohydrometallurgy. Miner Eng 76:38–46. https://doi.org/10.1016/j.mineng.2014.12.029

Li J, Jiang Y, Xu Z (2017a) Eddy current separation technology for recycling printed circuit boards from crushed cell phones. J Clean Prod 141:1316–1323. https://doi.org/10.1016/j.jclepro.2016.09.144

Li L, Liu G, Pan D, Wang W, Wu Y, Zuo T (2017b) Overview of the recycling technology for copper-containing cables. Resour Conserv Recycl 126:132–140

Li X et al (2010) Smartphone evolution and reuse: establishing a more sustainable model. In: 39th international conference on parallel processing workshops

Lim S, Kang D, Ogunseitan OA, Schoenung JM (2013) Potential environmental impacts from the metals in incandescent, compact fluorescent lamp (CFL), and light-emitting diode (LED) bulbs. Environ Sci Technol 47:1040–1047. https://doi.org/10.1021/es302886m

Lu Y, Xu Z (2016) Precious metals recovery from waste printed circuit boards: a review for current status and perspective. Resour Conserv Recycl 113:28–39

Maneesuwannarat S, Teamkao P, Vangnai AS, Yamashita M, Thiavetyan P (2016a) Possible mechanism of gallium bioleaching fromgallium nitride (GAN) by Arthrobacter creatinolyticus: role of aminoacids/peptides/proteins bindings with gallium. Process Saf Environ Prot I03:35–45. https://doi.org/10.1016/j.psep.2016.06.036

Maneesuwannarat S, Vangnai AS, Yamashita M, Thiavetyan P (2016b) Bioleaching of gallium from gallium arsenide by Cellulosimicrobium funkei and its application to semiconductor/electronic wastes. Process Saf Environ Prot 99:80–87. https://doi.org/10.1016/j.psep.2015.10.008

Maneesuwannarat S, Kudpeng K, Yingchutrakul Y, Roytrakul S, Vangnai AS, Yamashita M, Thiravetyan P (2019) A possible protein model involved in gallium arsenide leaching by Cellulosimicrobium funkei. Miner Eng 137:207–216. https://doi.org/10.1016/j.mineng.2019.04.002

Matozinhos R (2012) Fios e cabos elétricos para edificações. Minas Gerais. Disponível em https://sindusconpr.com.br/principal/pub/Image/20130206090223Qualimat-Cabos-e-Fios-2012-Internet-Reduzido.pdf. Acesso em 08 nov. 2018

Mejame PPM, Kim YM, Lee DS, Lim SR (2016) Effect of technology development on potential environmental impacts from heavy metals in waste smartphones. J Mater Cycles Waste Manag 20:100–109. https://doi.org/10.1007/s10163-016-0548-2

Moksin V, Striska V, Tetsman I (2011) Method of removal of the plastic insulator from waste cables by passing them between two rotating rolls that have different surface temperatures. In: The 8th international conference "environmental engineering": selected papers, vol 8, pp 228–232. May 19–20, 2011, Vilnius, Lithuania

Monteiro MR, Moreira DGG, Chinelatto MA, Nascente PAP, Alcântara NG (2007) Characterization and recycling of polymeric components present in cell phones. J Polym Environ 15:195–199. https://doi.org/10.1007/s10924-007-0060-9

Moraes VT, Espinosa DCR, Lucena LL (2014) Tecnologias de tratamento para resíduos de equipamentos eletroeletrônicos. In: Carvalho TCMB, Xavier LH (Org) Gestão de Resíduos eletroeletrônicos – uma abordagem prática para a sustentabilidade, 1 edn. Elsevier, Rio de Janeiro, pp 113–128

Nagy S, Bokányi L, Gomnkötö I, Magyar T (2017) Recycling of gallium from end-of-life light emitting diodes. Arch Metall Mater 62:1161–1166. https://doi.org/10.1515/amm-2017-0170; ISSN 1733-3490

Nancharaiah YV, Mohan SV, Lens PNL (2016) Biological and bioelectrochemical recovery of critical and scarce metals. Trends Biotechnol 34(2):137–155

Nicolai FNP (2016) Mineração urbana: avaliação da economicidade da recuperação de componentes ricos em Au a partir de resíduo eletrônico (e-waste). 2016. 242f. Tese (doutorado em em Engenharia de Materiais) – Universidade Federal de Ouro Preto, Belo Horizonte

Nurul A, Roslan FBM, Suah Norita M (2017) The use of an electrogenerative process as a greener method for recovery of gold (III) from the E-waste. Seper Purifi Techno 1821–8 S1383586616321487 https://doi.org/10.1016/j.seppur.2017.03.032

O'dea S (2021a) Smartphones—Estatísticas e fatos. In: Statista. Disponível em https://www.statista.com/topics/840/smartphones/#dossierSummary__chapter1. Acesso em 17 out 2021

O'dea S (2021b) Número de smartphones vendidos para usuários finais em todo o mundo de 2007 a 2021 (em milhões de unidades). In: Statista. Disponivel em https://www.statista.com/statistics/263437/global-smartphone-sales-to-end-users-since-2007/#statisticContainer. Acesso em 17 out 2021

Oliveira PCF (2012) Valorização de Placas de Circuito Impresso por Hidrometalúrgia. Lisboa 320:2012. Tese (doutorado)—Universidade Técnica de Lisboa

Oliveira RP, Benvenuti J, Espinosa DCR (2021) A review of the current progress in recycling technologies for gallium and rare earth elements from light-emitting diodes. Renew Sustain Energy Rev 145. https://doi.org/10.1016/j.rser.2021.111090

Panda R, Jadhao PR, Pant KK, Naik SN, Bhaskar T (2020) Eco-friendly recovery of metals from waste mobile printed circuit boards using low temperature roasting. J Hazard Mater 395. https://doi.org/10.1016/j.jhazmat.2020.122642

Peiró LT, Méndez GV, Ayres RU (2011) Rare and critical metals as by-products and the implications for future supply. INSEAD, Social Innovation Centre. Disponível em https://sites.insead.edu/facultyresearch/research/doc.cfm?did=48916. Acesso em 10 jul 2019

Pieroni MC, Leonel J, Fillmann G (2017) Retardantes de Chama Bromados: Uma Revisão. Quim. Nova 40(3):317–326. https://doi.org/10.21577/0100-4042.20160176

Pita F, Castilho A (2018) Separação de cobre de resíduos de cabos elétricos com base em métodos de processamento de minerais: um estudo de caso. Minerals 8(11)

Pode R (2020) Organic light emitting diode devices: An energy efficient solid state lighting for applications. Renew Sustain Energy Rev 133(2020):110043. https://doi.org/10.1016/j.rser.2020.110043

Pourhossein F, Mousavi SM (2018) Enhancement of copper, nickel, and gallium recovery from LED waste by adaptation of Acidithiobacillus ferrooxidans. Waste Manage 79:98–108. https://doi.org/10.1016/j.wasman.2018.07.010

Pourhossein F, Mousavi SM (2019) A novel step-wise indirect bioleaching using biogenic ferric agent for enhancement recovery of valuable metals from waste light emitting diode (WLED). J Hazard Mater 378. https://doi.org/10.1016/j.jhazmat.2019.05.041

Pourhossein F, Rezaei O, Mousavi SM, Beolchini F (2021) Biolixiviação de metaisessenciais de telas detoque OLED residuaisusando bactériasacidofflicas adaptadas. J Environ Health Sci Eng 19:893–906. https://doi.org/10.1007/s40201-021-00657-2

Pradiian JK (2013) Environmental impact assessment and bioleaching of metals from electronic waste (E-waste). Tese (Doutorado em Filosofia e Biotecnologia). Jaypee University of Information Technology, Waknaghat, Índia

Priya A, Hait S (2018) Comprehensive characterization of printed circuit boards of various end-of-life electrical and electronic equipment for beneficiation investigation. Waste Manage 75:103–123. https://doi.org/10.1016/j.wasman.2018.02.014

PVC4Cables (2017) PVC wires and cables: bringing essential benefits to society [Online]. Acesso em https://www.pvc4cables.org/en/media-en/news/item/115-pvc-wires-and-cables-bringing-essential-benefits-to-society. Acesso em 17 out 2021

Quinet P, Proost J, Lierde AV (2005) Recovery of precious metals from electronic scrap by hydrometallurgical processing routes. Min Metall Explor 22:17–22. https://doi.org/10.1007/BF03403191

Rahimi A, García JM (2017) Chemical recycling of waste plastics for new materials production. Nat Rev Chem 1(6):0046. https://doi.org/10.1038/s41570-017-0046

Rahman SMM, Kim J, Lerondel G, Bouzidi Y, Nomenyo K, Clerget L (2017) Missing research focus in end-of-life management of light-emitting diode (LED) lamps. Resour Conserv Recycl 127:256–258. https://doi.org/10.1016/j.resconrec.2017.04.013

Rebello RZ, Lima MTWDC, Yamane LH, Siman RR (2018) Composição Gravimétrica de Lâmpadas de LED Inservíveis. In: 18.º ENASB/18.º SILUBESA, Porto

Rebello RZ, Lima MTWDC, Yamane LH, Siman RR (2020) Characterization of end-of-life LED lamps for the recovery of precious metals and rare earth elements. Resour Conserv Recycl 153

Rene ER, Sethurajan M, Ponnusamy VK, Kumar G, Dung TNB, Brindhadevi K, Pugazhendhi A (2021) Electronic waste generation recycling and resource recovery: Technological perspectives and trends. J Hazard Mater 416125664-S0304389421006282 125664 https://doi.org/10.1016/j.jhazmat.2021.125664

München DD, Bernardes AM, Veit HM (2018) Evaluation of Neodymium and Praseodymium Leaching Efficiency from Post-consumer NdFeB Magnets. J Sustain Metal 4(2) 288–294 https://doi.org/10.1007/s40831-018-0180-6

Nurbaidah F, Soetrisno JM, Delgado S (2020) Chronic exposure to heavy metals from informal e-waste recycling plants and children's attention executive function and academic performance. Sci Total Environ 717137099-S0048969720306094 137099 https://doi.org/10.1016/j.scitotenv.2020.137099

Sanchez Junior O (2018) Avaliação do Potencial de Mitigação de Impactos Ambientais Associados à Reciclagem de LEDs. In: International Workshop Advances in Cleaner Production – Academic Work, 7, 2018, Barranquilla, Colombia. Anais eletrônicos... Universidad de La Costa, Barranquilla, Colombia. Disponível em http://www.advancesincleanerproduction.net/7th/files/sessoes/6B/3/sanchez_jr_o_academic.pdf. Acesso em 20 out. 2019

Santos CTS (2016) Obtenção de blendas poliméricas a partir da reciclagem de sucatas de aparelhos celulares e polietileno de baixa densidade reciclado. 2016. 115f. Dissertação (Mestrado em Engenharia Química)—Universidade Federal Rural do Rio De Janeiro, Rio de Janeiro

Santos ECA (2021) Avaliação Técnica, Econômica e Ambiental Para Recuperação de Materiais a Partir de Lâmpadas LED Pós-Consumo. 2021. 177f. Dissertação (Mestrado em Engenharia Civil)—Universidade do Vale do Rio dos Sinos, São Leopoldo—RS—Brsil

Santos ECA, Camacho ALD, Rauber LD, Moraes CAM (2019) Desmontagem e caracterização de Lâmpadas LED para Recuperação de Materiais. In: Tullio L (Org) Gestão de Resíduos Sólidos 2. Atena Editora, Ponta Grossa (PR). E-book. Disponível em https://www.atenaeditora.com.br/wp-content/uploads/2019b/03/E-book-Gest%C3%A3o-de-Res%C3%ADduos-S%C3%B3lidos-2-2.pdf. Acesso em 17 out. 2021

Santos ECA, Colling AV, Schaab A, Moraes CAM (2021a) Avaliação da Influência do Design na Desmontagem de Lâmpadas LED do Tipo Bulbo Para Posterior Reciclagem. In: Puppim R (Org) Design: Contribuições significativas para o desenvolvimento sustentável. Atena Editora, Ponta Grossa (PR). E-book. Disponível em https://doi.org/10.22533/at.ed.9282114104. Acesso em 15 out 2021a

Santos ECA, Silva JLC, Calheiro D, Moraes CAM (2021b) Beneficiamento de Lâmpadas LED Pós-Consumo Em Uma Cooperativa de Reciclagem de Resíduos Eletroeletrônicos. Mix Sustentável, Florianópolis—SC [No prelo]

Sarath P, Bonda S, Mohanty S, Nayak SK (2015) Mobile phone waste management and recycling: view and trends. Waste Manage 46:536–545. https://doi.org/10.1016/j.wasman.2015.09.013

Silveira TA (2019) Caracterização e Valorização de Placas de Circuito Impresso de Aparelhos Smartphones. 2019.128f. Dissertação (Mestrado em Engenharia Civil)—Universidade do Vale do Rio dos Sinos, São Leopoldo—RS—Brasil

Silveira AVM, Fuchs MS, Meili L, Bertuol DA (2012) Caracterização e Processamento de Telas de Lcd De Celulares Visando A Reciclagem. Rev. Elet. em Gestão, Educação e Tecnologia Ambiental 8(8):1785–1793. https://doi.org/10.5902/223611707314

Silveira AVM, Fuchs MS, Pinheiro DK, Tanabe EH, Bertuol DA (2015) Recovery of indium from LCD screens of discarded cell phones. Waste Manage 45:334–342. https://doi.org/10.1016/j.wasman.2015.04.007

Silveira TA, Santos ECA, Moraes CAM (2019) O Ecodesign e a Geração de Resíduos: Uma Abordagem Sobre os Eletroeletrônicos. In: Aguilera JG, Zuffo AM (Org) A Preservação do Meio Ambiente e o Desenvolvimento Sustentável 2. Atena Editora, Ponta Grossa, PR. E-book. Disponível em https://www.atenaeditora.com.br/wp-content/uploads/2019/08/Ebook-A-Preservacao-do-Meio-Ambiente-e-o-Desenvolvimento-Sustentavel-2-1.pdf. Acesso em 15 Nov 2019

Silveira TA, Dorneles KEO, Brehm FA, Moraes CAM (2020) Physical processing and analysis of materials content of small-size waste electric and electronic equipment (WEEE): a case of smartphones printed circuit boards. In: 11° Fórum Internacional de Resíduos Sólidos, 2020, Porto Alegre—RS—Brasil. Anais do 11° Fórum Internacional de Resíduos Sólidos, vol 11, pp 1–11. Instituto Venturi, Porto Alegre. Disponível em https://institutoventuri.org/ojs/index.php/FIRS/article/view/110. Acesso em 17 out 2021

Singh N, Hui D, Singh R, Ahuja I, Feo L, Fraternali F (2016) Recycling of plastic solid waste: a state of art review and future application. Composites B 115:409–422. https://doi.org/10.1016/j.compositesb.2016.09.013

Stuhlpfarrer P, Luidold S, Antrekowitsch H (2016) Recycling of waste printed circuit boards with simultaneous enrichment of special metals by using alkaline melts: a green and strategically advantageous solution. J Hazard Mater 307:17–25. https://doi.org/10.1016/j.jhazmat.2015.12.007

Suresh SS, Bonda S, Mohanty S, Nayak SK (2018) A review on computer waste with its special insight to toxic elements, segregation and recycling techniques. Proc Safe Environ Prot 116:477–493. https://doi.org/10.1016/j.psep.2018.03.003

Tabelin CB, Park I, Phengsaart T, Jeon S, Villacorte-Tabelin M, Alonzo D, Alonzo, Yoo K, Ito M, Hiroyoshi N (2021) Copper and critical metals production from porphyry ores and E-wastes: a review of resource availability processing/recycling challenges socio-environmental aspects and sustainability issues. Resour Conser Recyc 170105610-S0921344921002196 105610. https://doi.org/10.1016/j.resconrec.2021.105610

Suresh SS, Monhanty S, Nayak SK (2017) Composition analysis and characterization of waste polyvinyl chloride (PVC) recovered from data cables. Waste Manage 60:100–111. https://doi.org/10.1016/j.wasman.2016.08.033

Tiseo I (2021) Distribution of polyvinyl chloride consumption worldwide as of 2017, by application [Online]. Disponivel em https://www.statista.com/statistics/615454/distribution-of-pvc-consumption-worldwide-by-end-use/. Acesso em 17 out 2021

Top Cable (2020a) Parts of an electrical cable [Online]. Disponivel em https://www.topcable.com/blog-electric-cable/en/parts-of-an-electrical-cable/. Acesso em 17 out 2021

Top Cable (2020b) Electrical cable types, sizes, and installation [Online]. Disponivel em https://www.topcable.com/blog-electric-cable/en/electrical-cable-types/. Acesso em 17 out 2021

Tuenge JR, Hollomon B, Dillon HE, Snowden-Swan LJ (2013) Life-cycle assessment of energy and environmental impacts of LED lighting products, Part 3: LED environmental testing. United States. https://doi.org/10.2172/1074312

Umicore Brasil Ltda (2016). Processamento de sucatas eletrônicas. V Seminário Internacional Sobre Resíduos de Equipamentos Eletroeletrônicos – 18 e 19 de agosto de 2016, Recife, Pernambuco, Brasil

UNEP (2009) United Nations Environment Programme. Recycling—from e-waste to resources. Sustainable Innovation and Technology Transfer Industrial Sector Studies, July. Disponível em http://www.unep.fr/shared/publications/pdf/DTIx1192xPA-Recycling%20from%20ewaste%20to%20Resources.pdf. Acesso em 19 jul de 2018

van Eygen E, de Meester S, Tran HP, Dewulf J (2016) Resource savings by urban mining: the case of desktop and laptop computers in Belgium. Resour Conserv Recycl 107:53–64

Veit HM, Bernardes AM, Bertuol DA, Oliveira CT (2008) Utilização de processos mecânicos e eletroquímicos para reciclagem de cobre de sucatas eletrônicas. Metalurgia e Materiais 61(2):159–164

Venditti B (2021) Visualizing the critical metals in a smartphone. Disponível em https://elements.visualcapitalist.com/critical-metals-in-a-smartphone/. Acesso em 17 out 2021

Xavier LH, Carvalho TCMB (2014) Introdução à Gestão de Resíduos de Equipamentos Eletroeletrônicos. In: Carvalho TCMB, Xavier LH (Org) Gestão de Resíduos eletroeletrônicos – uma abordagem prática para a sustentabilidade, 1 edn. Elsevier, Rio de Janeiro, pp 1–18

Xavier LH, Lins FAF, Nascimento HFF, Bellan IO, Ribeiro F, Caldas MB, Silva LOS, Zomer B, Araujo RA, Filho OOD, Reinol PC, Fagundes RL, Gusmão ACF (2017) Manual para a destinação de resíduos eletroeletrônicos: orientação ao cidadão sobre como dispor adequadamente os resíduos eletroeletrônicos na cidade do Rio de Janeiro. 1ª Edição. Cetem, Rio de Janeiro

Xavier LH, Ottoni MSO, Gomes CF, Araujo RA, Bicov N, Nogueira M, Espinosa D, Tenório J (2020a) Guia de desmontagem de resíduos de equipamentos eletroeletrônicos. CETEM, Rio de Janeiro

Xavier LH, Ottoni MSO, Araujo RA, Cugula JS, Contador L, Petrungaro GN, Abreu LPP, Santos LA, Rebello RZ, Gomes CF, Sierpe R, Mançano MR, Cardoso ER, Romay KVM, Suemitsu W, Caldas MB (2020b) Manual de destinação de resíduos eletroeletrônicos. Orientações à sociedade sobre como dispor adequadamente os resíduos eletroeletrônicos no estado do Rio de Janeiro. 3ª Edição. CETEM/MCTI, Rio de Janeiro

Xiao S, Xiong W, Wang L, Ren Q (2015) The treatment technology of recycling scrap wire and cable. In: 4th international conference on sustainable energy and environmental engineering (ICSEEE 2015). https://doi.org/10.2991/icseee-15.2016.82

Xiu F, Qi Y, Zhang F (2015) Leaching of Au, Ag, and Pd from waste printed circuit boards of mobile phone by iodide lixiviant after supercritical water pre-treatment. Waste Manage 41:134–141. https://doi.org/10.1016/j.wasman.2015.02.020

Xu J, Kumagai S, Kameda T, Saito Y, Takahashi K, Hayashi H, Yoshioka T (2019) Separation of copper and polyvinyl chloride from thin waste electric cables: a combined PVC-swelling and centrifugal approach. Waste Manage 89:27–36. https://doi.org/10.1016/j.wasman.2019.03.049

Yamane LH, Moraes VT, Espinosa DCR, Tenório JAS (2011) Recycling of WEEE: characterization of spent printed circuit boards from mobile phones and computers. Waste Manage 31:2553–2558. https://doi.org/10.1016/j.wasman.2011.07.006

Zazycki MA, Tanabe EH, Bertuol DA, Dotto GL (2017) Adsorption of valuable metals from leachates of mobile phone wastes using biopolymers and activated carbon. J Environ Manage 188:18–25. https://doi.org/10.1016/j.jenvman.2016.11.078

Zhan L, Wang Z, Zhang Y, Xu Z (2020) Recycling of metals (Ga, In, As and Ag) from waste light-emitting diodes in sub/supercritical ethanol. Resour Conserv Recycl 155. https://doi.org/10.1016/j.resconrec.2020.104695

Zhan L, Xia F, Ye Q, Xiang X, Xie B (2015) Novel recycle technology for recovering rare metals (Ga, In) from waste light-emitting diodes. J Hazard Mater 299:388–394

Zhang W, Xu C, He W, Li G, Juang J (2018) A review on management of spent lithium ion batteries and strategy for resource recycling of all components from them. Waste Manage Res 36(2):99–112. https://doi.org/10.1177/0734242X17744655

Zhuang WQ, Fitts JP, Ajo-Franklin CM, Maes S, Alvarez-Cohen L, Hennebel T (2015) Recovery of critical metals using biometallurgy. Curr Opin Biotechnol 33:327–335

Zhang L, Xu Z (2016) A review of current progress of recycling technologies for metals from waste electrical and electronic equipment. J Clean Prod 127(2016)1:9–36

Sustainable Bioprospecting of Electronic Waste via Omics-Aided Biometallurgy

Chukwudi O. Onwosi, Victory C. Nnaemeka, Chioma L. Onyishi, Victor C. Igbokwe, and Flora N. Ezugworie

Abstract The tremendous use of electronics and digital gadgets in the modern world, combined with rapid innovations in technology, has resulted in a dramatic increase in the amount of electronic and electrical devices disposed of as waste. Although problematic, electronic waste (e-waste) has been identified as secondary sources of valuable metallic and/or non-metallic materials over the years. Thus, finding innovative and ecologically sustainable methods and technologies to recover these valuable materials, particularly metals, from e-waste have piqued researchers' interest as a means of supplementing the ever-increasing market demand for precious metals and rare earth elements. For leaching metals from their respective ores, biometallurgy is one of the most promising technologies in the metallurgical sector. Bioleaching of metals from e-waste, on the other hand, is still in its early phases. The performance of biometallurgy operations depends greatly on a comprehensive understanding of the microbial community during bioleaching. Omics technologies have opened new and exciting possibilities for discovering and probing complex microbial communities during bioleaching. Using Omics technologies and techniques in bioleaching can enable reliable predictions of how to keep microbial consortia active during the bioleaching process. Various Omics technologies essential to studying bioleaching microorganisms are addressed in this chapter. Omics technologies that have been leveraged towards bioprospecting for novel non-cultivable bioleaching bacteria and elucidate diverse metabolic pathways during bioleaching are also discussed.

C. O. Onwosi (✉) · V. C. Nnaemeka · C. L. Onyishi · F. N. Ezugworie
Department of Microbiology, Faculty of Biological Sciences, University of Nigeria, Nsukka, Enugu State, Nigeria
e-mail: chukwudi.onwosi@unn.edu.ng

C. O. Onwosi · V. C. Igbokwe · F. N. Ezugworie
Bioconversion and Renewable Energy Research Unit, University of Nigeria, Nsukka, Enugu State, Nigeria

V. C. Igbokwe
Bio-Inspired Materials, Department of Materials Science and Engineering, Université de Pau et des Pays de l'Adour, 64012 Pau Cedex, France

M. Jawaid and A. Khan (eds.), *Conversion of Electronic Waste in to Sustainable Products*, Sustainable Materials and Technology,
https://doi.org/10.1007/978-981-19-6541-8_2

1 Introduction

The consistent advent of electronic devices with smart designs, features, and technologies during the last two decades has resulted in the accelerated obsolescence of many electric and electronic equipment (Ding et al. 2019; Jothi et al. 2018). Accordingly, the lifespan of electronic devices has been considerably reduced because of variables such as customer satisfaction, marketing, and software and hardware compatibility constraints (Jothi et al. 2018). Concomitantly, the widespread use of electronics and digital gadgets in the modern world has led to a dramatic increase in the disposal of electronics and electrical devices as wastes (Li et al. 2019; Sovacool 2019). Proper disposal of electronic wastes (e-waste) is a critical problem that constitutes a real worldwide burden due to the heavy metal content and other hazardous components in e-waste. Heavy metals, for example, can potentially leak into landfill soils and eventually into groundwater supplies (Jothi et al. 2018). Hence, e-waste now maintains a spot as one of the most rapid-growing environmental problems in the world, majorly caused by the rudimentary technique of recycling and disposal methods of the harmful compounds they contain (Asante et al. 2019). Besides the heavy metal content, e-waste is also one of the most significant sources of persistent organic pollutants in the environment, and if not carefully managed, they can lead to serious ecosystem deterioration in the medium to long term. As a result, proper collection of e-waste, recovering precious metals and other valuable components from e-waste, and recycling the precious metals and valuable components to develop new products is an eco-friendly approach to e-waste management (Jothi et al. 2018).

Excessive metals waste is no longer acceptable in a time when nonrenewable raw material scarcity is imminent (Bindschedler et al. 2017). Considering most e-waste contains copious amounts of precious metals such as palladium, platinum, gold, tantalum, and selenium, recovering these precious metals is an ambitious e-waste management approach (Ding et al. 2019). As a result, easily accessing these precious metals and other valuable components in e-waste is critical to their use as a means of economic sustainability in the green manufacture of electrical and electronic equipment (Debnath et al. 2018). The quantities of valuable metals contained in e-waste can be argued to exceed those found in mineral ores (Li et al. 2019). However, these valuable metals are typically present alongside heavy metals (lead, mercury, cadmium, etc.) and brominated flame retardants in e-waste, which are all potent hazards. Thus, it is critical to adopt an effective management solution to combat the e-waste menace to avoid threats to human and environmental health (Onwosi et al. 2020). Mining metals from e-waste is more cost-effective than extracting them from their natural sources (Li et al. 2019). Meanwhile, mining raw materials from the earth's crust is usually associated with both environmental and socioeconomic concerns. As a result of the environmental and social impact of ore mining, numerous methods for efficiently recycling and/or appreciating the various metal components in e-waste have been developed (Bindschedler et al. 2017). Extracting metals from e-waste has added environmental benefits including mineral supply stabilization and reduced extractive mining (Li et al. 2019). When comparing the carbon footprint of value recovery from various e-waste to primary metal production, the global warming potential of value recovery from various e-waste showed a considerable reduction in carbon footprint (Iannicelli-Zubiani et al. 2017).

Physical, chemical, and biological processes have been developed to extract useful metals from e-waste. Hydrometallurgy (the use of an aqueous solution to remove metals from trash), pyrometallurgy (the use of heat to melt equipment and recover metal), and biohydrometallurgy (bioleaching using adapted microorganisms) are examples of effective metal recovery technologies that have been adopted hitherto (Onwosi et al. 2020). Physical and chemical recyclings are frequently associated with a variety of environmental issues; hence biological techniques are favored (Bindschedler et al. 2017).

Biotechnology is thought to be one of the most promising technologies in the metallurgical processing field. For several years, bioleaching has been employed to recover precious metals and rare earth elements from their ores (Cui and Zhang 2008). Bioleaching is well suited to the processing of tiny deposits of heavy metals. This approach will reduce high capital costs associated with processing these metals in remote deposits where manufacturing and exporting concentrates are not viable options. This is also applicable for low-grade ores that cannot be processed cost-effectively with conventional physicochemical methods, and complicated ores (Brierley 2010). In the domain of biomining, prokaryotes (bacteria and archaea) have customarily been more established and developed on an industrial scale than eukaryotes (algae or fungi). Extracting metals from sulfidic low-grade ores by prokaryotic bioleaching is one of the most prominent microbial techniques in biomining. Recent studies, however, have shown that both prokaryotes and eukaryotes can recover metals from a variety of solid matrices including e-waste (Bindschedler et al. 2017). Due to its cost-effectiveness and environmental friendliness, biohydrometallurgy has been considered a viable method of recovering metal from e-waste. It is also widely accepted to save energy owing to its simplicity of use when compared to alternative recovery methods (Onwosi et al. 2020).

The holistic understanding of the microbial species or consortia present in the environmental medium is critical to the biometallurgy operations' performance. During bioleaching, for example, different microorganisms are isolated and can be identified as the dominant species. However, in the bioleaching process, these dominating species are not always the most essential. This is mainly because various microorganisms involved in the bioleaching process cannot be isolated or studied using traditional selective and enrichment culture methods. The identification and isolation of bioleaching microorganisms can be made more efficient with the understanding of these microbial species and their consortia. Omics technologies have opened new and intriguing opportunities for identifying and investigating complex microbial communities in a variety of media. Besides identifying and studying microbial communities' bioleaching, Omics technologies also allow for the robust prediction of how to keep the microbial consortia healthy and active during the bioleaching process.

2 Global Distribution of E-waste

E-waste, as the name implies, is a waste from electrical and electronic materials that have reached their end-of-life, become damaged, outdated, or unwanted (Ghimire and Ariya 2020). These wastes originate mostly from electrical and electronic equipment (EEE), which have reached their end-of-life. According to the United Nations, around 50 million tons of e-waste are generated each year (Sovacool 2019) and is expected to exceed 52.2 million metric tons by 2021 (Baldé et al. 2017). As reported in 2020 by the Global E-waste Monitor, 53.6 million metric tons of e-waste were generated in 2019, and this amount is projected to surpass 74.7 million metric tons by 2030 (Forti et al. 2020). Surprisingly, the amount of e-waste generated in 2019 surpassed the estimated quantity for 2021 (Forti et al. 2020). These estimates represent a worldwide scenario (Fig. 1). In a recent UNEDP report, the annual volume of e-waste is anticipated to increase by almost 500% by the next decade, particularly in developing countries (Luhar and Luhar 2019). Remarkably, the rate of e-waste generation and the global population growth rate from 2015 to 2019 were calculated to be 2.9% and 0.9%, respectively. This implies that the worldwide e-waste generation is outpacing the global population (Ghimire and Ariya 2020).

The Asian continent generates the highest volume of e-waste, with a total of 24.9 million tons of e-waste in 2019, and China and India have the highest contributions (ITU 2020). America occupies second place in global e-waste generation having generated approximately 13.1 million tons of e-waste in 2019 while Africa and Oceania had lesser e-waste production of 2.9 million tons and 0.7 million tons, respectively (ITU 2020).

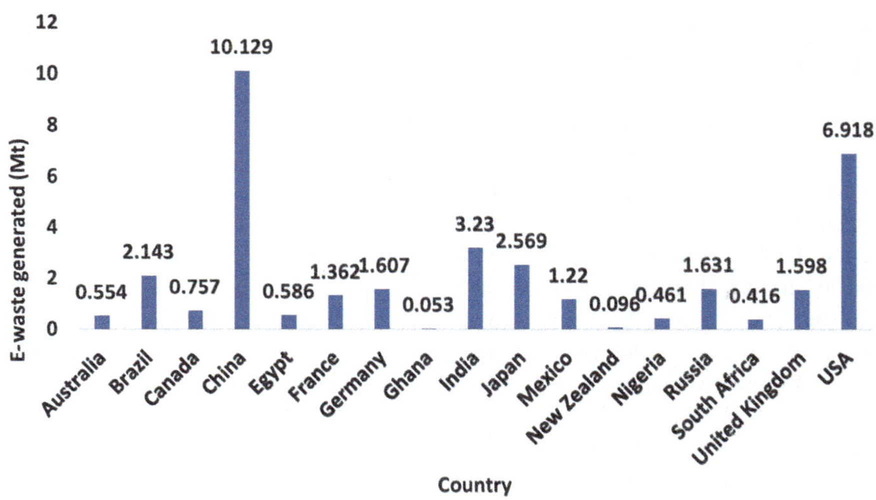

Fig. 1 Global e-waste generated in 2019. The data source is Global E-Waste Monitor (https://www.invest-data.com/eWebEditor/uploadfile/2020071100243938206180.pdf)

Generally, about 20% of the total e-waste generated globally is subjected to recycling processes with the rest 80% being unaccounted for (Daum 2017; Forti et al. 2020). The Global E-waste Monitor reported in 2020 that formally collected/recycled e-waste increased by 17.4% between 2014 and 2019. However, it is estimated that a large chunk of e-waste is in general not accounted for, primarily because they are illegally traded, incinerated, or dumped in landfills (Ghimire and Ariya 2020). Dumping e-waste in landfills raises major environmental concerns on a local and worldwide scale (Jothi et al. 2018). E-waste is typically disposed of in landfills or may even be imported from other countries as in the case of developing nations. Asia generated the most e-waste in 2019, with about 24.9 million metric tons (Mt), followed by Europe (12.0 Mt), the Americas (13.1 Mt), Africa (2.9 Mt), and Oceania (2.9 Mt) (0.7 Mt) (Teye and Tetteh 2021).

E-waste has also been reported to be moved illegally from developed countries to developing countries under the pretext of bridging the global divide. Shipping e-waste to developing countries is fueled by the strict e-waste policies and the higher cost of recycling harmful substances obtainable in developed countries rather than transporting them to developing countries (Ghimire and Ariya 2020). The burden of e-waste in Africa is largely due to the heavy importation of electronic devices, especially fairly used electronic devices, from developed countries and not from the ones produced in the continent (Baldé et al. 2017).

Citizens of a country also contribute to the quantity of e-waste generated in their countries specifically through the disposal of domestic e-waste that is no longer needed (Tran and Salhofer 2018). Similarly, electronic industries contribute to e-waste generation. E-waste generated from such industrial processes is seen as hazardous (Tran and Salhofer 2018). The sources of e-waste have been divided into six categories namely: temperature exchange equipment, screens and monitors, lamps, large equipment, small equipment, and small IT and telecommunication equipment (Forti et al. 2020).

2.1 E-waste Scenario in West Africa

Earlier studies have shown that the most prevalent continents where e-waste is found are Africa and Asia. The study by Diaz-Barriga (2013) reported that out of the 20–50 million tons of e-waste generated annually, more than 75% is dumped in Asia and Africa in the guise of recycling. A large quantity of e-waste is exported from developed countries that no longer have use for them (Odeyingbo et al. 2017). This e-waste then ends up mixed with second-hand electronic gadgets and shipped to low-income countries where these gadgets contribute more to the e-waste as they soon reach their half-life after being imported (Orisakwe et al. 2019).

With the remarkable growth in information and communication technology, Nigeria has recorded an increase in the demand for electronic gadgets, however, like in other low-income countries, citizens are not able to afford brand new gadgets in the markets but rather subscribe to secondhand alternatives (Mokolo 2019). This has afforded importers and broke traders the opportunity of business. In 2010, over 600,000 pieces of used electrical and electronic equipment (EEE) were imported into Nigeria. In addition, about 30% of second-hand imports into the country were deemed non-functional and hence classified as e-waste (Teye and Tetteh 2021). Agbogbloshie area in Accra, Ghana, the Alaba International Market, and Ikeja Computer Village in Lagos, Nigeria are notably the largest dumping sites for electronic wastes in Africa (Asante et al. 2019). With eight countries taking the lead as major exporters of used electronic gadgets to Nigeria, Germany having 20.0% ranks as the largest exporter of e-waste to Nigeria, followed by the United Kingdom (19.5%), Belgium (9.4%), Netherlands (8.2%), Spain (7.35%), China (7.33%), United States of America (7.33%), and Ireland contributing the least with 6.2% (Odeyingbo et al. 2019).

In Ghana, the amount of e-waste generated increased from 63,000 tons per year in 2003 to 169,000 tons in 2008, and then to 215,000 tons in 2009. Only 30% of the total electrical appliances that arrived in Ghana in 2009 were deemed to be new, with the rest being used equipment. About 15% of this used equipment was either malfunctioning or obsolete and hence could not be sold, and thus they ended up in the informal recycling industry (Teye and Tetteh 2021). Another African country with significant e-waste in South Africa produced up to 346 metric kilotons of e-waste in 2014 (Asante et al. 2019). Egypt on its own has been reported as one of the top three African countries that have the highest e-waste production with a whopping 0.37 million tons of e-waste generated in 2016 (Abdelbasir et al. 2018) to 0.585 million tons generated in 2019 (Forti et al. 2020) (Fig. 2).

For a long time, the trans-border flow of e-waste from developed to developing countries has been a major source of concern. In developing countries, there is still a lot of work to be done in terms of e-waste management. However, there are still debates that e-waste generation in African countries and indeed some developing countries do not require critical intervention (Onwosi et al. 2020). This is mainly due to the extended half-life of electronic items in these countries because of financial limitations, both on national and regional levels. In the case of developed countries, however, the opposite is the case (Onwosi et al. 2020).

3 Composition and Nature of E-waste

From the perspective of material composition, e-waste is defined as a combination of various metals, primarily copper, aluminum, and steel, bonded to, covered with, or fused with various types of plastics and ceramics (Cui and Zhang 2008). E-waste is generally heterogeneous and complex (Debnath et al. 2018). Depending on the age of the e-waste, the type, and the categories of EEEs, e-waste contains a wide range of chemicals and materials as shown in Table 1. A mobile phone, for example, may contain more than 40 reusable elements. Typically, e-waste contains about 50% iron-steel, 13% non-ferrous metals, 21% plastics, and about 16% of other components such as glass, rubber, ceramics, concrete, wood, etc. (Pant et al. 2012; Song et al. 2012). Precious metals (Au, Ag, Pd, Pt), base metals (Cu, Al, Ni, Sn, Zn, Fe, etc.),

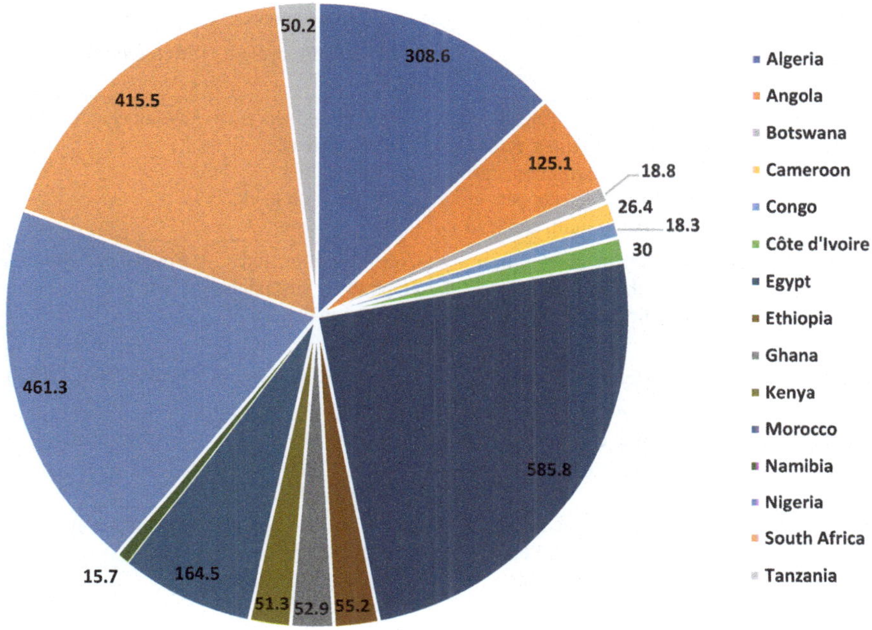

Fig. 2 E-waste generated (in kt) in Africa in 2019. The data source is Global E-Waste Monitor (https://www.invest-data.com/eWebEditor/uploadfile/2020071100243938206180.pdf)

metals of concern (Hg, Be, In, Pb, Cd, As, Sb, etc.), halogens (bromine, fluorine, chlorine), and combustibles are among the substances found in e-waste (plastics, organic fluids, etc.) (Ghimire and Ariya 2020). All these materials are generally classified into hazardous and non-hazardous elements (Omole et al. 2015). Plastic wastes generated by the disassembly of e-waste are one of the world's fastest-growing waste streams (Luhar and Luhar 2019).

A large portion of e-waste is toxic substances that have potential intolerable risks for human, animal, and environmental health (Murugappan and Karthikeyan 2021; Alabi et al. 2012). Metals in e-waste are categorized into ferrous and non-ferrous metals (Priya and Hait 2017). Toxic metals in e-waste include lead, cadmium, mercury, and nickel (Omole et al. 2015). Some monitors and cathode ray televisions contain about 8% lead by weight, they also contain quantifiable amounts of mercury, cadmium, and beryllium (Omole et al. 2015). Copper has been listed as the most common metal used in electrical and electronic equipment with an average weight of 9 g (Hagelüken and Corti 2010). Other components of e-waste are useful and recyclable elements such as precious metals and rare earth metals (Priya and Hait 2017).

Table 1 Value-added products isolated from different e-waste sources

E-waste sources	Composition	Value added/recycled product	References
End of life vehicles	Rubber, plastic, fluids, glass, textiles, tyres, steel, aluminum, cast iron, zinc, copper, magnesium, and lead	Building products such as roofing sheets, insulators, filler materials, piping, etc.	Wong et al. (2018)
Lithium-ion batteries	Cobalt, copper, iron, lithium, manganese, tin, nickel, silicon, graphite	Cobalt, lithium, and graphite for sustainable lithium-ion battery production	Pinegar and Smith (2019)
Fluorescent lamps	Mercury, barium	Mercury extraction for lighting	Kadam et al. (2019)
End of life flat panel displays	Iron, copper, aluminum, lead, nickel, tin, silver	Indium for use in producing liquid crystal displays (LCDs)	Fontana et al. (2020)
Mobile phones	Cd, Fe, Cu, Al, Pb, Ni, Sn, Ag, Au, Pd, glass, plastic materials, batteries, displays	Plastic casing used as supplementary cementing material for cement mortars	Fontana et al. (2019)
Cathode ray tubes (CRT)	Lead, barium, strontium, glass	Radiation shielding glass from glass of CRT Building materials such as fly ash, clay bricks, and roof tiles substitute	Shehada et al. (2018) Shi et al. (2019)
Printed circuit board	Copper, zinc, silver, tin, gold, nickel, lead, aluminum, glass fiber, polymers, ceramics	Copper, zinc, nickel, lead, aluminum	Nekouei et al. (2019)

3.1 Precious Metals

Among the various compositions of e-waste are precious metals. The precious metals are members of the transition elements and include silver, gold, platinum, and other platinum-group metals such as ruthenium, palladium, iridium, rhodium, and osmium (Mokolo 2019; Lucier and Gareau 2019). Gold and silver are generally employed as contacts, for bonding wires, and as switches in the electronics industry, whereas palladium is used in computer hard disk drives (Ding et al. 2019). Precious metals possess specific characteristics that make them important including great economic value, beauty, and properties such as high corrosion resistance, high chemical stability, low electron affinity, electrical conductivity, and high density (Cayumil et al. 2016; Mokolo 2019). Owing to these superior properties of precious metals, they are in high demand in so many industries. However, precious metal reserves are insufficient to sustain world demand. Hence their recovery from e-waste is critical from both an economic and environmental perspective (Ding et al. 2019).

Printed circuit boards (PCBs) are considered the core component of e-waste (Priya and Hait 2017). Some precious metals are found in much higher quantities in e-waste, such as PCBs than they are found in their corresponding ores (Li et al. 2019), thus, making e-waste, like PCBs, important secondary reserves for the extraction of precious metals (Cayumil et al. 2016). Due to a large growth in the number of electronic devices being made, demand for precious metals for use in electrical and electronic equipment has risen significantly (Schluep et al. 2009). Therefore, their recovery from e-waste will have a major impact on the economy (Cayumil et al. 2016; Ding et al. 2019). For example, precious metals in mobile phones are just in very small amounts but their value is up to 80% of the total worth of the entire device (Hagelüken and Corti 2010). Other importance of precious metal recovery from e-waste includes the minimization of wastes being landfilled, reduction in energy that would have been expended in primary production, and decline in gaseous emissions (Cayumil et al. 2016).

3.2 Rare Earth Elements (REE)

Rare earth elements (REE) are another set of valuable elements found within electronic devices (Lucier and Gareau 2019). They include members of the lanthanide family with atomic numbers 57–71, scandium, and yttrium (Qu and Lian 2013). These 17 elements are however divided into two categories: heavy REE and light REE. The heavy REEs are terbium (Tb), gadolinium (Gd), europium (Eu), yttrium (Y), erbium (Er), thulium (Tm), dysprosium (Dy), lutetium (Lu), holmium (Ho) and ytterbium (Yb) while the light REEs include lanthanum (La), neodymium (Nd), cerium (Ce), samarium (Sm), scandium (Sc), promethium (Pm), and praseodymium (Pr) (Fathollahzadeh et al. 2018).

REEs can be found in great quantity worldwide; however, there is a limitation to their extraction as this could lead to global environmental challenges of which radioactive contamination is one of (Lucier and Gareau 2019). Apparently, the usefulness of these rare earth elements, especially yttrium, neodymium, europium, terbium, and dysprosium, in advanced technology has led to their increasing demand (Dev et al. 2020) and investment of funds and researches towards extracting from industrial, mining and electronic wastes (Onal and Binnemans 2019; Wang et al. 2019). Though used in relatively small quantities, their properties have a direct effect on the functioning of most advanced technologies (Gradael et al. 2015). The importance of REEs in communication, health, security, and defense has caused a rise of interest of industrialized countries in finding alternative ways of extracting REEs especially after China, the largest producer of REEs, enforced quotas on the quantities of these elements that are to be exported (U.S. Department of Energy 2010; Jowitt et al. 2018). Cost-effective technologies for the recovery and recycling of secondary rare earth elements have now been developed (Harler 2018).

Some of the technologies in which REEs find application include:

- Electronics with LED lighting
- Electronics with touch screen
- Magnets
- Flat-screen televisions
- Smartphones
- Solar panels
- Satellite systems
- Electric cars
- Biomedical devices.

3.2.1 Magnets

Magnets, which contain around 20–30% REEs, are a significant component of e-waste (Venkatesan et al. 2018). They are found in most end-of-life consumer products such as hard drives, speakers, headphones, and spectrometers (Gergoric et al. 2018). The recoverable REEs in NdFeB permanent magnets are Nd, Pr, and Dy (Kim et al. 2014) with concentrations of 259.5, 3.4, and 42.1 ppm, respectively (Hoogerstraete et al. 2014). Efficient hydrometallurgical procedures for recovering REEs from NdFeB magnets include leaching, solvent extraction, and precipitation (Gergoric et al. 2018). Normally, the elements in the magnet are dissolved in an aqueous solution, followed by a solvent extraction of the leachate and a reprocessing into a new aqueous solution (Gergoric et al. 2018). Auerbach et al. (2019a, b) conducted bioleaching tests on waste magnets using different acidophilic microbial strains and found that *Leptospirillum ferrooxidans* had the highest extraction rate of Nd at 91.3%. The three microbial strains with the best results, *Acidithiobacillus thiooxidans*, *Acidithiobacillus ferrooxidans*, and *L. ferrooxidans*, were able to recover 100% of Dy and Pr (Auerbach et al. 2019a, b).

3.2.2 NiMH Batteries

To boost the battery's hydrogen storage capacity, NiMH batteries include roughly 10% REEs (Jowitt et al. 2018). NiMH batteries have been found to contain 237, 67, and 36 ppm of La, Ce, and Nd, respectively (Maroufi et al. 2018). Numerous hydrometallurgical and pyrometallurgical procedures have been used to extract REEs from NiMH batteries (Kivanc et al. 2018). However, there is still a need to investigate the biotechnological approach. *Aspergillus niger* has been used for the biorecovery of certain metals, such as Ni, Cu, Mn, Al, and Cu from NiMH batteries (Horeh et al. 2016; Bahaloo-Horeh and Mousavi 2017). Though these metals are not REEs, the success of the microbial strain in extracting these metals from NiMH batteries (Horeh et al. 2016; Bahaloo-Horeh and Mousavi 2017) and REEs from monazite sand (Brisson et al. 2016) show their great potential for use in the extraction of REEs from NiMH batteries.

3.2.3 Lithium-Ion Batteries

Lithium-ion batteries possess properties that have made them begin to gain popularity and preference over NiMH batteries (Horeh et al. 2016; Dev et al. 2020). These properties include a longer lifespan, superior safety features, higher energy density, high voltage, low discharge, and compatibility with a varying range of temperatures (Horeh et al. 2016). It has been reported that doping lithium-ion batteries cathodes with La, Ce, Pr, Gd, and Dy improves the electrochemical performance and structural strength of the batteries (Ram et al. 2016). *Aspergillus niger*, *Acidithiobacillus* spp., and *Sulfobacillus* spp. are some of the microorganisms that are promising in the biorecovery of REEs from lithium-ion batteries (Horeh et al. 2016; Baniasadi et al. 2019).

3.2.4 Phosphors

Products such as fluorescent lights, semiconductor light-emitting diodes, LCDs, and cathode ray tubes are all made with phosphor (Liang et al. 2016; Jha et al. 2016). The rapid growth of the electronic information business has made the production and use of luminous materials more frequent. As a result, waste phosphor powders are becoming increasingly abundant in the environment (Liang et al. 2016). Phosphor-contain high number of REEs (Jowitt et al. 2018) must be recycled to obtain the REEs, thereby reducing the strain on resources and limiting environmental risks. Innocenzi et al. (2018) used solvent extraction process to extract REEs from the fluorescent lamp and reported 2.5 2.7, 3.8, 4.4, 4.9, and 112.1 ppm as the concentrations of Gd, Tb, La, Eu, Ce, and Y, respectively in the fluorescent lamp. Furthermore, different microbial strains, such as *Lactobacillus casei*, *Gluconobacter oxydans*, *Yarrowia lipolytica*, and *Komagataeibacter xylinus*, have been used to bioleach REEs from phosphors (Reed et al. 2016) showing that biorecovery of REEs from electronic wastes containing phosphor is possible using these microorganisms.

4 Biometallurgy

Biometallurgy also referred to as biohydrometallurgy is a term that refers to technologies that employ biological systems, particularly microbes, in recovering metals from ores and waste products ((Bindschedler et al. 2017; Das et al. 2015). It is a unique approach for recovering metals in a manner that is energy efficient and non-threatening to the environment (Priya and Hait 2017). Biometallurgy employs the mechanism of converting metals into a soluble extractable form that can then be recovered from the solution by taking advantage of the bioleaching abilities of certain microorganisms (Priya and Hait 2017). Leaching helps the formation of metal ions into complexes or precipitates, which are then removed from the culture broth for immediate use or subsequent refinement (Ning et al. 2017).

Chemolithoautotrophic bacteria such as *L. ferrooxidans*, *A. ferooxidans*, *A. thiooxidans*, and *Sulfolobus* spp. are among the most used microorganisms in the bioleaching process (Priya and Hait 2017). The use of autotrophs in this process makes it advantageous over conventional techniques like pyrometallurgy as large quantities of CO_2 are not emitted. Other advantages of biometallurgy over other recovery techniques are the ability to generate its own heat for the exothermal reaction, hence, no external heat or energy is required, the ability to perform at atmospheric pressure, and adaptation to temperatures ranging from 20 to 80 °C (Das et al. 2015). Apart from chemolithoautotrophic bacteria, other microorganisms involved in the bioleaching process include fungi (e.g. *Penicillium* spp., *Aspergillus niger*), cyanogenic bacteria (e.g. *Pseudomonas fluorescens*, *Chromobacterium violaceum*), acidophilic and thermophilic bacteria (Debnath et al. 2018; Ghimire and Ariya 2020). The major issues of the biometallurgy process include the relatively slow rate of leaching, the possibility of introducing microbial contaminants into the recovered metal, toxic effects of metals on microbes, the sensitivity of the process, and the selection of microorganisms capable of operating under varying conditions (Pollmann et al. 2018).

Biohydrometallurgy is a reformed hydrometallurgical procedure that employs a variety of microorganisms to improve metal solubility and mobilization from wastes and ores (Ilyas et al. 2013; Priya and Hait 2017) (Fig. 3).

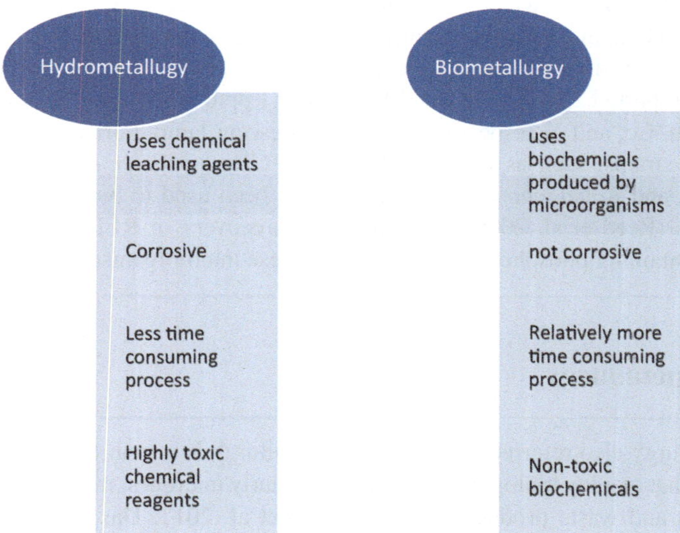

Fig. 3 Differences between biometallurgy and hydrometallurgy

4.1 Approaches Adopted in Biometallurgy

4.1.1 Direct Bioleaching

This mechanism is also known as the one-step process (Rasoulnia et al. 2020). Here, both the microbial culture and the e-waste are added and incubated together in the bioreactor (Qu and Lian 2013). The microbial-e-waste complex is formed, and the metals are recovered (Priya and Hait 2017).

4.1.2 Indirect Bioleaching

Indirect bioleaching is a two-step process (Rasoulnia et al. 2020). First, the microorganisms are cultured without the e-waste. During incubation, biochemical products are released into the media by microorganisms. Then, e-waste is added to form the biochemical-e-waste complex. Metals are recovered from this complex (Priya and Hait 2017).

4.2 The Biometallurgical Process

In biohydrometallurgy, microorganisms solubilize metals from the solid matrix (bioleaching) after which the microbial biomass sorbs the metals from the aqueous solution (biosorption). The mobilization of REEs from solution is the initial stage in recovering REEs from either e-waste or low-grade ores. Acidolysis, redoxolysis, and complexolysis reactions facilitate the mobilization of REEs from the solution. Biosorption and bioaccumulation take place afterward allowing the REE to be eventually recovered from the aqueous phase of the bioleaching media (Marra et al. 2018).

4.2.1 Mobilization

Mobilization is a process of recovery of metals from the solid phase. It involves biochemical processes that include acidolysis, redoxolysis, and complexolysis.

Acidolysis

Acidolysis can also be referred to as proton-induced metal solubilization (Palumbo-Roe et al. 2014). Changes in metal mobility are caused by protons secreted by sulfur-oxidizing and phosphate-solubilizing bacteria (Dev et al. 2020). Sulfide is oxidized to sulfuric acid by sulfur-oxidizing bacteria (Eq. 1) while the phosphate-solubilizing bacteria liberate phosphate from mineral elements and in both processes, REEs are solubilized. *A. thiooxidans*, *A. ferrooxidans*, and *Alicyclobacillus disulfidooxidans* are some sulfur-oxidizing bacteria that can carry out acid-mediated dissolution of REEs from the solid matrix (Baker and Banfield 2003; Plyatsuk et al. 2020). Phosphate solubilizing microorganisms (PSMs) that have been reported to dissolve mineral phosphate for REE solubilization include *Erwinia*, *Rhizobium*, *Pseudomonas*, *Bacillus*, *Enterobacter*, *Micrococcus*, *Acidithiobacillus*, *Acetobacter*, *Streptomyces*, and *Klebsiella*. However, some of these PSMs (including *Enterobacter aerogenes* and *Pseudomonas putida*) produce phosphatase enzyme that also facilitates the solubilization of REEs (Fathollahzadeh et al. 2019; Corbett et al. 2017).

Acidolysis is an indirect leaching process in which oxygen atoms from the organic acids, produced by the microbes, are protonated. By weakening bonds and removing metal ions from the matrix surface, these protons bind and react to the metal-ore or metal-matrix surface. When protonated oxygen reacts with water, it causes the metal oxide to separate from the solid metal surface (Dusengemungu et al. 2021).

$$S^0 + H_2O + 1.5O_2 \xrightarrow{\text{sulphur-oxidizing bacteria}} H_2SO_4 \tag{1}$$

$$E^0 + H_2SO_4 + 0.5O_2 \rightarrow E^{2+} + SO_4^{2-} + H_2O \tag{2}$$

where E is the earth element.

In Eq. 1, a sulfur oxidizing bacteria such as *A. thiooxidans* catalyzes the oxidation of sulfur (S^0) to H_2SO_4, which is then responsible for the solubilization of REEs. During the process, a balance occurs between the H^+ produced by the oxidation of S^0 to H_2SO_4 and the H^+ consumed via the oxidation of metals (Marra et al. 2018).

During acidolysis, organic acids such as citric acid, gluconic acid, acetic acid, oxalic acid, and malic acid are secreted by the participating microorganisms such as PSMs depending on the source of the REE which could be a metal-ore or e-waste. The type of organic carbon utilized for their growth influences the secretion of organic acids by these PSMs for the dissolution of phosphate minerals (Corbett et al. 2017; Carmo et al. 2019). Furthermore, the rate of mineral or solid matrix solubilization is dependent on the microbial strain used, the composition of media, an organic acid produced, and the properties of the REE source (Corbett et al. 2017; Brisson et al. 2016). For instance, when glucose was used as a carbon source, *Acinetobacter* spp. produced gluconic acid but malic acid was produced when mannitol was used as a carbon source instead (Marra et al. 2019). Also, *Pseudomonas aeruginosa* produced pyruvate, citrate, and gluconate when phosphate was part of the medium composition but produced only gluconate in the absence of phosphate in the medium (Buch et al. 2008).

The organic acids produced during acidolysis serve several important functions the first of which is to keep the pH reasonably low during bioleaching. This is crucial for an efficient bioleaching process. Also, through the proton translocating ATPase present on the plasma membrane of the microbial cells, the organic acids supply core sources of protons that reduce the access of anions to the cations in a metal compound reaction. Preventing the association of anions and cations in a metal compound improves the solubility of metal ions in the solution. Furthermore, the hydrogen ions (H^+) produced by organic acids promote metal ion mobilization while also stabilizing metal chelation (Dusengemungu et al. 2021).

Redoxolysis

The redoxolysis reaction involves the oxidation or reduction of metals by microorganisms during bioleaching (Palumbo-Roe et al. 2014). This reaction increases metal mobility in solution, subject to the type and oxidation state of the metal (Palumbo-Roe et al. 2014). Firstly, electrons are transferred directly from mineral elements to microorganisms by the oxidation of Fe^{2+} to Fe^{3+} (Eq. 3). The oxidative breakdown of solid-phase REEs then occurs because of the Fe^{3+} formation (Dev et al. 2020). Equation 4 proposes the reductive leaching of REEs by anaerobic autotrophs in the presence of elemental sulfur, which acts as the electron donor (Barnett et al. 2020).

$$4Fe^{2+} + 4H^{+} + O_2 \xrightarrow{\text{aerobic bacteria}} 4Fe^{3+} + 2H^2O \tag{3}$$

$$6Fe^{3+} + S^0 + 2H_2O \xrightarrow{\text{anaerobic bacteria}} 6Fe^{2+} + 8H^{+} + SO_4^{2-} \tag{4}$$

Microorganisms can interact with the minerals in two ways: contact mode or non-contact mode. It is also possible for interaction to follow both ways (Vera et al. 2013). In the contact mode, microorganisms in contact with the minerals produce extracellular polymeric substances (EPS), which are made up of polysaccharides, proteins, and nucleic acids (Park and Liang 2019). Microbial oxidation of Fe^{2+} to Fe^{3+}, followed by oxidation of sulfides to sulfuric acid, results in the dissolution of REE in the EPS. Dissolution of REE and reduction in pH can also occur in the presence of organic acids secreted by the microbes (Park and Liang 2019). For the non-contact mode, non-attached microbes oxidize dissolved Fe^{2+} to Fe^{3+}, resulting in sulfide minerals oxidation to sulfuric acid and subsequent dissolution of REEs (Jia et al. 2019). Some microorganisms that have been used extensively in this process include *A. thiooxidans*, *A. ferrooxidans,* and *L. ferrooxidans* (Auerbach et al. 2019a, b; Sethurajan et al. 2018). These microorganisms were reported by Auerbach et al. (2019a, b) to leach 100% Dy and Pr from the magnet.

Complexolysis

Following acidolysis and redoxolysis, the metal ions released into the solution are stabilized via complexolysis. Complexolysis is a process that involves the microbial-assisted formation of complexing or chelating agents such as siderophores and organic acids (Dev et al. 2020). Organic acid(s) are natural chelating agents that stabilize metal ions by complexing them in solution. By stabilizing metal complexes in solutions, the potentially toxic effects of these metal ions on bioleaching microbes are minimized, and their metabolic activities are not impaired (Dusengemungu et al. 2021). These agents (organic acids and siderophores) alter the environmental pH and facilitate the formation of metal ion(s)-organic acids or metal ions(s)-siderophores complexes thereby increasing the mobilization of REEs from the solid matrix (Rasoulnia et al. 2020). This is because compared to the lattice bonds existing between solid particles and metal ions, the bonds formed between metal ions and organic acids (ligands) are stronger, strengthening the solid particle bioleaching process (Dusengemungu et al. 2021).

Some studies have examined supplementing the bioleaching medium with organic acids to enhance the level of metal ion complexes to improve the bioleaching process. Qu et al. (2019) reported extractions of more than 50% Y and Sc from red mud by *Acetobacter* spp. via REE-organic acid complex formation. Likewise, *Gluconobacter oxylans* NRRL B85 produced gluconic acid which was capable of extracting Y, Sm, Eu, Yb, Ce, and Nd from synthetic phosphogypsum (Antonick et al. 2019).

Siderophores from Actinobacteria, *Streptomyces* spp., *Aspergillus niger* have all been implicated in bioleaching of a considerable amount of REEs (Chao et al. 2016; Zhang et al. 2018; Osman et al. 2019). Due to their role in the overexpression of genes that code for the siderophore biosynthetic enzymes, substrates like glucose, glycerol, and NH_4Mo^{2+} are said to promote siderophore synthesis (Dev et al. 2020). Therefore, the inclusion of one or all these substrates in the bioleaching process could considerably increase the extraction of REEs (Dev et al. 2020).

The REE metal ion-ligand complex is recovered from the aqueous phase of the bioleaching media through biosorption, bioaccumulation, and bioprecipitation.

4.2.2 Biosorption

Biosorption is one of the mechanisms involved in the recovery of REEs from aqueous solution after mobilization or formation of metal ion complex. Biosorption occurs through the passive binding of the metal ions onto active or dead microbial biomass. The REE metal ion-biomass interaction is determined by the chemical characteristics of biosorbent cell coatings rather than their biological activity (Beni and Esmaeili 2020). Though active biomass can be used, metabolically inactive biosorbents are more beneficial as they permit REEs recovery and reuse and are not limited by toxic elements in the solution (Giese 2020).

Biosorption offers different advantages including negligible sludge production, high recovery competency, simple operation, economically and environmentally friendly (Gupta et al. 2019). Electrostatic interactions, ion exchange, surface complex formation, and precipitation are all involved in the adsorption of REEs on the surface of a microbial cell used. The functional groups on the surface of the cell enable adsorption of the REEs to the cell (Zhuang et al. 2015). The pH of the solution plays a role in determining which functional group binds to the REEs (Takahashi et al. 2010). For instance, at pH 3–4, a functional group $-PO_4^{2-}$ binds to REEs while at pH 6–7, it is the –COOH groups that bind to REEs. In addition, the pH of the solution also affects the adsorption of different molecular weights. Light molecular weight REEs (Nd and La) adsorbed easily at pH > 4, whereas medium and high molecular weight REEs (Gd and Sm) adsorbed heavily below pH 4 (Dev et al. 2020). Because adsorption is a surface process, a biosorbent with a wide surface area promotes metal ion adsorption by providing a broad area of contact for metal ions to interact (Beni and Esmaeili 2020). Engineering microorganisms to express a lanthanide binding tag on their cell surface can improve the overall biosorption of REEs. This is confirmed in the study by Park et al. (2017) where the biosorption ability of engineered *Escherichia coli* (having 16 lanthanide binding sites) to Y, La, Pr, Nd, and Ce improved by 56.1, 87, 72.4, 58.8, and 63.6%, respectively, when compared to the control.

Two of the known disadvantages of the biosorption process are slow recovery rate and that it has not been well developed for recycling more complicated metallic e-waste (Ghimire and Ariya 2020).

4.2.3 Bioaccumulation

The intracellular uptake of REEs by metabolically active cells is referred to as bioaccumulation (Dev et al. 2020). When soluble metal ions are conveyed between the cell membrane and the cell (metabolism-dependent intracellular uptake), solid particles accumulate or precipitate in vacuoles, resulting in bioaccumulation (Dusengemungu et al. 2021). The REEs are translocated into the intracellular space by the membrane lipid bilayer immediately after they are adsorbed on the cell surface (Maleke et al. 2019). Thereafter, intracellular proteins and peptide ligands sequester the REEs (Lederer et al. 2017).

Bacillus cereus is one of the microorganisms that have been reported to bioaccumulate REEs. Emmanuel et al. (2011) reported that B. cereus was able to bioaccumulate Ce and Nd from soil enriched with REEs. The red algae, *Galdieria phlegrea* has also been able to recover REEs from luminophores using bioaccumulation with yttrium being the most bioaccumulated element (Čížková et al. 2019). Other microorganisms that have been used to recover REEs by bioaccumulation include *Galdieria sulphuraria* (wastewater) (Minoda et al. 2015), *Desmodesmus quadricauda* (waste luminophores) (Čížková et al. 2019), and *Gracilaria gracilis* (wastewater) (Jacinto et al. 2018). Though bioaccumulation has the advantage of being able to selectively recover REEs from a mixed metal solution, one notable disadvantage of this recovery process is that the solubility of REEs in lipids can be relatively low (Ochsner et al. 2019). However, this shortfall can be eliminated by catechol-type siderophore producing *Arthrobacter luteolus* (Emmanuel et al. 2012).

The catechol-type siderophore forms a complex with the REEs, this complex is then easily accumulated through the membrane (Emmanuel et al. 2012). Interestingly, research has shown that some microorganisms that can accumulate REEs use them as co-factors for microbial metabolism as seen in *Methylacidiphilum fumariolicum* (Nancharaiah et al. 2016).

4.2.4 Bioprecipitation

Bioprecipitation is an enzyme-assisted method of precipitating precious metals. This mechanism involves changing the positive valence state of the precious metals to a zero valence state. Metal bioreduction can occur because of direct or indirect interaction with the microbial cell surface, through electron mediators. Secondary metabolites (also known as metallophores) ensure adherence of the metal surfaces to the microbes (Rana et al. 2020).

In the case of low-grade ores as secondary sources of REE, the phosphatase activity of microorganisms is used to hydrolytically liberate inorganic phosphate resulting in the precipitation of dissolved REEs as phosphates (Crocket et al. 2018). *Serratia* spp. immobilized on polyurethane foam were able to recover over 90% of Nd and 85% of Eu through phosphate-based precipitation (Deplanche et al. 2012). Another study also reported the recovery of Dy as $DyPO_4$ by *Penidiella* spp. using the same mechanism earlier stated (Horiike et al. 2018). Though greatly influenced by pH, bioprecipitation process can selectively recover a specific REE and this is enabled by the selectivity of bioprecipitation towards distinct REEs (Dev et al. 2020). The co-precipitation process that occurs during passive treatment of AMD, a rich source of REEs, has been reported as a novel REE recovery strategy (Ayora et al. 2016; Dominguez-Benetton et al. 2018). A large proportion of REEs is sequestered in the sludge formed during the passive treatment procedure. The substance, baaluminite, formed during passive treatment of AMD contained an estimated 0.04–0.3 tons/year of yttrium (Naidu et al. 2019) (Table 2).

5 Bioprospecting of Metal Recovery Microorganisms Using Omics Technologies

The pursuit of unifying separate biological events, such as the human genome project in 2003 or the mapping of various amino acid metabolic pathways in the 1950s, is not new, and it cannot be accomplished using traditional biological technologies, which embraces a reductionist approach that is unsatisfactory in demystifying a biological system (Karahalil 2016). System biology, rather than using a reductionist approach, takes an integrated approach to better understand the entire processes occurring in a biological system. It leverages the beauty of the fundamental dogma of molecular

Table 2 REEs that have been recovered from various materials

Recovery process	Material	REEs	Microorganisms	pH	References
Biosorption	Aqueous solution of REEs oxides Aqueous solution of REEs salts	La, Nd, Eu, Gd La, Ce, Eu, Yb	*Sargassum* spp. *Turbinaria conoides*	5.0 4.9	Oliveira and Garcia (2009) Vijayaraghavan et al. (2010)
Complexolysis	Red mud Synthetic phosphogypsum	Sc, Y, La, Nd, Ce, Eu, Tb, Y, Ce, Nd, Sm, Eu, Yb	*Penicillium tricolor* RM-10 *Gluconobacter oxydans* NRRL B85	9–10.4 2.1	Qu and Lian (2013) Antonick et al. (2019)
Complexolysis/acidolysis	Waste phosphors and cracking catalysts Coal fly ash	Tb, Eu, Ce, La, Y Yb, Er, Sc, Y, La, Dy, Gd, Eu	*Gluconobacter oxydans* *Candida bombicola*	3.3 3.3–3.5	Reed et al. (2016) Park and Liang (2019)
Redoxolysis	Monazite	Ce, La Ce, La	*Acidithiobacillus ferrooxidans* *Acidithiobacillus thiooxidans*	1.8 0.9/1.2	Nancucheo et al. (2019)
Redoxolysis/acidolysis	Incineration slag	Ce, La, Er Ce, La, Er	*Acidithiobacillus thiooxidans*, *Acidithiobacillus ferrooxidans* *Leptospirillum ferrooxidans*	1.8 3.2	Auerbach et al. (2019a, b)
Acidolysis	WEEE shredding dust Coal ash	Ce, Eu, Nd, La, Y Sc, Y, La	*Acidithiobacillus thiooxidans* *Acidithiobacillus thiooxidans*, *Acidithiobacillus ferrooxidans*	1.2–1.5 2.0–4.0	Marra et al. (2018) Muravyov et al. (2015)

WEEE Waste Electrical and Electronic Equipmen

biology, along with the first rule of thermodynamics which elucidates that biological information is interconvertible primarily in the order of DNA to RNA, and then to proteins. Thus, mapping these different forms of biological information from an integral perspective would readily deliver the intricate features of the biological system under study. Omics technology is the vehicle that provides us with this whole picture of the components of the central dogma of molecular biology. Data obtained from genomics, proteomics, metabolomics, epigenomics, transcriptomics, phenomics, lipidomics, interactomics, and other Omics approaches (Chandran et al. 2020; Jerez 2017) (Fig. 4) enables the global demystification of the physiological processes in a biological system and the informed prediction of how these organisms operate in their communities and change over time under varying conditions.

Consortia of various microorganisms are responsible for the cascades of biological processes involved in the extraction of REEs and various precious metals from e-waste. These microorganisms that are mostly extremophiles are not amenable to traditional biological approaches due to the dearth of suitable culture media that can recover these organisms from bioleaching samples (Jerez 2017). As a result, Omics approaches are ideal for studying these bioleaching populations.

The biochemical processes involved in iron and sulfur compound oxidation, quorum sensing, metal adhesion, biofilm development, and a variety of adaptive resistance mechanisms are all important aspects of metal bioleaching that must be properly understood in order to improve the process metrics for pure metal extraction (Martinez et al. 2015). These biochemical processes are revealed by a subtle evaluation of the Omics sequences of various bioleaching microorganisms, and as more of these sequences are mapped or accessed, the researchers would have sufficient information that would enable a quality understanding of the adaptive processes that allow these organisms to thrive in their environment (Jerez 2017). Furthermore, these sequences serve as a boon for recombinant DNA technology in the production of avid strains via tools like the CRISPR-Cas9 technologies to produce suitable transgenic bioleaching microorganisms (Brar et al. 2021).

5.1 Application of Omics in Studying Bioleaching Communities

The micro-environment of organisms involved in biomining is dynamic due to the production of several metabolites resulting from the solubilization of metals. It is, therefore, not sufficient to know only the predominant species in this hostile environment but also to map out the mechanisms they utilize to persist in the dynamic bio-leaching environment. The 16S ribosomal ribonucleic acid and ribosomal DNA profiles are commonly considered as the standard template sequence for Omics research in the study of bacterial diversity, although recent Omics discoveries have exposed the pitfalls of totally depending on this conserved region (rDNA) for bioprospecting. Moreover, for downstream processing of a bioleaching sample

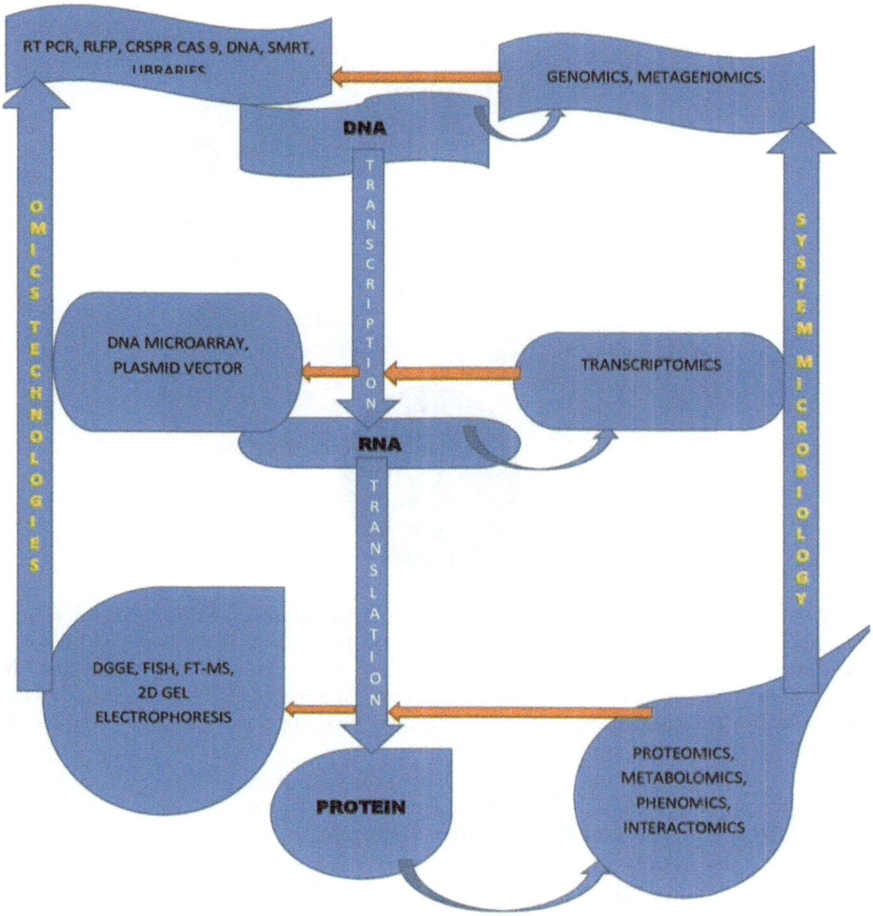

Fig. 4 Various Omics approaches used in system biology with the corresponding Omics technologies such as: fluorescent in situ hybridization (FISH), denaturing gradient gel electrophoresis (DGGE), two-dimensional polyacrylamide gel electrophoresis (2PAGE), real-time polymerase chain reaction (PCR) and clustered regularly interspaced short palindromic repeat, Caspase 9 (CISPR CAS 9)

into useful data, several molecular techniques and software are used including those mentioned in (Fig. 5) (Chandran et al. 2020; Jerez 2008). Over 91 full genome sequences of microorganisms found in biomining environments have been mapped using these molecular techniques (Wheaton et al. 2015; Siezen and Wilson 2009; Cardenas et al. 2010). The study of bioleaching communities can be done using various Omics approaches.

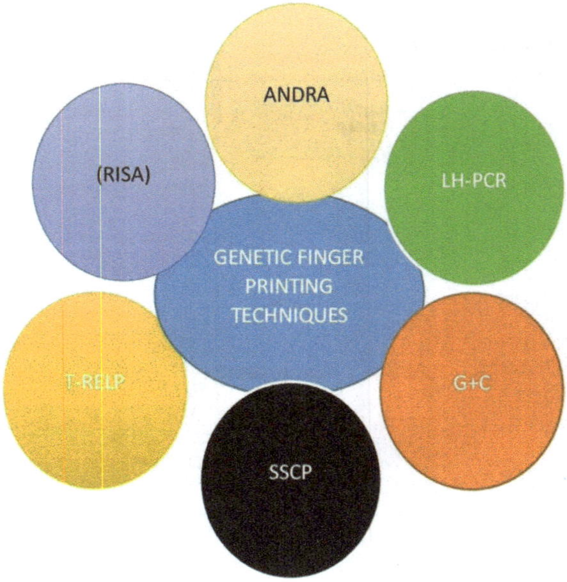

Fig. 5 Techniques for determining the structure, function, and dynamics of microbial communities in bioleaching systems: Single-strand conformation polymorphism (SSCP), amplified ribosomal DNA restriction analysis (ANDRA), length heterogeneity PCR (LH-PCR), polymorphism of the terminal restriction fragment length (T-RELP) microbial lipid analysis DNA microarrays, content separation of G + C guanine and cytosine, temperature gradient gel electrophoresis (TGGE) and ribosomal intergenic spacer analysis (RISA)

5.1.1 Metagenomics and Comparative Genomics

Metagenomics is the study of all the genomes of a microbial community without necessarily passing through the traditional cultural protocols (Oulas et al. 2015; Bilal et al. 2018). It is interested in the collective genomes of the microorganisms present in an environmental sample (Levett et al. 2021). Metagenomics studies of a metal solubilizing environment are majorly aimed at identifying predominant species in the sample of interest. Under ideal conditions, metagenomics is suitable for environments with low microbial diversity to permit a non-complicated synthesis of the constituent genomes during downstream processing (Jerez 2008). Direct (shotgun) sequencing and indirect (library-based) metagenomics are the two most common sequencing techniques used in metagenomics research (Levett et al. 2021).

In direct shotgun sequencing, DNA is randomly fragmented by either physical or chemical processes before being read and generally excludes a pre-cloning stage of genetic materials (Chandran et al. 2020). This technique is famously used for environmental characterization, and it is a technique that is increasingly available for metagenomic studies (Levett et al. 2021). The second method in metagenomics is library-based targeted metagenomics, which involves recovering environmental genomes using specialized DNA extraction and purification techniques. The fragments obtained are cloned into a plasmid vector and, thereafter, introduced into a host organism to create a library, which is then screened for specific genetic features before being sequenced. The cost of sequencing has dropped by five orders of magnitude with the advent of second and third-generation sequencing technologies. This has made it possible to conduct more extensive environmental metagenomic research (Chandran et al. 2020; Bouhajja et al. 2016). As a result, the efficiency or cost of the sequencing process is no longer a key concern; rather, the bottleneck is now the downstream reconstruction of genomes of specific species in complex situations. However, with the introduction of new machine learning algorithms, in silico engineering, and bioinformatics, recovering almost complete genomes from increasingly vast metagenomic datasets has become much feasible (Albertsen et al. 2013; Kaksonen et al. 2020).

A. ferrooxidans is the first metal-solubilizing bacterium whose genome has been entirely sequenced (Jerez 2017). This microorganism has been used as a model species to explore the genes involved in metal-microorganism interactions (Levett et al. 2021). Genomic data from this organism has also helped in the identification of genes involved in critical physiological functions such as solubilization of metals, metal resistance, biofilm formation, cell adsorption on the surface of these metals, biomarker identification for organisms that can potentially utilize metal, amino acid biosynthesis and the formation of extracellular polymeric substances (Chadran et al. 2020). Specifically, genomics helps in the unraveling of paralogues, orthologues, operons, regulons, and even stimulons that coordinate critical bioleaching functions such as the formation of biofilms. A good example of the significance of genomics is in the unraveling of the ability of these organisms to live in a highly acidic environment while their cytoplasmic pH is paradoxically near neutral. Research using comparative genomics shows that *Acidithiobacillus* spp. has a gene-sequence similarity with microorganisms that have Clusters of Orthologous Groups that produce proteins including K-dependent transporting ATPase, and proton antiporters (Brar et al. 2020). The reverse membrane potential generated by K-transporting ATPases produces an osmotic field that limits the influx of hydrogen ions into the cell, and additional secondary transporters (Fig. 6) are important for internal pH homeostasis (Baker-Austin and Dopson 2007). *Acidithiobacillus* spp. also has its metal resistance mechanisms, such as the Mer and CzcD operons for mercury, and copper-zinc-cadmium respectively. In addition, genes for biosynthesis of isoprenoid lipids and a variety of proton efflux pumps were discovered, giving more insight into survival strategies in acid mine environments, as some bacteria have been shown to use these resistant mechanisms including phospholipid head group alteration (Torres et al. 2011).

With the genomic sequences of over 1000 microorganisms completed and over 2400 in process, comparative genomics has become a useful tool for improving the identification of genes and predicting their metabolic capacity and regulations (Valdes et al. 2010). Comparative genomics, as the name suggests, is genomic research that makes use of microbial community data from several different metagenomics investigations. This wealth of information is revealing important details about the

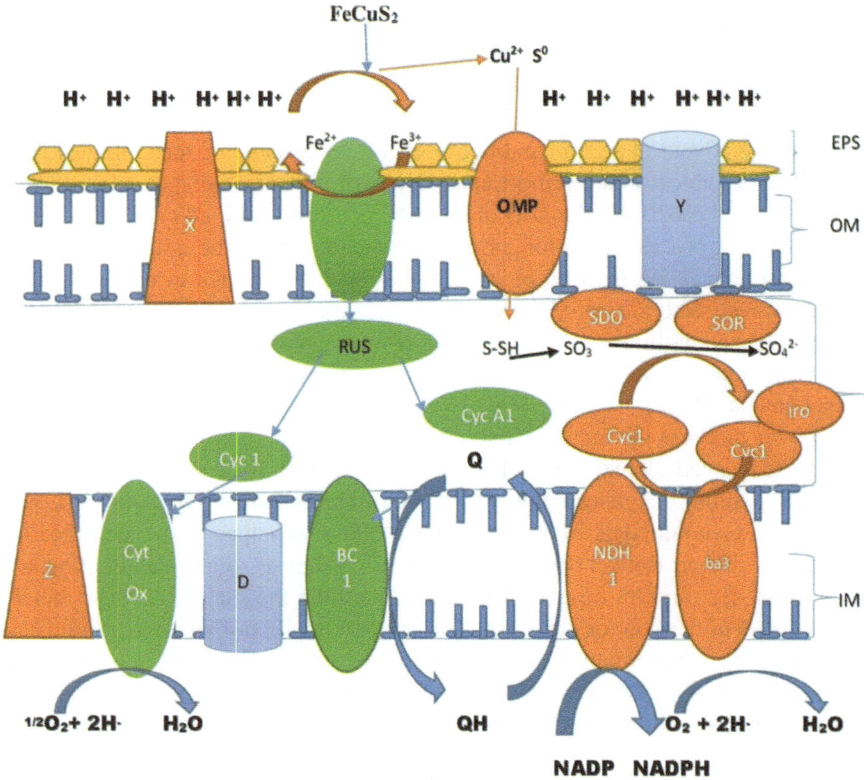

Fig. 6 Schematic depiction of the periplasmic proteins involved in oxidative bioleaching in *Acidothiobacillus* spp. The first layer is the exopolymeric substances (EPS). Positively basic acid-tolerant transmembrane proteins protrude through it, supposedly preventing hydrogen ions influx. The outer membrane (OM) is next, followed by the periplasmic space (PPS), and finally the inner membrane (IM). Proteins Cys 2, C-type cytochrome; RUS, rusticyanin; Cyc 1, C552; Cyt Ox, aa3 cytochrome oxidase; OMP, outer membrane proteins; SDO, sulfur dioxygenase; SOR, sulfite oxidoreductase involved in iron and sulfur oxidation from left to right. The alphabetized putative proteins (X, Y, D) are those that have no known functions and were described as unfamiliar proteins

metabolic capabilities and interactions that aid in the development and maintenance of bioleaching microbial consortia and ecophysiology (Jerez 2017). Comparative genomics reveals the elements of genome evolution, suggesting that bioleaching bacteria are more varied than previously thought.

There is a developing interest among environmental scientists in the validation of the sensitivity and specificity of rDNA typing in bioprospecting of bioleaching acidophiles. While rDNA, on the other hand, gives a quick summary of the species present, it is possible, according to Valdes et al. (2010), that solely depending on the sequence of this region may underrepresent the real metabolic variance of the extremophiles in a heap. For example, microorganisms with 98% similarity in the 16S rDNA sequence are thought to belong to the same species according to conventional molecular understanding as the degree of similarity is frequently equated with the genomic content. Valdes et al. (2010) went on to show that even when two species of *A. ferrooxidans* have 100% rDNA homology and are thought to be the same strains, comparative genomics have revealed substantial variations in their gene content. In his observation, these strains exhibit two significant divergent genome segments (Indels), close to 300 kb in one strain and 200 kb in another respectively. These indels account for roughly 16% of the variation in gene content of the organisms (Valdes et al. 2010). Furthermore, genes known to be involved in horizontal gene transfer events, including phage integrase gene, conjugation genes, transposases, and numerous genes involved in DNA mobilization, flanks the indels of these organisms. The presence of a CRISPR locus in the indel of *A. ferrooxidans* suggests that this strain has virus-resistant adaptive mechanisms. These findings, along with the discovery of further potential genome deletions, insertions, and rearrangements, suggest that *Acidithiobacillus* spp. has a great deal of genomic plasticity. This adds to the idea of a pan-genome, in which organisms have a core or conserved genome that encodes specific activities, as opposed to a transient or epigenetic genome that encodes auxiliary functions like metal resistance, which is desirable in a bioleaching environment (Valdes et al. 2010).

Comparative genomics is emerging to highlight the inadequacies of relying solely on rDNA type to measure microbial diversity, necessitating the identification of a set of gene markers that can provide a better way of assessing the exact metabolic potentials of acidophiles. In addition, genomics research has contributed significantly to our understanding of how acidophiles deal with reactive oxygen species. A disproportion between the production of reactive oxygen species (ROS) and their inhibition by antioxidant defense systems causes oxidative stress. Cardenas et al. (2012) used comparative genomic analysis to rebuild individual oxidative stress responses and conserved mechanisms in 44 sequenced genomes of acidophiles. They discovered that acidophiles lack genes encoding conventional oxidative stress response regulators (SoxR, OhrR, OxyR), as well as a sparse glutathione/glutaredoxin system component and an underrepresentation of conventional reactive oxygen species consumption enzymes like catalases. These characteristics distinguish acidophilic organisms from well-studied neutrophilic species. Alternative systems, such as the hydrogen peroxide-scavenging rubrerythrin and the Fur family of regulators, have been suggested to perform the majority of these functions in acidophiles. Furthermore, the large number of thioredoxins and peroxiredoxins seen in acidophiles implies that they have developed a robust response to attenuate the impacts of reactive oxygen species on macromolecules, complementing the relatively weak direct reactive oxygen species removal mechanisms.

5.1.2 Transcriptomics

The study of differential genome expression at the level of messenger RNA production is known as transcriptomics or functional genomics. Transcriptomics relies heavily on DNA microarrays, also known as biochips, to determine a sample's transcription profile. A biochip is a collection of tiny DNA patches that have been covalently or non-covalently deposited in a two or three-dimensional array on a solid surface. However, knowing the genome of the organism to be examined necessitates the use of microarrays, and the extracted RNA must be converted to complementary DNA before analysis. The use of DNA chips that is impregnated with the whole genomic sequences of the organisms will allow for a comprehensive view of the expressed genes of members of the microbial community under different bioleaching conditions.

Early transcriptomics studies were carried out by Valenzuela et al. (2006) as they assayed the relative changes in mRNA abundance of certain genes associated with sulfur metabolism in *A. ferrooxidans*. The effect of different oxidizable substrates on the physiology of the consortia of bioleaching organisms was studied using a pilot DNA microarray containing 70 distinct genes.

A genome-wide microarray impregnated with 3000 genes was later used for transcript profiling study of the *A. ferrrooxidans* ATCC 23270 (Quatrini et al. 2006; Appia-Ayme et al. 2006). The genes that were upregulated in ferrous iron or sulfur growth conditions were investigated. The findings confirmed and expanded models of iron and sulfur oxidation, as well as a possible alternative electron route and the possibility that iron and sulfur oxidation may be coordinated (Jerez 2017). In a transcriptomic study by Auernik et al. (2008) on the genome of *Metallosphaera sedula*, 88 open reading frames were up-regulated by two-fold when ferrous sulfate was added to its medium. These included genes predicted to be involved in the metabolism of sulfur as well as constituents of terminal oxidase clusters suggested to be involved in the oxidation of sulfur.

Furthermore, transcriptomic studies can be used to study the genes upregulated when exposed to high concentrations of organic solvent. Most times, in order to speed up extraction, most e-waste extraction systems use high concentrations of organic solvents and biosurfactants, but these compounds have been shown to harm the bioleaching microbial community (Wijte et al. 2011). This is because they disrupt the cell membrane's integrity and permeability (Jafari et al. 2019). The rpoH, OxyR, and NrdR regulon which improves acidophile bioleaching performance under adverse conditions has been shown to be overexpressed in the presence of alcohols, reactive oxygen species, and aromatic compounds according to transcriptomics studies (Rutherford et al. 2010; Horinouchi et al. 2010). Furthermore, Jerez (2017) determined the active microorganisms during bioleaching cycles using a microarray designed for acidophiles. During the first period of the leaching cycle, *A. ferrooxidans* was predominant while the population of *Ferroplasma acidiphilum* and *Leptospirillum ferriphilum* increased across the period while that of *A. thiooxidans*, was steady throughout the entire leaching process. In summary, the Alphaproteobacteria genus was active in all samples, and the sulfur-oxidizing bacteria genus was active in older samples, according to the prokaryotic acidophile microarray analysis. Similar transcriptomics studies show that changes in pH and the ratio of Fe^{2+} to Fe^{3+}, for example, were found to be primary factors shaping the dynamics of the bioleaching microbiome (Cárdenas et al. 2010; Remonsellez et al. 2009; Orell et al. 2010).

5.1.3 Proteomics

Proteomic studies further help to establish facts regarding genomics and transcriptomics studies because the proteome of an organism is the product of its genome. Its goal is the correlation of identified proteins to the corresponding gene(s). Most proteins exist as protein complexes with a specific interacting moiety. To understand the physiology of proteins, the effect of their interacting partners must be determined thus the reason for mapping out all the constituents of a protein complex in proteomic studies (Mann et al. 2001). Identifying protein complexes involved in oxidative reactions is a top priority for a bioleaching bacterium like *A. ferrooxidans*.

In general, proteomic studies provide a more detailed look at proteins such as (NADH-quinone oxidoreductase); nitrogen fixation (nitrogenase complex); CO_2 fixation (ribulose bisphosphate carboxylase); pH homeostasis (kdpD-encoded sensor and transport proteins); heavy metal resistance (efflux pumps and metal reductases); chemotaxis and motility (flagellar motor switch proteins); and biofilm formation (UDP-galactose-4-epimerase, UDP-galactopyranose mutase) quorum sensing (c-di-GMP phosphodiesterase) (Brar et al. 2021). The most commonly used method for the resolution of the aforementioned proteins as well as comparative bacterial proteomics study is 2-D PAGE in tandem with mass spectrometry but other states-of-the-art technologies like LC–MS, MALDI-ToF, and ICPL are used (Marques 2018; Chandran et al. 2020). In a single run, 2-D PAGE may resolve about 103 protein locations (Jerez 2017). One of the drawbacks of 2-D PAGE is that it can only detect the predominant proteins in the cell. To improve resolution, it is also ideal to examine localized regions in a cell such as a sub proteomic cell fraction. Proteins involved in energetics, intracellular and extracellular logistics may be found in the proteomic fractions of acidophilic species, which is of particular interest. The full collection of proteins in the periplasmic fraction of *A. ferrooxidans* cultured in sulfur, copper, and iron-enriched medium has been determined experimentally using high-throughput proteomics (Chi et al. 2007). About 26.1% of these proteins have no homologs in databases (Chi et al. 2007), suggesting that they are unique to this bacterium and may play essential roles that have yet to be discovered.

It is considerably efficient and easier to detect proteins under the different environmental conditions in bioleaching microorganisms because of the development and advancement of quantitative proteomics, which employs radio-labeling of the bioleaching samples to be comparatively assayed. Quantitative proteomics using isotope-coded protein labeling was recently used to assess the behavior of *A. ferrooxidans* to high copper concentrations. Putative proteins overexpressed were those involved in copper resistance, efflux pumps, and enzymes involved in the repair of copper-damaged disulfide bonds whereas the outer membrane proteins were down-regulated, indicating an overall reduction in the inflow of the metal into the cell (Martnez-Bussenius et al. 2016). The majority of energy-metabolizing proteins found in the periplasm revealed that most of them have basic isoelectric values higher than 7 (Jerez 2017). The proteins found in this fraction were found to have high basic acid-tolerant properties like rusticyanin. This was not the case with neutrophils like *Escherichia coli* or alkalophiles, which had a larger proportion of acidic periplasmic proteins on the contrary. According to Chi et al. (2007), the basic nature of the proteins strongly suggests that acidophiles like *A. ferrooxidans* have adapted to their acidic environment by providing a positively charged barrier that prevents protons from entering the cytoplasm (Fig. 3), preventing excessive periplasmic acidification. Proteins involved in iron oxidation (cytochromes c4, cytochrome, and rusticyanin) are found in the periplasm of *A. ferrooxidans*, as well as the putative high redox potential iron-sulfur protein or iron oxidase, which acts as an electron acceptor in several Fe^{2+}-oxygen transfer chains (Brar et al. 2021). Some of the proteins involved in the metabolism of sulfur, such as sulfur dioxygenase, sulphite oxidoreductase, tetrathionate hydrolase, and many cytochromes involved in both iron and sulfur oxidation, are also shown in Fig. 3.

5.1.4 Metabolomics

The metabolome is an organism's total metabolites, and metabolomics is the study of a cell's metabolite profile under specific conditions (Beale et al. 2017). This is an important aspect of Omics, but it is frequently overlooked due to a dearth of research interest in microbial metabolites generated by acidophilic species. When a cell is stressed, it creates a variety of primary and secondary metabolites (Malla et al. 2018). Metabolomics helps to better understand the metabolic dynamics of biomining acidophiles in harsh environments. Metabolites identified during proteomics studies can be utilized as important biomarkers for the activity of an acidophile during bioleaching. Capillary electrophoresis and mass spectrometry are commonly utilized techniques for the separation and identification of metabolites respectively, in metabolomics studies (Lu et al. 2019; Chandran et al. 2020).

The goal of metabolomics research is to find out what metabolites are generated when acidophilic organisms are exposed to various energy sources. When *A. thiooxidans* and *A. ferrooxidans* were cultured in the presence of sulfur, metabolites such as glutamate and aspartate were overexpressed, a phenomenon not observed when these microorganisms were grown in a copper and iron-enriched medium (Martinez et al. 2013). Similar amino acids have been linked to biofilm development in *Bacillus subtilis*, therefore these amino acids were thought to be involved in the production of exopolymeric substances (Morikawa et al. 2006). Sugars such as dihydroxyacetone phosphate and sedoheptulose-7-phosphate, both of which are overexpressed when the same organism is cultured on sulfur media, are also extremely intriguing metabolites. These sugars are also linked to the production of exopolymeric compounds, which is also corroborated by earlier genomics research.

There have been reports of a variety of polycationic compounds also known as polyamines such as putrescine, spermidine, and spermine (Terui et al. 2005) being produced differentially under various conditions. Studies by Martinez et al. (2016) showed that the concentration of spermidine in *A. ferrooxidans* strain grown in an iron-enriched medium is 33% higher than in sulfur enriched medium. A lot of hypotheses have been made regarding the role of spermidine in the activity of acidophiles and there is consensus in its role as a signaling molecule for adherence and biofilm formation and serve as a biomarker for sulfur-oxidization in bioleaching processes (Martinez et al. 2015; Karatan et al. 2005; Sturgill and Rather 2004). Metabolomics research is without a doubt at its early stage of application in Omics studies. However, the identification of new metabolites and their role in bioleaching activities highlights its prospects in industrial application.

5.2 *Omics in Rewiring the Metabolic Pathways*

The importance of Omics research heavily lies in leveraging the global knowledge gained to manipulate the organism and bioleaching conditions in order to expedite the process metrics involved in metal extraction. Although there are still ethical concerns about introducing modified microbes into the bioleaching environment, there have been several strains produced that have shown great potentials and demonstrated that the value outweighs the danger (Kaksonen et al. 2020). Thus the transformation of these organisms to knock in or knock out certain genes is pivotal to generating robust strains that would leach metals efficiently. Unfortunately, there is a paucity of effective transformation and conjugation systems needed to produce strains with good adherence qualities and upregulated chemolithotrophic ability. But the novel molecular tool known as the CRISPR-Cas9 has the potentials to overturn this uphill gene editing task for bioleaching acidophiles. CRISPR-Cas9 acts like the humoral immunity for bacteria; it is the important antiviral tool that protects microorganisms from phage invasion (Cardenas et al. 2016). Relevant viruses and other transposable elements disrupt the host's genomic stability and microbial activity (Dellas et al. 2014; Wang et al. 2010; Garrett et al. 2015). There is, therefore, a rising interest by environmental scientists tailored towards adopting CRISPR-Cas9 as a gene-editing (insertion/knockout) technology because of its precision and low cost. Access to an array of viral metagenomic and CRISPR spacer sequences will allow scientists to better understand the nexus between viruses and their hosts, as well as control or prevent viral infection by gene-knockout or modification that enhances the endurance of bioleaching communities. In addition, a molecular method such as homologous recombination, metabolic modeling (Wang et al. 2012; Yu et al. 2014) can be utilized to alter microbial surfaces to enhance the availability of carboxylic groups in the exopolymeric layer (as shown in Fig. 3) and stimulate its strong interaction with metallic compounds. As a result, further study in this field is necessary to enhance ion accumulation rates by controlling the electrochemical environment. A protein called licanantase, which was first purified from *A. thiooxidans,* for example, has the potential to enhance the recovery of copper (Bobadilla-Fazzini et al. 2011). In addition, recombinant strains of *A. ferrooxidans* with an upregulated tetH gene have been shown to effectively oxidize sulfur and various tetrathionate energy sources compared to its wild strain (Yu et al. 2014). Studies also showed that the overexpression of genes implicated in Kreb's cycle like those of α-ketoglutarate dehydrogenase and succinate dehydrogenase in *Acidithiobacillus caldus,* expedited oxidation, and bioleaching performance. Similar findings were seen in an auxotrophic strain of *A.*

ferrooxidans with a 93% increase in the generation of isobutyric acid from carbon dioxide and the oxidation of Fe^{2+} to Fe^{3+} (Kernan et al. 2016).

The accessibility of the meta-omics dataset would simplify the task of relating physicochemical characteristics to the bioleaching flora, and genome-editing for the regulation of interest genes to make bioleaching strains more powerful will be possible (TerAvest and Ajo-Franklin 2016).

6 Conclusion

Bioleaching to recover precious and rare earth elements from e-waste is a brilliant approach to meeting market demands for metal raw materials for the manufacture of new electrical and electronic products. Biometallurgical processes have benefited from the use of Omics technology and techniques. Knowing which microbes are present in a bioleaching operation is critical, but also being able to monitor their activities throughout the bioleaching process, determining the dominant species and how they adapt in response to the changing environment as the metals are solubilized, is crucial. There is an expanding generation of Omics sequences of bioleaching communities due to the development of inexpensive and high throughput Omics techniques like CRISPR-Cas9, 2-D PAGE, next-generation sequencing, in silico engineering, and so on. This big dataset will be crucial in elucidating the exact microbial complexity in the biomining environment, as well as pinpointing factors that can enhance the bioleaching environment and develop robust recombinant strains that can efficiently leach metals from e-waste.

References

Abdelbasir SM, Hassan SSM, Kamel AH, El-Nasr RS (2018) Status of electronic waste recycling techniques: a review. Environ Sci Pollut Control Ser 2018. https://doi.org/10.1007/s11356-018-2136-6

Alabi OA, Bakare AA, Xu X, Li B, Zhang Y, Huo X (2012) Comparative evaluation of environmental contamination and DNA damage induced by electronic-waste in Nigeria and China. Sci Total Environ 423:62–72

Albertsen M, Hugenholtz P, Skarshewski A (2013) Genome sequences of rare, uncultured bacteria obtained by differential coverage binning of multiple metagenomes. Nat Biotechnol 31:533–538. https://doi.org/10.1038/nbt.2579

Antonick PJ, Hu Z, Fujita Y, Reed DW, Das G, Wu L, Shivaramaiah R, Kim P, Eslamimanesh A, Lencka MM, Jiao Y, Anderko A, Navrotsky A, Riman RE (2019) Bio- and mineral acid leaching of rare earth elements from synthetic phosphogypsum. J Chem Thermodyn 132:491–496

Appia-Ayme C, Quatrini R, Denis Y, Denizot F, Silver S, Roberto F, Veloso F, Valdes J, Cardenas JP, Esparza M, Orellana O, Jedlicki E, Bonnefoy V, Holmes DS (2006) Microarray and bioinformatic analyses suggest models for carbon metabolism in the autotroph *Acidithiobacillus ferrooxidans*. Hydrometallurgy 83:273–280

Asante KA, Amoyaw-Osei Y, Agusa T (2019) E-waste recycling in Africa: risks and opportunities. Curr Opin Green Sustain Chem 18:109–117

Auerbach R, Bokelmann K, Stauber R, Gutfleisch O, Schnell S, Ratering S (2019a) Critical raw materials—advanced recycling technologies and processes: recycling of rare earth metals out of end-of-life magnets by bioleaching with various bacteria as an example of an intelligent recycling strategy. Miner Eng 134:104–117

Auerbach R, Ratering S, Bokelmann K, Gellermann C, Bramer T, Baumann R, Schnell S (2019b) Bioleaching of valuable and hazardous metals from dry discharged incineration slag. An approach for metal recycling and pollutant elimination. J Environ Manag 232:428–437

Auernik KS, Maezato Y, Blue PH, Kelly RM (2008) Genome sequence of the metalmobilizing, extremely thermoacidophilic archaeon *Metallosphaera sedula* provides insights into bioleaching metabolism. Appl Environ Microbiol 74:682–692

Ayora C, Macias F, Torres E, Lozano A, Carrero S, Nieto J-M, Perez-Lopez R, Fernandez-Martinez A, Castillo-Michel H (2016) Recovery of rare earth elements and yttrium from passive-remediation systems of acid mine drainage. Environ Sci Technol 50(15):8255–8262

Bahaloo-Horeh N, Mousavi SM (2017) Enhanced recovery of valuable metals from spent lithium-ion batteries through optimization of organic acids produced by *Aspergillus niger*. Waste Manag 60:666–679

Baker BJ, Banfield JF (2003) Microbial communities in acid mine drainage. FEMS Microbiol Ecol 44:139–152

Baker-Austin C, Dopson M (2007) Life in acid: pH homeostasis in acidophiles. Trends Microbiol 15(4):165–171. https://doi.org/10.1016/j.tim.2007.02.005

Brar KK, Magdouli S, Etteieb S, Zolfaghari M, Fathollahzadeh H, Calugaru L, Komtchou SP, Tanabene R, Brar SK (2021) Integrated bioleaching-electrometallurgy for copper recovery-A critical review. J Clean Produ 291:125257. https://doi.org/10.1016/j.jclepro.2020.125257

Baldé CP, Wang F, Kuehr R, Huisman J (2015) The global e-waste monitor-2014. United Nations University, IAS-SCYCLW, Bonn, Germany. 978-92-808-4556-3

Baldé CP, Forti V, Gray V, Kuehr R, Stegmann P (2017) The global E-waste monitor, 2017. Quantities, flows, and resources. United Nations University (UNU), International Telecommunication Union (ITU) and International Solid Waste Association (ISWA), Bonn/Geneva/Vienna

Baniasadi M, Vakilchap F, Bahaloo-Horeh N, Mousavi SM, Farnaud S (2019) Advances in bioleaching as a sustainable method for metal recovery from e-waste: a review. J Ind Eng Chem 76:75–90

Barnett MJ, Palumbo-Roe B, Deady EA, Gregory SP (2020) Comparison of three approaches for bioleaching of rare earth elements from bauxite. Minerals 10(8):649

Beale DJ, Karpe AV, Ahmed W, Cook S, Morrison PD, Staley C, Sadowsky MJ, Palombo EA (2017) A community multi-omics approach towards the assessment of surfacewater quality in an urban river system. Int J Environ Res Public Health 14:E303. https://doi.org/10.3390/ijerph14030303

Beni AA, Esmaeili A (2020) Biosorption, an efficient method for removing heavy metals from industrial effluents: a review. Environ Technol Innov 17:100503. https://doi.org/10.1016/j.eti.2019.100503

Bilal T, Malik B, Hakeem KR (2018) Metagenomic analysis of uncultured microorganisms and their enzymatic attributes. J Microbiol Methods 155:65–69. https://doi.org/10.1016/j.mimet.2018.11.014

Bindschedler S, Bouquet TQTV, Job D, Joseph E, Junier P (2017) Fungal biorecovery of gold from e-waste. Adv Appl Microbiol 99:53–81

Bobadilla-Fazzini RA, Levican G, Parada P (2011) Acidithiobacillus thiooxidans secretome containing a newly described lipoprotein Licanantase enhances chalcopyrite bioleaching rate. Appl Microbiol Biotechnol 89(3):771–780. https://doi.org/10.1007/s00253-010-3063-8

Bouhajja E, Agathos SN, George IF (2016) Metagenomics: probing pollutant fate in natural and engineered ecosystems. Biotechnol Adv 34(8):1413–1426

Brierley CL (2010) Biohydrometallurgical prospects. Hydrometallurgy 104(3–4):324–328

Brisson VL, Zhuang WQ, Alvarez-Cohen L (2016) Bioleaching of rare earth elements from monazite sand. Biotechnol Bioeng 113(2):339–348

Buch A, Archana G, Naresh Kumar G (2008) Metabolic channeling of glucose towards gluconate in phosphate-solubilizing *Pseudomonas aeruginosa* P4 under phosphorus deficiency. Res Microbiol 159:635–642

Cardenas JP, Moya F, Covarrubias P, Shmaryahu A, Levicán G, Holmes DS, Quatrini R (2012) Comparative genomics of the oxidative stress response in bioleaching microorganisms. Hydrometallurgy 127–128:162–167. https://doi.org/10.1016/j.hydromet.2012.07.014

Cardenas JP, Quatrini R, Holmes DS (2016) Genomic and metagenomic challenges and opportunities for bioleaching: a mini-review. Res Microbiol 167:529–538. https://doi.org/10.1016/j.resmic.2016.06.007

Cárdenas JP, Valdés J, Quatrini R (2010) Lessons from genomes of extremely acidophilic bacteria and archaea with special emphasis on bioleaching microorganisms. Appl Microbiol Biotechnol 88:605–620

Cayumil R, Khanna R, Rajarao R, Mukherjee PS, Sahajwalla V (2016) Concentration of precious metals during their recovery from electronic waste. Waste Manag 57:121–130

Chao Y, Liu W, Chen Y, Chen W, Zhao L, Ding O (2016) Structure, variation, and co-occurrence of soil microbial communities in abandoned sites of a rare earth elements mine. Environ Sci Technol 50(21):11481–11490

Chandran H, Meena M, Sharma K (2020) Microbial biodiversity and bioremediation assessment through omics approaches. Frontiers Environ Chem 1:570326. https://doi.org/10.3389/fenvc.2020.570326

Chi A, Valenzuela L, Beard S, Mackey AJ, Shabanowitz J, Hunt DF, Jerez CA (2007) Periplasmic proteins of the extremophile *Acidithiobacillus ferrooxidans*: a high throughput proteomic analysis. Mol Cell Proteomics 6(12):2239–2251

Čížková M, Bisova K, Zachleder V, Mezricky D, Rucki M, Vitova M (2019) Recovery of rare earth elements from luminophores using red alga Galdieria. In: 16th international conference on environmental science and technology, pp 1–3

Corbett MK, Eksteen JJ, Niu X-Z, Croue J-P, Watkin ELJ (2017) Interactions of phosphate solubilizing microorganisms with natural rare-earth phosphate minerals: a study utilizing Western Australian monazite. Bioproc Biosys Eng 40(6):929–942

Crocket KC, Hill E, Abell RE, Johnson C, Gary SF, Brand T, Hathorne EC (2018) Rare earth elements distribution in the NE Atlantic: evidence for benthic sources, longevity of the seawater signal, and biogeochemical cycling. Front Mar Sci 5:147

Cui J, Zhang L (2008) Metallurgical recovery of metals from electronic waste: a review. J Hazard Mater 158(2–3):228–256

Das AP, Ghosh S, Mohanty S, Sukla LB (2015) Biometallurgy: greener technology for mineral recovery from wastes. In: Sukla LB, Panda S, Esther J (eds) Applied and industrial biotechnology. J Appl Environ Biotechnol Innov, Bhubaneswar

Daum K (2017) Toward a more sustainable trajectory for E-Waste policy: a review of a decade of E-Waste research in Accra, Ghana. Int J Environ Res Public Health 14:135

Debnath B, Chowdhury R, Ghosh SK (2018) Sustainability of metal recovery from E-waste. Front Environ Sci Eng 12:2. https://doi.org/10.1007/s11783-018-1044-9

Dellas N, Snyder JC, Bolduc B, Young MJ (2014) Archaeal viruses: diversity, replication, and structure. Annu Rev Virol 1:399–426. https://doi.org/10.1146/annurev-virology-031413-085357

Deplanche K, Merroun ML, Casadesus M, Tran DT, Mikheenko IP, Bennett JA, Zhu J, Jones IP, Attard GA, Wood J, Selenska-Pobell S, Macaskie LE (2012) Microbial synthesis of core/shell gold/palladium nanoparticles for applications in green chemistry. J R Soc Interface 9:1705–1712. https://doi.org/10.1098/rsif.2012.0003

Dev S, Sachan A, Dehghani F, Ghosh T, Briggs BR (2020) Mechanisms of biological recovery of rare-earth elements from industrial and electronic wastes: a review. Chem Eng J 397:124596. https://doi.org/10.1016/j.cej.2020.124596

Diaz-Barriga F (2013) Evidence-based intervention programs to reduce children's exposure to chemicals in e-waste sites. Discussion Paper for WHO Working Meeting on E-waste and Children's Health, pp 1–90

Ding Y, Zhang S, Liu B, Zheng H, Chang CC, Ekberg C (2019) Recovery of precious metals from electronic waste and spent catalysts: a review. Resour Conserv Recycl 141:284–298. https://doi.org/10.1016/j.resconrec.2018.10.041

do Carmo TS, Moreira FS, Cabral BV, Dantas RCC, de Resende MM, Cardoso VL, Ribeiro EJ (2019) Phosphorus recovery from phosphate rocks using phosphate-solubilizing bacteria. Geomicrobiol J 36:195–203

Dominguez-Benetton X, Varia JC, Pozo G, Modin O, Heijne AT, Fransaer J, Rabaey K (2018) Metal recovery by microbial electro-metallurgy. Prog Mater Sci 94:435–461

Dusengemungu L, Kasali G, Gwanama C, Mubemba B (2021) Overview of fungal bioleaching of metals. Environ Adv 5:100083. https://doi.org/10.1016/j.envadv.2021.100083

Emmanuel ESC, Vignesh V, Anandkumar B, Maruthamuthu S (2011) Bioaccumulation of cerium and neodymium by *Bacillus cereus* isolated from rare earth environments of Chavara and Manavalakurichi, India. Indian J Microbiol 51(4):488–495

Emmanuel EC, Ananthi T, Anandkumar B, Maruthamuthu S (2012) Accumulation of rare earth elements by siderophore-forming *Arthrobacter luteolus* isolated from rare earth environment of Chavara, India. J Biosci 37(1):25–31

Fathollahzadeh H, Becker T, Eksteen JJ, Kaksonen AH, Watkin ELJ (2018) Microbial contact enhances bioleaching of rare earth elements. Bioresour Technol Rep 3:102–108

Fathollahzadeh H, Eksteen JJ, Kaksonen AH, Watkin ELJ (2019) Role of microorganisms in bioleaching of rare earth elements from primary and secondary resources. Appl Microbiol Biotechnol 103(3):1043–1057. https://doi.org/10.1007/s00253-018-9526-z

Fontana D, Pietrantonio M, Pucciarmati S, Rao C, Forte F (2019) A comprehensive characterization of end-of-life mobile phones for secondary material resources identification. Waste Manag 99:22–30. https://doi.org/10.1016/j.wasman.2019.08.011

Fontana D, Forte F, Pietrantonio M, Pucciarmati S (2020) Recent developments on recycling end-of-life flat panel displays: a comprehensive review focused on indium. Crit Rev Environ Sci Technol 51(5):429–456. https://doi.org/10.1080/10643389.2020.1729073

Forti V, Baldé CP, Ruediger K, Garam B (2020) The global E-waste monitor 2020: quantities, flows, and the circular economy potential. https://globalewaste.org/?fbclid=

Garrett RA, Shah SA, Erdmann S, Liu G, Mousaei M, Leon-Sobrino C, Peng W, Gudbergsdottir S, Deng L, Vestergaard G, Peng X (2015) CRISPR-Cas adaptive immune systems of the sulfolobales: unravelling their complexity and diversity. Life (Basel) 5:783–817. https://doi.org/10.3390/life5010783

Gergoric M, Ravaux C, Steenari B, Espegren F (2018) Leaching and recovery of rare-earth elements from neodymium magnet waste using organic acids. Metals 8(9):721. https://doi.org/10.3390/met8090721

Ghimire H, Ariya PA (2020) E-wastes: bridging the knowledge gaps in global production budgets, composition, recycling and sustainability implications. Sustain Chem 1(2):154–182. https://doi.org/10.3390/suschem1020012

Giese EC (2020) Biosorption as green technology for the recovery and separation of rare earth elements. World J Microbiol Biotechnol 36(4):52. https://doi.org/10.1007/s11274-020-02821-6

Gradael TE, Harper EM, Nassar NT, Reck BK (2015) On the materials basis of modern society. Proc Natl Acad Sci USA 112(20):6295–6300. https://doi.org/10.1073/pnas.131275211

Gupta NK, Gupta A, Ramteke P, Sahoo H, Sengupta A (2019) Biosorption—a green method for the preconcentration of rare earth elements (REE) from waste solutions: a review. J Mol Liq 274:148–164. https://doi.org/10.1016/j.molliq.2018.10.134

Hagelüken C, Corti WC (2010) Recycling of gold from electronics: cost-effective use through 'design for recycling.' Gold Bull 43:209–220

Harler K (2018) Rare opportunity to recycle rare earths. Recycl Today. www.recyclingtoday.com/article/rare-earth-metals-recycling/. Accessed 18 Jan 2022

Hoogerstraete TV, Blanpain B, Gerven TV, Binnemans K (2014) From NdFeB magnets towards the rare-earth oxides: a recycling process consuming only oxalic acid. RSC Adv 4:64099–64111

Horeh NB, Mousavi SM, Shojaosadati SA (2016) Bioleaching of valuable metals from spent lithium-ion mobile phone batteries using *Aspergillus niger*. J Power Sources 320:257–266
Horiike T, Kiyono H, Yamashita M (2018) Dysprosium biomineralization by *Penidiella* sp. Strain T9. In: Endo K, Kogure T, Nagasawa H (eds) Biomineralization. Springer Singapore, Singapore, pp 251–257
Horinouchi T, Tamaoka K, Furusawa C, Ono N, Suzuki S, Hirasawa T, Yomo T, Shimizu H (2010) Transcriptome analysis of parallel-evolved *Escherichia coli* strains under ethanol stress. BMC Genom 11:579. https://doi.org/10.1186/1471-2164-11-579
Iannicelli-Zubiani EM, Giani MI, Recanati F, Dotelli G, Puricelli S, Cristiani C (2017) Environmental impacts of a hydrometallurgical process for electronic waste treatment: a life cycle assessment case study. J Clean Prod 140:1204–1216
Ilyas S, Lee JC, Chi R (2013) Bioleaching of metals from electronic scrap and its potential for commercial exploitation. Hydrometallurgy 131:138–143
Innocenzi V, Ippolito NM, Pietrelli L, Centofanti M, Piga L, Veglio F (2018) Application of solvent extraction operation to recover rare earths from fluorescent lamps. J Clean Prod 172:2840–2852
ITU (2020) The global E-waste statistics partnership: country and regional sheets. https://globalewaste.org/country-sheets/. Accessed 5 July 2020
Jacinto J, Henriques B, Duarte AC, Vale C, Pereira E (2018) Removal and recovery of critical rare elements from contaminated waters by living *Gracilaria gracilis*. J Hazard Mater 344:531–538
Jafari M, Abdollahi H, Shafaei S, Gharabaghi M, Jafari H, Akcil A, Panda S (2019) Acidophilic bioleaching: a review on the process and effect of organic-inorganic reagents and materials on its efficiency. Min Proc Ext Met Rev 40:87–107. https://doi.org/10.1080/08827508.2018.1481063
Jerez CA (2008) The use of genomics, proteomics and other OMICS technologies for the global understanding of biomining microorganisms. Hydrometallurgy 94(1):162–169
Jerez CA (2017) Biomining of metals: how to access and exploit natural resource sustainably. Microb Biotechnol 10(5):1191–1193. https://doi.org/10.1111/1751-7915.12792
Jha MK, Kumari A, Panda R, Kumar JR, Yoo K, Lee JY (2016) Review on hydrometallurgical recovery of rare earth metals. Hydrometallurgy 165:2–26
Jia Y, Tan Q, Sun H, Zhang Y, Gao H, Ruan R (2019) Sulfide mineral dissolution microbes: community structure and function in industrial bioleaching heaps. Green Energy Environ 4:29–37
Jothi VR, Bose R, Rajan H, Jung C, Yi SC (2018) Harvesting electronic waste for the development of highly efficient eco-design electrodes for electrocatalytic water splitting. Adv Energy Mater 8(34):1802615
Jowitt SM, Werner TT, Weng Z, Mudd GM (2018) Recycling of the rare earth elements. Curr Opin Green Sustain Chem 13:1–7
Kadam AR, Nair GB, Dhoble SJ (2019) Insights into the extraction of mercury from fluorescent lamps: a review. J Environ Chem Eng 7(4):103279. https://doi.org/10.1016/j.jece.2019.103279
Kaksonen AH, Deng X, Bohu T (2020) Prospective directions for biohydrometallurgy. Hydrometallurgy 195:105376. https://doi.org/10.1016/j.hydromet.2020.105376
Karahalil G (2016) Overview of systems biology and omics technologies. Curr Med Chem 23(37):4221–4230. https://doi.org/10.2174/0929867323666160926150617
Karatan E, Duncan TR, Watnick PI (2005) NspS, a predicted polyamine sensor, mediates activation of *Vibrio cholerae* biofilm formation by norspermidine. J Bacteriol 187:7434–7443
Kernan T, Majumdar S, Li X, Guan J, West AC, Banta S (2016) Engineering the iron-oxidizing chemolithoautotroph Acidithiobacillus ferrooxidans for biochemical production. Biotechnol Bioeng 113(1):189–197. https://doi.org/10.1002/bit.25703
Kim CJ, Yoon HS, Chung KW, Lee JY, Kim SD, Shin SM, Lee SJ, Joe AR, Lee SI, Yoo SJ, Kim SH (2014) Leaching kinetics of lanthanum in sulfuric acid from rare earth element (REE) slag. Hydrometallurgy 146:133–137
Kivanc K, Alemrajabi M, Rasmuson AC, Forsberg KM (2018) Sustainable hydrometallurgical recovery of valuable elements from spent nickel-metal hydride HEV batteries. Metals 8(12):1062

Lederer FL, Curtis SB, Bachmann S, Dunbar WS, MacGillivray RTA (2017) Identification of lanthanum-specific peptides for future recycling of rare earth elements from compact fluorescent lamps. Biotechnol Bioeng 114:1016–1024

Levett A, Gleeson SA, Kallmeyer J (2021) From exploration to remediation: a microbial perspective for innovation in mining. Earth-Sci Rev 216:103563. https://doi.org/10.1016/j.earscirev.2021.103563

Li H, Zhu F, He S (2019) The degradation of decabromodiphenyl ether in the e-waste site by biochar supported nanoscale zero-valent iron/persulfate. Ecotoxicol Environ Saf 183:109540

Liang Y, Liu Y, Lin R, Guo D, Liao C (2016) Leaching of rare earth elements from waste lamp phosphor mixtures by reduced alkali fusion followed by acid leaching. Hydrometallurgy 163:99–103. https://doi.org/10.1016/j.hydromet.2016.03.020

Lu H, Que Y, Wu X, Guan T, Guo H (2019) Metabolomics deciphered metabolic reprogramming required for biofilm formation. Sci Rep 9:13160. https://doi.org/10.1038/s41598-019-49603-1

Lucier CA, Gareau BJ (2019) Electronic waste recycling and disposal: an overview. In: Assessment and management of radioactive and electronic wastes technology. IntechOpen

Luhar S, Luhar I (2019) Potential application of E-wastes in construction industry: a review. Constr Build Mater 203:222–240

Maleke M, Valverde A, Vermeulen J-G, Cason E, Gomez-Arias A, Moloantoa K, Coetsee-Hugo L, Swart H, van Heerden E, Castillo J (2019) Biomineralization and bioaccumulation of europium by a thermophilic metal resistant bacterium. Front Microbiol 10:81. https://doi.org/10.3389/fmicb.2019.00081

Malla MA, Dubey A, Yadav S, Kumar A, Hashem A, Abd Allah EF (2018) Understanding and designing the strategies for the microbe-mediated remediation of environmental contaminants using omics approaches. Front Microbiol 9:1132. https://doi.org/10.3389/fmicb.2018.01132

Mann M, Hendrickson RC, Pandey A (2001) Analysis of proteins and proteomes by mass spectrometry. Annu Rev Biochem 70(1):437–473. https://doi.org/10.1146/annurev.biochem.70.1.437

Maroufi S, Nekouei RK, Hossain RM, Assefi M, Sahajwalla R (2018) Recovery of rare earth (i.e., La, Ce, Nd, and Pr) oxides from end-of-life Ni-MH battery via thermal isolation. ACS Sustain Chem Eng 6:11811–11818

Marra A, Cesaro A, Rene ER, Belgiorno V, Lens PN (2018) Bioleaching of metals from WEEE shredding dust. J Environ Manag 210:180–190

Marra A, Cesaro A, Belgiorno V (2019) Recovery opportunities of valuable and critical elements from WEEE treatment residues by hydrometallurgical processes. Environ Sci Pollut Res 26(19):19897–19905

Martínez-Bussenius C, Navarro CA, Orellana L (2016) Global response of *Acidithiobacillus ferrooxidans* ATCC 53993 to high concentrations of copper: a quantitative proteomic approach. J Proteomics 145:37–45

Martínez P, Gálvez S, Ohtsuka N, Budinich M, Cortés MP, Serpell C. Nakahigashi K, Hirayama A, Tomita M, Soga T, Martínez S, Maass A, Parada P (2013) Metabolomic study of Chilean biomining bacteria Acidithiobacillus ferrooxidans strain Wenelen and Acidithiobacillus thiooxidans strain Licanantay. Metabolomics 9:247–257. https://doi.org/10.1007/s11306-012-0443-3

Martinez P, Vera M, Bobadilla-Fazzini RA (2015) Omics on bioleaching: current and future impacts. Appl Microbiol Biotechnol 99(20):8337–8350. https://doi.org/10.1007/s00253-015-6903-8

Marques CR (2018) Extremophilic microfactories: applications in metal and radionuclide bioremediation. Front Microbiol 9:1191. https://doi.org/10.3389/fmicb.2018.01191

Minoda A, Sawada H, Suzuki S, Miyashita S-I, Inagaki K, Yamamoto T, Tsuzuki M (2015) Recovery of rare earth elements from the sulfothermophilic red alga *Galdieria sulphuraria* using aqueous acid. Appl Microbiol Biotechnol 99:1513–1519

Mokolo SC (2019) Management of computer and electronic waste in Nigeria for industrialization and greener environment in Nigeria. Submitted to the Computer engineering programme, Department of Computer and Electrical and Electronics Engineering, College of Engineering Afe Babalola University, Ado-Ekiti, Ekiti State, Nigeria

Morikawa M, Kagihiro S, Haruki M, Takano K, Branda S, Kolter R, Kanaya S (2006) Biofilm formation by a *Bacillus subtilis* strain that produces gamma-polyglutamate. Microbiology 152:2801–2807

Muravyov MI, Bulaev AG, Melamud VS, Kondrateva TF (2015) Leaching of rare earth elements from coal ashes using acidophilic chemolithotrophic microbial communities. Microbiology 84:194–201

Murugappan RM, Karthikeyan M (2021) Microbe-assisted management and recovery of heavy metals from electronic wastes. In: Hussain CM (ed) Environmental management of waste electrical and electronic equipment. Elsevier, pp 65–88

Naidu TS, Sheridan CM, Van Dyk LD, Higginson A (2019) Acid mine drainage pilot remediation system using waste products from the steel manufacturing and sugar industries. In: Wolkersdorfer C, Khayrulina E, Polyakova S, Bogush A (eds) IMWA 2019: mine water: technological and ecological challenges, pp 121–126

Nancharaiah Y, Mohan SV, Lens P (2016) Biological and bioelectrochemical recovery of critical and scarce metals. Trends Biotechnol 34:137–155

Nancucheo I, Oliveira G, Lopes M, Johnson DB (2019) Bioreductive dissolution as a pretreatment for recalcitrant rare-earth phosphate minerals associated with lateritic ores. Minerals 9:136

National inventory of dams (NID). https://nid-test.sec.usace.army.mil/. Accessed 12 Sept 2021

Nekouei RK, Pahlevani F, Asse M, Maroufi S, Sahajwalla V (2019) Selective isolation of heavy metals from spent electronic waste solution by macroporous ion-exchange resins. J Hazard Mater 371:389–396. https://doi.org/10.1016/j.jhazmat.2019.03.013

Ning C, Lin CSK, Hui DCW, McKay G (2017) Waste printed circuit board (PCB) recycling techniques. Top Curr Chem 375:331

Ochsner AM, Hemmerle L, Vonderach T, Nussli R, Bortfeld-Miller M, Hattendorf B, Vorholt JA (2019) Use of rare-earth elements in the phyllosphere colonizer *Methylobacterium extorquens* PA1. Mol Microbiol 111:1152–1166

Odeyingbo O, Nnorom I, Deubzer O (2017) Person in the port project—assessing import of used electrical and electronic equipment into Nigeria. http://collections.unu.edu/eserv/UNU:6349/PiP_Report.pdf

Odeyingbo AO, Nnorom IC, Deubzer OK (2019) Used and waste electronics flows into Nigeria: assessment of the quantities, types, sources, and functionality status. Sci Total Environ 666:103–113

Oliveira RC, Garcia O Jr (2009) Study of biosorption of rare earth metals (La, Nd, Eu, Gd) by *Sargassum* sp. biomass in batch systems: physicochemical evaluation of kinetics and adsorption models. Adv Mater Res 71:605–608. https://doi.org/10.4028/www.scientific.net/AMR.71-73.605

Omole DO, Tenebe IT, Emenike CP, Umoh AS, Badejo AA (2015) Causes, impacts and management of electronic wastes: case study of some Nigerian communities. ARPN J Eng Appl Sci 10(18):7876–7884

Onal MAR, Binnemans K (2019) Recovery of rare earths from waste cathode ray tube (CRT) phosphor powder by selective sulfation roasting and water leaching. Hydrometallurgy 183:60–70

Onwosi CO, Igbokwe VC, Nwagu TN, Odimba JN, Nwuche CO (2020) E-waste management from macroscopic to microscopic scale. In: Khan A, Inamuddin AA (eds) E-waste recycling and management. Environmental chemistry for a sustainable world, vol 33. Springer, Cham, pp 143–157

Orell A, Navarro CA, Arancibia R (2010) Life in blue: copper resistance mechanisms of bacteria and archaea used in industrial biomining of minerals. Biotechnol Adv 28:839–848

Orisakwe OE, Frazzoli C, Ilo CE, Oritsemuelebi B (2019) Public health burden of e-waste in Africa. J Health Pollut 9(22):190610. https://doi.org/10.5696/2156-9614-9.22.190610

Osman Y, Gebreil A, Mowafy AM, Anan TI, Hamed SM (2019) Characterization of *Aspergillus niger* siderophore that mediates bioleaching of rare earth elements from phosphorites. World J Microbiol Biotechnol 35(6):93. https://doi.org/10.1007/s11274-019-2666-1

Oulas A, Pavloudi C, Polymenakou P, Pavlopoulos GA, Papanikolaou N, Kotoulas G, Arvanitidis C, Iliopoulos I (2015) Metagenomics: tools and insights for analyzing next-generation sequencing

data derived from biodiversity studies. Bioinform Biol Insights 9:75–88. https //doi.org/10.4137/BBI.S12462
Palumbo-Roe B, Gregory S, Milodowski A, West J, Wragg J, Banwart S, Romero Gonzalez M, Huang W, Wharfe E, Harding J, Freeman C, Hall S (2014) Using microbes to recover rare earths with low environmental impact? [Poster] In: ERES 2014—1st conference on European rare earth resources, Milos, Greece, 4–7 September, 2014
Pant D, Joshi D, Upreti MK, Kotnala RK (2012) Chemical and biological extraction of metals present in e-waste: a hybrid technology. Waste Manag 32(5):979–990
Park S, Liang Y (2019) Bioleaching of trace elements and rare earth elements from coal fly ash. Int J Coal Sci Technol 6(1):74–83. https://doi.org/10.1007/s40789-019-0238-5
Park DM, Brewer A, Reed DW, Lammers LN, Jiao Y (2017) Recovery of rare earth elements from low-grade feedstock leachates using engineered bacteria. Environ Sci Technol 51:13471–13480
Pinegar H, Smith YR (2019) Recycling of end-of-life lithium ion batteries, Part I: Commercial processes. J Sustain Metall 5:402–416. https://doi.org/10.1007/s40831-019-00235-9
Plyatsuk L, Balintova M, Chernysh Y, Ablieieva I, Ablieiev O (2020) The process of environmentally safe biochemical recycling of phosphogypsum. In: Ivanov V et al (eds) Advances in design, simulation and manufacturing II. DSMIE 2019. Lecture notes in mechanical engineering. Springer International Publishing, Cham, pp 843–852
Pollmann K, Kutschke S, Matys S, Raff J, Hlawacek G, Lederer FL (2018) Bio-recycling of metals: recycling of technical products using biological applications. Biotechnol Adv 36(4):1048–1062. https://doi.org/10.1016/j.biotechadv.2018.03.006
Priya A, Hait S (2017) Comparative assessment of metallurgical recovery of metals from electronic waste with special emphasis on bioleaching. Environ Sci Pollut Res 24:6989–7008
Qu Y, Lian B (2013) Bioleaching of rare earth and radioactive elements from red mud using *Penicillium tricolor* RM-10. Bioresour Technol 136:16–23
Qu Y, Li H, Wang X, Tian W, Shi B, Yao M, Zhang Y (2019) Bioleaching of major, rare earth, and radioactive element from red mud by using indigenous chemoheterotrophic bacterium *Acetobacter* sp. Minerals 9(2):67
Quatrini R, Appia-Ayme C, Dennis Y, Ratouchniak J, Veloso F, Valdes J, Lefimil C, Silver S, Roberto F, Orellana O, Denizot F, Jedlicki E, Holmes D, Bonnefoy V (2006) Insights into the iron and sulfur energetic metabolism of *Acidithiobacillus ferrooxidans* by microarray transcriptome profiling. Hydrometallurgy 83:263–272
Ram P, Goren A, Ferdov S, Silva MM, Singhal R, Costa CM, Sharma RK, Lanceros-Mendez S (2016) Improved performance of rare earth doped $LiMn_2O_4$ cathodes for lithium-ion battery applications. New J Chem 40(7):6244–6252
Rana S, Mishra P, Ab Wahid Z, Thakur S, Pant D, Singh L (2020) Microbe-mediated sustainable bio-recovery of gold from low-grade precious solid waste: a microbiological overview. J Environ Sci 89:47–64
Rasoulnia P, Barthen R, Lakaniemi A (2020) A critical review of bioleaching of rare earth elements: the mechanisms and effect of process parameters. Crit Rev Environ Sci Technol 51(4):378–427. https://doi.org/10.1080/10643389.2020.1727718
Reed DW, Fujita Y, Daubaras DL, Jiao Y, Thompson VS (2016) Bioleaching of rare earth elements from waste phosphors and cracking catalysts. Hydrometallurgy 166:34–40
Remonsellez F, Galleguillos F, Moreno-Paz M, Parro V, Acosta M, Demergasso C (2009) Dynamic of active microorganisms inhabiting a bioleaching industrial heap of low-grade copper sulfide ore monitored by real-time PCR and oligonucleotide prokaryotic acidophile microarray. Microb Biotechnol 2:613–624
Rutherford BJ, Dahl RH, Price RE, Szmidt HL, Benke PI, Mukhopadhyay A, Keasling JD (2010) Functional genomic study of exogenous n-butanol stress in *Escherichia coli*. Appl Environ Microbiol 76:1935–1945. https://doi.org/10.1128/AEM.02323-09
Schluep M, Hagelüken C, Meskers C, Magalini F, Wang F, Mueller E, Kuehr R, Maurer C, Sonnemann G (2009) Market potential of innovative e-waste recycling technologies in developing

countries. R'09 Twin World Congress on Resource Management and Technology for Material and Energy Efficiency at Davos Switzerland, September 2009
Sethurajan M, van Hullebusch ED, Nancharaiah YV (2018) Biotechnology in the management and resource recovery from metal bearing solid wastes: recent advances. J Environ Manag 211:138–153
Shehada AM, Kharita MH, Yousef S (2018) Studying characteristics of radiation shielding glass made of glass of cathode ray tubes and heavy metal oxides. Int J Electr Electron Sci 5(1):14–32
Shi J, Miao X, Zhang Z (2019) Effect of different particle sizes of waste cathode-ray tube glass sand on the properties of concrete. Bull Chin Ceram Soc 38(4):1191–1195
Siezen RJ, Wilson G (2009) Bioleaching genomics. Microb Biotechnol 2:297–303
Song Q, Wang Z, Li J, Duan H (2012) Sustainability evaluation of an e-waste treatment enterprise based on emergy analysis in China. Ecol Eng 42:223–231
Sovacool BK (2019) Toxic transitions in the lifecycle externalities of a digital society: The complex afterlives of electronic waste in Ghana. Resour Policy 64:101459. https://doi.org/10.1016/j.resourpol.2019.101459
Sturgill G, Rather PN (2004) Evidence that putrescine acts as an extracellular signal required for swarming in *Proteus mirabilis*. Mol Microbiol 51:437–446
Takahashi Y, Yamamoto M, Yamamoto Y, Tanaka K (2010) EXAFS study on the cause of enrichment of heavy REEs on bacterial cell surfaces. Geochim Cosmochim Acta 74:5443–5462
Terui Y, Ohnuma M, Hiraga K, Kawashima E, Oshima T (2005) Stabilization of nucleic acids by unusual polyamines produced by an extreme thermophile, Thermus thermophilus. Biochem J 388(2):427–433. https://doi.org/10.1042/BJ20041778
TerAvest MA, Ajo-Franklin CM (2016) Transforming exoelectrogens for biotechnology using synthetic biology. Biotechnol Bioeng 113(4):687–697. https://doi.org/10.1002/bit.25723
Teye AK, Tetteh IK (2021) Assessment of heavy metal pollution resulting from informal E-wastes recycling in the Greater Accra Region of Ghana
Torres S, Pandey A, Castro GR (2011) Organic solvent adaptation of Gram-positive bacteria: applications and biotechnological potentials. Biotechnol Adv 29:442–452. https://doi.org/10.1016/j.biotechadv.2011.04.002.g
Tran CD, Salhofer SP (2018) Analysis of recycling structures for e-waste in Vietnam. J Mater Cycles Waste Manag 20(1):110–126. https://doi.org/10.1007/s10163-016-0549-1
U. S. Department of Energy (2010) Report on rare earth elements from coal and coal byproducts. Report to Congress. https://www.energy.gov/sites/prod/files/2018/01/f47/EXEC-2014-000442%20-
Valenzuela L, Chi A, Beard S (2006) Genomics, metagenomics and proteomics in biomining microorganisms. Biotechnol Adv 24:197–211
Valdes J, Cárdenas JP, Quatrini R, Esparza M, Osorio H, Duarte F, Lefimil C, Sepulveda R, Jedlicki E, Holmes DS (2010) Comparative genomics begins to unravel the ecophysiology of bioleaching. Hydrometallurgy 104(3–4):471–476.
Venkatesan P, Vander Hoogerstraete T, Binnemans K, Sun Z, Sietsma J, Yang Y (2018) Selective extraction of rare-earth elements from NdFeB magnets by a room-temperature electrolysis pretreatment step. ACS Sustain Chem Eng 6:9375–9382
Vera M, Schippers A, Sand W (2013) Progress in bioleaching: fundamentals and mechanisms of bacterial metal sulfide oxidation—Part A. Appl Microbiol Biotechnol 97:7529–7541
Vijayaraghavan K, Sathishkumar M, Balasubramanian R (2010) Biosorption of lanthanum, cerium, europium, and ytterbium by a brown marine alga, *Turbinaria conoidesa*. Ind Eng Chem Res 49(9):4405–4411. https://doi.org/10.1021/ie1000373
Wang Q, Deng J, Liu X, Zhang Q, Sun S, Jiang C, Zhou F (2010) Discovery of the REE minerals and its geological significance in the Quyang bauxite deposit, West Guangxi, China. J Asian Earth Sci 39(6):701–712. https://doi.org/10.1016/j.jseaes.2010.05.005
Wang H, Liu X, Liu S, Yu Y, Lin J, Lin J, Pang X, Zhao J (2012) Development of a markerless gene replacement system for *Acidithiobacillus ferrooxidans* and construction of a pfkB mutant. Appl Environ Microbiol 78:1826–1835. https://doi.org/10.1128/AEM.07230-11

Wang Z, Dai S, Zou J, French D, Graham IT (2019) Rare earth elements and yttrium in coal ash from the Luzhou power plant in Sichuan, Southwest China: concentration, characterization and optimized extraction. Int J Coal Geol 203:1–14

Wheaton G, Counts J, Mukherjee A (2015) The confluence of heavy metal biooxidation and heavy metal resistance: implications for bioleaching by extreme thermoacidophiles. Minerals 5:397–451

Wijte D, van Baar BL, Heck AJ, Altelaar AF (2011) Probing the proteome response to toluene exposure in the solvent tolerant *Pseudomonas putida* S12. J Proteome Res 10:394e403. https://doi.org/10.1021/pr100401n

Wong YC, Al-Obaidi KM, Mahyuddin N (2018) Recycling of end-of-life vehicles (ELVs) for building products: concept of processing framework from automotive to construction industries in Malaysia. J Clean Prod 190:285–302

Yu Y, Liu X, Wang H, Li X, Lin J (2014) Construction and characterization of tetH overexpression and knockout strains of *Acidithiobacillus ferrooxidans*. J Bacteriol 196:2255–2264. https://doi.org/10.1128/JB.01472-13

Zhang L, Dong H, Liu Y, Bian L, Wang X, Zhou Z, Huang Y (2018) Bioleaching of rare earth elements from bastnaesite-bearing rock by actinobacteria. Chem Geol 483:544–557

Zhuang W-Q, Fitts JP, Ajo-Franklin CM, Maes S, Alvarez-Cohen L, Hennebel T (2015) Recovery of critical metals using biometallurgy. Curr Opin Biotechnol 33:327–335

Diverse Technological Initiatives for E-Waste Management and Its Impact on Ecosystem

Sujit Das, Bishal Gupta, and Abhijit Sarkar

Abstract In the current scenario, e-waste is a growing concern for every nation in the world; because every year, its production gives a shocking outcome. Here, we have taken an attempt to determine the link between business opportunities and recycling of e-waste by applying microbial activities. To build such a research hypothesis, the advances in the biotechnology-driven recycling process for metal extraction from e-waste have been analyzed in brief. E-waste contains metals, toxic substances, plastic, and halogenated compounds in it. The toxicity of e-waste is unsatisfactory; because it pollutes soil, water, and air by releasing different toxic compounds. Not only human beings it also affects plants and animals adversely in the landfills areas. To tackle a huge amount of e-waste, every country follows some regulating approaches. But, its management is a serious issue owing to informal recycling activities, lack of infrastructure, lack of awareness, and sensitization to handle different types of e-waste's components in developing countries. Proper management can reduce e-waste and develop a sustainable environment. Besides this, e-waste can boost a country's economy if proper management is going on; as it contains valuable materials; recovering such materials can help to earn money, creates job, and also reduce raw materials. So, it is serious as well as beneficial for a country.

Keywords Electrical and electronic equipment (EEE) · E-wastes · Biotechnological recycling · Toxicity · Management · Sustainable environment

1 Introduction

In recent times, e-waste is a rising global concern that is increasing day by day in developed and developing countries (Kumar et al. 2017; Ikhlayel 2018; Li and Achal 2020; Adam et al. 2021). Generally, it means electronic waste when electrical and electronic equipments come to their end life (Kiddee et al. 2013; Lu et al. 2015;

S. Das · B. Gupta · A. Sarkar (✉)
Laboratory of Applied Stress Biology, Department of Botany, University of Gour Banga, Malda, West Bengal, India
e-mail: abhijitbhu@gmail.com; abbhijitbot@ugb.ac.in

M. Jawaid and A. Khan (eds.), *Conversion of Electronic Waste in to Sustainable Products*, Sustainable Materials and Technology,
https://doi.org/10.1007/978-981-19-6541-8_3

Sankhla et al. 2016; Ilankoon et al. 2018). Directive 2012/19/EU of the European Parliament of the Council defines e-waste as "Electrical and electronic equipment including all components, sub-assemblies and consumables which are part of the product at the time of discarding" (EU 2012). Examples such as phones, laptops, desktops, projectors, electronic toys, electric cables, audio cassettes, Compact Discs (CDs), Digital Video Discs (DVDs), printed circuit boards, refrigerators, air condition, etc. (Nnorom and Osibanjo 2008; Pariatamby and Victor 2013; Ikhlayel 2018; Ahirwar and Tripathi 2021). E-waste increases day by day with the significant amount of upgradation of electrical devices. With the upgradation in technology constantly people are renewing their electronic devices and gadgets which makes a huge number of discarded products very quickly which increases e-waste to the solid waste stream (Perkins et al. 2014). The electronic Industrial revolution increases the market penetration of the product in developing countries and advanced technology during the last century has changed people's lives all over the world (Halim and Suharyanti 2020; Murthy 2012). For example, 80 percent of phones have been upgraded every 2 years which makes a major waste stream in the world. As per the current situation, e-waste is growing almost three times more globally. In 2019, 54 million tons of e-waste generates globally; and in India, it is 3.2 million tons. But, it is projected to increase by 75 million tons in 2030 worldwide (Forti et al. 2020). E-waste contains valuable metals such as gold, silver, palladium, and platinum. Apart from these, it also contains a wide range of toxic substances. When these substances release in such a concentration from dumping sites and or during the recycling of e-waste, exceed their standards; and create a severe impact on human health as well as on the environment (Namias 2013; Li and Achal 2020). Therefore, its management should be performed in a scientific way. But, due to a lack of knowledge in e-waste management, today many developing countries face serious issues. This country not only generates a huge amount of e-waste due to its fast consumption; and also provides a huge amount of data all over the world as data saves in storage which makes their personal information unsafe (RACGP e-health). If it is managed scientifically, it will reduce the number of e-wastes in landfills. E-waste contsins Some valuable materials such as gold, platinum, silver, copper, and aluminum which are existent in e-waste (Fedorenko 2016). If we reclaim such valuable materials, it also helps to reduce the production of new materials; if we reused or donate discarded electronic devices can also reduce e-waste. So, e-waste can make circular money and job in a country if proper management is imposed (Heacock et al. 2016). If its management is not maintained scientifically, it will become a serious environmental concern (Song and Li 2015). Keeping it in mind, the present study discusses the detailed views of e-waste and its biotechnological recycling, benefits, impact, management, and overall sustainability.

2 Categories of E-Waste

Any electrical and electronic equipment or device is unable to play their work properly, that is discarded, known as e-waste. A suitable definition is very crucial for the development of e-waste policies and disposal standards. Balde et al. (2017) in their study have reported that e-wastes are divided into six categories: (1) temperature exchange instruments such as a heater, air conditioner, cooler, freezer, and refrigerator; (2) screen and monitors such as television, laptop, desktop, notebook, and tablets; (3) small household equipments which include toaster, vacuum cleaner, micro oven, trimmer, camera, medical devices, etc.; (4) large household equipments that include washing machine, cloth dryer, electric stove, etc.; (5) lighting equipments such as lamps and LED lamps; (6) small IT telecommunication equipment includes mobile phone, GPS, telephones, routers, printers, etc.

3 Existence Elements in E-Waste

E-wastes mostly contain the valuable as well as most toxic materials (Table 1); and this composition depends on what kind of electronic device it is (Mmereki et al. 2016). For example, a cell phone comprises almost 40 elements such as metals like copper (Cu), tin (Sn), lithium (Li), cobalt (co), and antimony (Sb) and valuable metals like silver (Ag), gold (Au), and platinum (Mmereki et al. 2016). Cathode ray tubes (CRTs) are present in television, laptop, and desktop monitors and contain barium (Ba), zinc (Zn), lead (Pb), cadmium (Cd), and Cu (Osibanjo and Nnorom 2007; Sing et al. 2016). Printed circuit boards are found in many electronic devices such as stabilizers and lights; and contain mercury (Hg), arsenic (Sb), chromium (Cr), Cd and Pb, etc. (Kumar and Singh 2014; Annamalai 2015; Mmerek et al. 2016; Hembrom et al. 2020). Many e-wastes such as air conditioner freeze and refrigerators contain toxic substances like chlorofluorocarbons (CFCs) (Mmerek et al. 2016). According to Purchase et al. (2020), plastic is a significant component of e-waste, consisting primarily of thermoplastics that may be reprocessed and repurposed. In the production of electrical and electronic equipment (EEE), different types of engineering plastics such as acrylonitrile butadiene styrene (ABS), polypropylene (PP), polyamide (PA), styrene-acrylonitrile (SAN) and polyurethane (PU) are used; and release dioxin identified. Therefore, Mmereki et al. (2016) reported that specific contents such as plastics, metal plastic mixture, and cables are present in waste electrical and electronic equipment (Fig. 1).

Halogenated compounds mean the addition of a halogen to an organic compound. But, e-waste also contains such compounds like polychlorinated biphenyls (PCBs), polychlorinated diphenyl ethers (PCDE), Polybrominated diphenyl (PBD), and polybrominated diphenyl ether (PBDE) mainly in the form of flame retardants (FRs); that are the mixture of plastic and other components. Such compounds are used in printed

Table 1 some waste from electrical and electronic equipments and their existence elements

Waste from electrical and electronic equipment (WEEE)	Existence elements	References
Printed circuit boards in mobile phones	Copper (Cd), tin (Sn), cobalt (Co), gold (Au), silver (Ag), etc.	Chi et al. (2011), Mmereki et al. (2016)
Cathode ray tubes found in television and computer monitors	Lead (Pb), Barium (Ba), cobalt (Co), zinc (Zn), copper (Cu), etc.	Osibanjo and Nnorom (2007), Sing et al. (2016)
Circuit boards found in stabilizers and lights	Mercury (Hg), arsenic (Sb), chromium (Cr), cadmium (Cd), etc.	Kumar and Singh (2014), Annamalai (2015), Mmerek et al. (2016), Hembrom et al. (2020)
Plastics in electrical and electronic equipment (EEE)	Dioxins, Polybrominated diphenyl (PBD) and polybrominated diphenyl ether (PBDE)	Purchase et al. (2020)

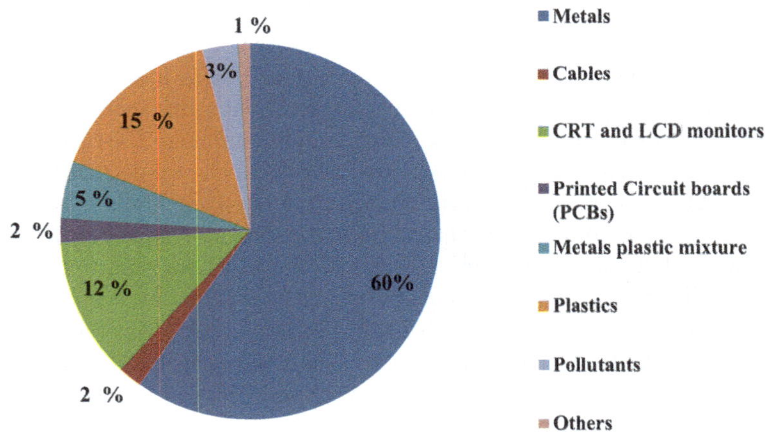

Fig. 1 Waste from electrical and electronic equipment (WEEE) and its specific components

circuit Boards (PCBs), plastic housing, cables, chargers, keyboards, and liquid crystals in LCD (Purchase et al. 2020). Organic chemicals such as carbon, hydrogen, oxygen, and fluorine make up the majority of liquid crystals found in LCDs. Although liquid crystals used as mixtures in display technology have been determined to be non-toxic, uncontrolled burning or cremation could produce hydrogen fluoride and organofluorine compounds (Ikhlayel 2017, 2018; Purchase et al. 2020).

4 Generation of E-Waste at Global Level

Nowadays, technology is increasing fastly that makes the buyers to upgrade their electronic equipment; and this increases consumption, which brings more waste to the environment as the older devices have no more use (Murthy 2012; Sivaramanan 2013). Forti et al. (2020) in their study have reported that worldwide approximately 53.6 million tons (Mt) of e-waste are generated during the year of 2019. Furthermore, it is expected to increase and reach 74 Mt in 2030 worldwide. Although, Asia takes place first position for e-waste production as it contributes a total of 24.9 Mt e-waste production; followed by America (13.1 Mt) and Europe (12 Mt). Nevertheless, China, India, Japan, and Indonesia contribute a total of 10.1, 3.23, 2.57, and 1.62 Mt of e-waste production in 2019; they share totally 70.36% of e-waste production; therefore, these countries are identified as the biggest e-waste generator in Asian continent during the year of 2019.

5 Recovering of Valuable Metals from E-Waste

In the USA, approximately a total of 29,000 tons of metal including 4500 tons of aluminum, 19,900 tons of steel, 4600 tons of copper, and 1 ton of valuable metals (gold, silver, platinum, and palladium) are recovered from e-waste by recycling process during the period of 1998. After sorting, e-wastes are transferred to metal recovery facilities. If metal recovery facilities are small or intermediate, e-wastes are sent to scrape dealers. Then, the scrape dealer sells them to the overseas market or sends them to the smelter for further processing. Kang and Schoenung (2005) have depicted that the extraction of precious metals such as gold, silver, platinum, and palladium from e-waste is a step-by-step process (Fig. 2). At first, e-wastes are cut into small pieces. Then, these are transferred to the conveyer belt in the magnetic separation system. Due to magnetic attraction, ferrous metals are attached to the belt; and non-ferrous such as aluminum and copper are deposited in a bin by gravity. Non-ferrous metals (Aluminum and copper) are passed into the eddy current separator where magnates inside the shell rotate at high speed. As a result, a magnetic field is formed around these metals. But, the polarity of this magnetic field and the rotating magnet is the same. Therefore, non-ferrous metals such as aluminum and copper are repelled away. The anode slime of copper is leached by pressure from the copper electrolysis process. Thereafter, these are dried; and then some fluxes are added with them; and transferred to the furnace for smelting. At the time of smelting, selenium is recovered. Then, high-intensity electrolytic refining process is performed; consequently, high-purity cathode and anode slime is formed. Then, gold is leached from anode slime and silver is leached from cathode slime and precipitated as sludge.

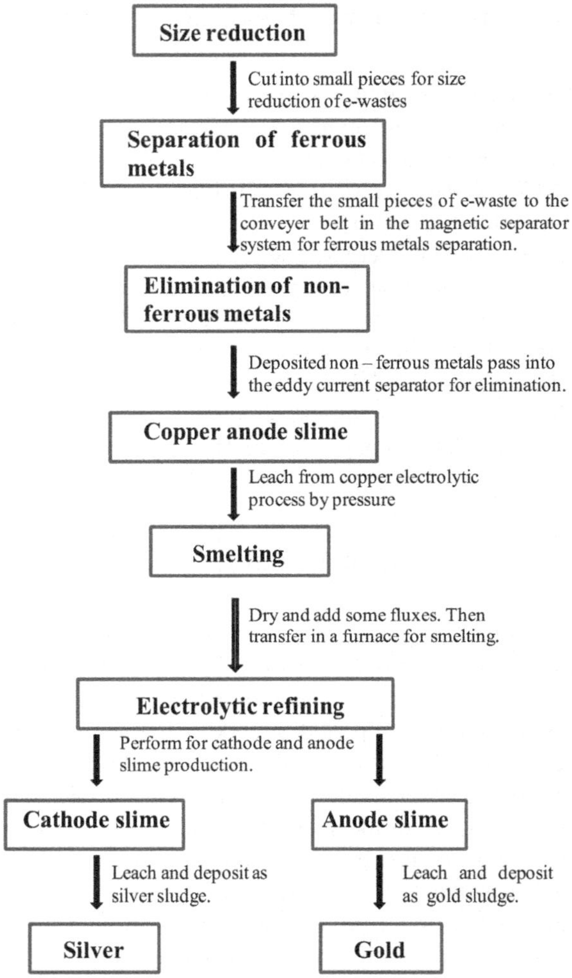

Fig. 2 Simplified flow chart for precious metals such as gold and silver extraction from e-waste

Therefore, it has been determined that recycled precious metals such as gold and silver improve economic conditions worldwide in developed and developing countries.

6 Biotechnological Recycling of E-Wastes

Generally, e-waste recycling is smelting-based process that needs high energy intensive and emits high concentration of greenhouse gases which has adverse human health impacts (Rocchetti et al. 2013). Other hand, biotechnology-driven e-waste recycling is the most eco-friendly process. Here instead of smelting, micro-organisms

Fig. 3 Simplified flow chart of the biotechnological recycling process for Copper (Cu) and Gold (Au) extraction from Printed Circuit Boards (PCBs) of mobile phones

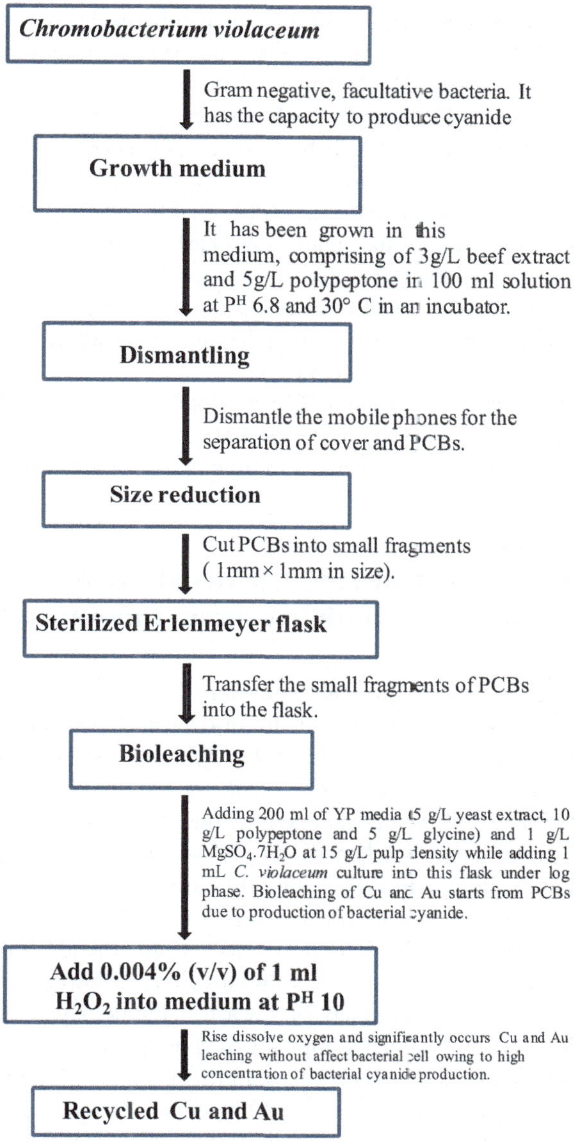

are used to extract metals from e-waste. Therefore, the emission of greenhouse gases decreases and also makes a sustainable environment (Brandl et al. 2001; Yang et al. 2014; Ilyas et al. 2021). Chi et al. (2011) in their study have depicted that *Chromobacterium violaceum* (C. violaceum) is a cyanide-generating mesophilic, facultative gram-negative bacterium, which has been used to extract gold and copper from the waste mobile phone printed circuit boards (Fig. 3). At first, these bacteria are grown in a media comprising of 3 g/L beef extract and 5 g/L polypeptone in 100 ml solution at pH 6.8 and 30° C in an incubator. Thereafter, mobile phones are dismantled for the separation of covers and printed circuit boards (PCBs). Then, the size of PCBs is reduced by cutting into small pieces in the size of 1 mm × 1 mm. After that, a sterilized Erlenmeyer flask is taken and transferred the small fragments of PCBs into the flask; and finally bioleaching of Cu and Au occurs owing to bacterial cyanide production after adding 200 ml of YP media (5 g/L yeast extract, 10 g/L polypeptone, and 5 g/L glycine) and 1 g/L $MgSO_4.7H_2O$ at 15 g/L pulp density while adding 1 mL *C. violaceum* culture under log phase. Thereafter, the flask is incubated at 30 °C; and also added 0.004 percent (v/v) of 1 ml H_2O_2 into the flask containing medium at pH 10. As a result, dissolved oxygen (DO) increases, which significantly enhance Au and Cu leaching without affecting bacterial cell; due to the high concentration of bacterial cyanide production.

Pradhan and Kumar (2012) supported this biotechnological e-waste recycling process and also they reported that *C. violaceum* extracts 79% of Cu and 69% of Au, respectively, from printed circuit boards of mobile phones.

7 Tokyo Olympic 2020 Medals Made from Electronic Scrap: An Initiative for an Innovative Future for All

The Olympic Games are held every two years as a global competition where more than 200 countries participate in over 400 events representing international unity. Each and every medal for Tokyo Olympic 2020 are made from the metals; which are extracted from e-wastes by the recycling process. It has Japan's goal of making Olympics 2020 the "most environmentally friendly and sustainable Games so far." The initiative is taken by the Tokyo Organizing Committee of the Olympic and Paralympic Games (TOCOG) to craft sustainable medals in an effort to build "an innovative future for all" (Hernandez 2021).

The "Tokyo 2020 Medal Project" is an official "Tokyo (2020) Nationwide Participation Programme." This initiative has followed the tradition of making medals from recycled products of e-wastes. Previously, it has implemented in the 2010 Vancouver Winter Olympics which adopted a similar method of reusing e-wastes for medals (Hernandez 2021). In Rio Summer Olympics 2016, the silver and bronze medals are also made from 30 percent recycled materials. However, Tokyo (2020) Olympic medals are created entirely from recycled products of electronic waste. Tokyo (2020) Medal Project is unique in its scale. It is the first time in Olympic history that 100% of the required medals are made from recycled electronic wastes proactively involving the country's citizens in donating the obsolete electronic gadgets (Great Lakes Electronic Corporation). ReNet Japan Group, one of the leading companies whose fundamental thrust revolves around sustainability, is engaged in this project. Toshio Kamakura, Director of ReNet Japan Group, says, "We developed a waste management movement for the medal project with the cooperation of many stakeholders, from the Japanese government to local communities" (Shallow 2021).

The collection of electronic gadgets (e-wastes) is commenced on April 1, 2017 and closed on March 31, 2021. In the 2 years, 100% of the metals require to craft nearly 5000 gold, silver and bronze medals are reached. According to the International Olympic Committee (IOC), nearly 78,985 tons of used electronic devices are collected by 1621 municipal authorities across Japan. Besides, 90% of the 1741wards/cities/towns/villages across the country participate in this mega project. Approximately 6.21 million used cell phones and other gadgets like digital cameras, handheld games, and laptops are collected by NTT Docomo shops across Japan. To maximize the collection, yellow donation boxes are mounted in different post offices and street corners across the nation. TOCOG partner encourage the citizens to donate their old cell phones voluntarily at one of 2400 stores across the nation. The highly trained contractors are then employed to classify, dismantle, and smelt all collected e-wastes (Kamczyc 2021). Finally, about 32 kg of gold, 3500 kg of silver, and 2200 kg of bronze metals are collected to reach the goal within time. The Tokyo (2020) medal ribbon, too, is made in an eco-friendly manner using "chemically recycled polyester fibers; that produce less CO_2 during their manufacturing process" (DH Web desk 2021). To lessen the environmental footprint, the other sustainability practices in Tokyo Olympics 2020 include lightweight recycled cardboard beds for the athletes' sleeping quarters, mattresses, podiums, and electric cars.

The precious metals such as gold and silver which are used in electronic gadgets get thrown away each year worldwide. Thus the billion dollars get wasted simply by dumping or burning what could be collected and recycled. According to the United Nations, in 2019 a record of 53.6 million tons or 7.3 kg per person of e-waste is produced worldwide. In the last five years, e-waste production has grown by five times due to the growing demand for electronic gadgets. Less than a fifth of the e-waste is properly gathered and recycled, thus posing serious environmental and health risks (Marshall 2021). Recycling electronic wastes saves the environment in a number of ways and thus reaps benefits to the human health. Molding medals from e-waste recycling has obviously lessened both the greenhouse gas emissions that are primarily generated in the making of medals and energy consumption therein. Traditionally, medals are molded using raw materials. Leader et al. (2017) estimated that the metals are extracted from the recycled e-waste for Olympic medals instead of traditional raw materials; that saved approximately 4.5–5.1 TJ of energy and reduced 420 metric tons of CO_2. This reduction in CO_2 emissions along with energy savings has resulted in indirect health benefits. If not recycled, these wastes would go into landfills and as a result the noxious toxins would leach out into soils and groundwater. E-waste is comprised of brominated flame retardants, arsenic, cadmium, chromium, lead, and mercury, all of which have health risks (Leader et al. 2017). Therefore, e-waste recycling is a best option to build sustainable environment and better quality of life for all.

8 Impacts of E-Waste on the Ecosystem

Uncontrolled, informal dumping can lead to very devastating situations around the globe. When dumped in an informal way it can directly affect our ecosystem. Formal e-waste recycling focuses on the disassembly and mechanical processing of EEE in order to recover valuable elements. Informal e-waste recycling has the potential to unleash a huge amount of pollutants into recycling facilities and the environment, for example, it includes open burning to recover valuable materials (Purchase et al. 2020). Informal or formal both have very bad effects on their surrounding environment as they have toxic materials. Therefore, it has been depicted that these processes have become major environmental fretfulness (Sarkar et al. 2013). But, it has been reported that such type of concern minimizes owing to proper recycling of e-waste (Sharma et al. 2017).

8.1 Atmospheric Environment

Large volumes of fine particles, gases, and smoke are released into the environment by the formal or informal e-waste recycling sector, which includes physically breaking apart the components and open pit or barrel burning. Fine particles can travel long distances through the air from their point of origin (Das et al. 2021), thus impacting communities far away from where the pollution is generated. Metals, NOx, and other aqua regia gaseous by-products, such as NOCl, Cl_2, and organic contaminants, are all present in these, resulting in air pollution. Surface dust samples taken at e-waste recycling facilities have shown copper, lead, nickel, and Zinc. Open burning is frequently employed at informal recycling facilities to destroy the plastic components in WEEE, allowing the valuable materials within to be salvaged. As a result, the quantities of halogenated chemicals in these locations are of great interest and concern such as PBDEs and PCBs. For example, samples collected from Delhi, PCBs, cadmium (Cd), chromium, copper, Mercury, lead, and zinc (Zn) were detected at high concentrations in India's e-waste, with the distribution of metals in the dust varying according to the type of e-waste (purchase et al. 2020).

8.2 Terrestrial Environment

The availability and mobility of metals are influenced by the diverse physico-chemical and biological features of soils. Around informal recycling sites, large levels of metals, particularly copper, lead, and zinc, have been observed. For example, Luo et al. (2011) found metal levels (cadmium, copper, lead, and zinc) that were higher in an old e-waste incinerator in the Chinese province of Guangdong than the Dutch standard's actionable values. Antimony, cadmium, copper, lead, nickel, tin, and zinc were detected in significant concentrations in soil samples collected from an acid leaching area in an informal recycling site in Guiyu, China. Metal levels at e-waste dumping and recycling facilities in the Mandoli industrial area in India reached huge levels of 6734 mg per kg, copper, 2645 mg per kg, lead, and 776 mg per kg zinc, according to reports (Purchase et al. 2020) Copper, lead, and zinc in high concentrations were also discovered in soil samples obtained from within and up to 12 m out from a site. Though the maximum levels of copper, lead, and zinc inside the recycling site were at least 10 times higher than those outside, the maximum levels of copper, lead, and zinc inside the recycling site were at least 10 times higher than those outside (Fujimori et al. 2016; Fujimori and Takigami 2014). Polychlorinated biphenyls were also detected inside the informal site. The influence of e-waste recycling on the biota (flora and fauna of a region) is a key worry of soil contamination from e-waste recycling. Contaminants generated by e-waste processes are transmitted directly to soils. TBBP-A level in biota at e-waste recycling and disposal sites in China ranged from 3.62 to 42.26 ng per gram wet mass in plant materials, 28.2–103.4 ng per gram dry mass in birds, and 0.98 ng/g wet mass to 1.52 ng/g dry mass in fish, indicating the potential for bioaccumulation (Malkoske et al. 2016). The pollutants present (e.g., PAHs, PBDEs, and metals) appeared to have a major impact on the soil microbial community at e-waste sites, but this did not appear to be related to the distance away from the pollution source (Jiang et al. 2017). When e-waste open-burning sites were compared to a control site, bacterial diversity did not decrease, probably due to the bacterial consortium's adaptability to the environmental contaminants (Zhang et al. 2010). Metals entering leaf and grain tissues via foliar uptake have been demonstrated to constitute a key route (Luo et al. 2011). Soil pH, redox potential, and soil composition, such as organic matter concentration, influence the mobility and bioavailability of contaminants, and chemical speciation dictates the toxicity of metals in the plant system In vegetables, rice, and wild plants taken from regions around residential gardens, paddy fields, and vacant land near a former e-waste incineration site in China, metal levels in the edible section of leafy vegetables were much greater than in the edible component of root vegetables (Luo et al. 2011; Huang et al. 2014). Bakare et al. (2013) gathered leachate (contaminate liquid) samples taken from wells 100–150 m away from an open e-waste dumpsite in Alaba International Market in Lagos State, Nigeria. In onions, the leachate appeared to cause chromosomal aberration, a lower mitotic index, and root development abnormalities (Allium cepa).

8.3 Aquatic Environment

Metals were transported from the surface soil to ponds, causing pond water to become considerably acidified and polluted with transition metals (particularly cadmium and copper) (Purchase et al. 2020). Metal concentrations (cadmium, copper, lead, nickel, and zinc) in groundwater collected from an e-waste recycling area in rural South China were estimated to be between 1.3 and 140 times higher than national groundwater guideline values, posing a major risk to families that drink and cook with it. Acid stripping with strong oxidizing acids, such as a result of e-waste recycling activities, soil acidification, and the production of leachate effluents that pollute water (Srivastava et al. 2017, 2021) and sediment have been recorded (Pradhan and Kumar 2012). During speciation and leaching investigations employing metal-contaminated soil from e-waste recycling facilities, copper and zinc were found in pore water and topsoil, leading the scientists to infer that the contaminants may be mobilized by rainfall irrigation and plant development (Cui et al. 2017; Sarkar et al. 2018). Metals released into the environment can upset the natural balance of aquatic ecosystems or settle in sediment, where aquatic plants and organisms might absorb them (Harguinteguy et al. 2014; Rezania et al. 2016). Persistent toxic substances (such as PBDEs) and metals (such as lead) can end up in the water and spread throughout the environment, causing bioaccumulation and biomagnification High amounts of BDE 47 (tetra bromodiphenyl ether) were identified in fish taken from a reservoir adjacent to several e-waste disassembly workshops.

Contaminants accumulate in the food chain and go up the food chain, which is an essential route for contaminants to travel through the ecological network. For example, Earth worms in an e-waste site have high amount of PCB and PBDEs (Shang et al. 2013). Birds examined at e-waste recycling regions in China had rather high levels of PBDEs (1000–5200 ng/g wet mass), PBBs (Polybrominated biphenyls) (110–340 ng per g wet mass), and polybutylene terephthalate (PBT; 0.5–1.2 ng 0 per g wet mass). Berries, soft fruits, and vegetables, as well as insects and small animals, are the main sources of food for these aquatic birds, this indicates that how bioaccumulation occurs in wildlife (Wu et al. 2016).

9 Impacts of E-Waste on Human Health

If the recycling process is carried out in a scientific manner, precious elements such as gold, copper, silver, palladium, and platinum are efficiently extracted and brought back into the economy, so that economic conditions can benefit from e-waste. In contrast, it has been determined that the primitive recycling process causes severe risk to the environment as well as human beings by releasing toxic substances (Ignatuschtschenko 2017); which enter into human body through inhalation, ingestion, and also with skin contact (Kampa and Castanas 2008). Subsequently, these toxic substances store in the fatty tissues and bring about a detrimental impact on human beings; who live around the informal e-waste site (Zeng et al. 2017; Zhang et al. 2017; Liu et al. 2018). Owing to the incineration of e-waste, fly ash (Singh et al. 2014, 2012) and particulate matters are released; and they can travel hundred and thousands of miles and deteriorate the air quality; and it also potent to bypass the body's defense mechanisms, increasing the risk for chronic diseases and cancers (Gu et al. 2010; Zeng et al. 2016). The hazardous substances are demonstrated in Table 2, with their routes of exposure as well as health effects being listed.

Table 2 some e-wastes, potential hazardous substances and their route of exposure and impact on human health

Metals	Uses in electrical and electronic equipment	Route of exposures	Impacts on human health	References
Nickel (Ni)	Used in batteries	Inhalation, ingestion, and skin contact	Asthma, chronic bronchitis, stomach aches; Increase protein quantity in urine; Damage blood cells	Li and Achal (2020), Bhutta et al. (2011)
Beryllium (Be)	Used in spring, relays, and coonection computer motherboard	Inhalation, ingestion, and skin contact	Carcinogenic, Inhalation can cause berylliosis	Kumar and Singh (2014), Perkins et al. (2014)
Lead (Pb)	Batteries, glass, stabilizers, lights, Printed circuit boards, etc.	Inhalation, ingestion, and skin contact	Damage central to peripheral nervous systems, blood systems, and kidney Headaches, pain	Kumar and Singh, (2014), Annamalai (2015), Hembrom et al. (2020)
Aluminum (Al)	Used in casing and frame also seen in metallization printed circuit board, computer motherboard, hard drives, etc.	Inhalation, ingestion, and skin contact	Slow growth, muscle weakness	Widmer et al. (2005), Anamalai (2015)
Silicon (Si)	Semiconductors like cathode ray tube, printed circuit board	Inhalation, ingestion, and skin contact	Irritation in eyes, lungs, and mucus membranes	Widmer et al. (2005), Anamalai (2015)
Tin (Sn)	Stabilizers	Inhalation, ingestion, and skin contact	Nausea, diarrhea	Matthews (1996), Widmer et al. (2005)
Scandium (Sc)	It's employed in lasers and high-intensity discharge lamps because of its photophysical properties	Inhalation, ingestion, and skin contact	Lung embolisms, threats to the liver	Kumar and Singh (2014), Perkins et al. (2014)

(continued)

Table 2 (continued)

Metals	Uses in electrical and electronic equipment	Route of exposures	Impacts on human health	References
Chromium (Cr)	Crucial components for stainless steel, smart card chips	Inhalation, ingestion	Asthma, cough, shortness of breath, damage to liver, kidney, etc.	Anamalai (2015)
Iron (Fe)	Used in transformers, and generators as Fe_2O_3 it is used as a hard drive and printed circuit board	Inhalation, ingestion, and skin contact	Stomach upset and pain, constipation or diarrhea, nausea, and vomiting	European Food Safety Authority, EFAS (2005)
Copper (Cu)	Used as wire, in smart card chips	Inhalation, ingestion, and skin contact	Irritation in nose, mouth, and eye causes headaches, and diarrhea	Widmer et al. (2005)
Zinc (Zn)	ZnO used in rubber, plastic, inks batteries, etc.	Inhalation, ingestion	Stomach cramps, nausea, and vomiting	Widmer et al. (2005)
Cadmium (Cd)	Used in switches	Inhalation, ingestion	Carcinogenic, severe damage to lungs, fragile bones, gene mutation, etc.	Kumar and Singh (2014), Annamalai (2015)
Mercury (Hg)	Used in vapor lamps, cold cathode fluorescent lamps also in switches	Inhalation, ingestion, and skin contact	Carcinogenic, permanent damage to the brain, tumors neurological disorder, etc.	Shamim et al. (2015), Annamalai (2015), Gangwar et al. (2019)
Arsenic (As)	Used as a semiconductor material, printed circuit boards and laser diodes	Inhalation, ingestion, and skin contact	Carcinogenic, skin lesions, nausea, blood vessel damage, diabetes	Gamble et al. (2018)
Titanium (Ti)	Used in the casing, circuits, and optical microdevices	Inhalation, ingestion, and skin contact	Dust inhalation may cause tightness and pain in the chest, coughing, and difficulty in breathing	Kumar and Singh (2014)

(continued)

Table 2 (continued)

Metals	Uses in electrical and electronic equipment	Route of exposures	Impacts on human health	References
Yttrium (Y)	Laser, lights, etc.	Inhalation, ingestion, and skin contact	Shortness of breath coughing chest pain, and cyanosis	Rim (2016)
Polychlorinated biphenyls (PCB) and Polychlorinated diphenyl ethers (PCDE)	PCB used as insulation material, transformer, capacitors, and PCBE used as fire retardants in plastic hydraulics, etc.	Inhalation, ingestion, and skin contact	Damage to the immune system, liver, skin, reproductive system, and thyroid gland	Tharappel et al. (2008), Shi et al. (2019)
Polybrominated diphenyl (PBD) and Polybrominated diphenyl ether (PBDE)	Both are used as fire retardants in plastic, plastic housing	Inhalation, ingestion, and skin contact	Causes cancer, damage central nervous system, liver, and kidneys	European Food Safety Authority, EFAS (2005), Kumar and Singh (2014)
Perfluo octane sulfonate (PFOS)	Used in photographic industry, semiconductor, photo resistance, etc.	Inhalation, ingestion, and skin contact	Causes bladder cancer, impairment in the reproductive system	European Food Safety Authority, EFAS (2005), Kumar and Singh (2014)
TetrabromobisphenolA (TBBP-A)	Used as thermoplastic components, cable insulation, widely in printed wiring boards, casing, etc.	Inhalation, ingestion, and skin contact	Health issue low, carcinogenic	European Food Safety Authority, EFAS (2005), Kumar and Singh (2014)

Some people who are directly involved with recycling activities without any suitable protection; they breathe contaminated fumes and dust to such an extent that they are adversely affected much faster than other individuals (Sepulveda et al. 2010; Chen et al. 2011; Kit et al. 2013). Apart from this, it has been reported that these hazardous substances can transport long distance and create secondary exposure health risk among the individuals in rural area (Peng et al. 2019). Furthermore, it has been reported that if this process is not maintained properly, these hazardous substance tends to enter into the human body through the food chain and cause various dangerous effect on the human health.

10 E-Waste Management Rules

To manage those different kinds of e-waste components every single country in the world needs some proper management ideas or laws against different valuable, hazardous elements that are covering the globe at a constant speed (e-waste). Proper management can reduce this type of waste. The Resource Conservation and Recovery Act (RCRA), implemented by the US government in 1976, was the first step in organizing e-waste. As a result of this law, e-waste has been dumped illegally in developing countries. Some common terms in e-waste management rules are given here; such as (1) Consumer—Any person who uses the electronic and electrical equipment (EEE); (2) Bulk consumer—The bulk users of EEE such as central state government, or central government departments public sectors; (3) Extended Producer Responsibility (EPR)—EPR is a policy strategy in which manufacturers are assigned significant financial and/or physical responsibility for the treatment or disposal of post-consumer products; (4) Producer—Any individual who manufactures and offers to sell electrical and electronic equipment and its components, consumables, parts, or spares under their own brand, regardless of the selling strategy utilized, such as dealer, retailer and e-retailer; (5) Recycler—Any individual who engages in the recycling and reprocessing of waste electrical and electronic equipment, assemblies, or components and has the facilities described in the rules is referred to as a recycler (Kahhat et al. 2008; Namias et al. 2013).

In industrialized countries, "EPR" or "Product Take Back" is the policy framework. WEEE directives offer a legal foundation for the EU's collection, recovery, and reuse/recycling goals (Zeng et al. 2017; Ferronato and Torretta 2019). In all EU countries, the creation of laws and compliance structures in accordance with EU regulations is ongoing. Member states must ensure a minimum level of collection, recovery, and reuse/recycling. The basic idea of the WEEE directive is "Extended Producer Responsibility," which means that producers are liable for WEEE/E-waste disposal. Japan, for example, has laws focusing on "Reuse, Recycling, and Recovery." Other countries, such as Canada and Australia, are creating systems based on the same principles as the United States. Electronics and electrical items, as well as their parts, are considered to have reached the end of their useful lives and can be discarded by the owner (customer), the service in charge for an area, a factory manager of a factory, or the manufacturer itself. The establishments are under the supervision of the administration manager. An item of machinery once the user or the above-mentioned agencies have permanently disposed it will be deed. There are four principles in e-waste regulation—protection of the environment, social responsibility, disposal and data protection (Nnorom and Osibanjo 2008; Cucchiella et al. 2015). Additional criteria, as required by RoHS (Restriction of Certain Hazardous Substances) standards or any other criteria updated by legislation, shall be added to the supplier requirements in how they fulfill the environmental requirements in regard to toxics reduction, design for end of life, material selection, life cycle extension, energy conservation (EU Directive 2011/65/EU).

11 Approaches for Regulating of E-Waste

There are several methods for regulating e-waste in formal &and informal ways (Yu et al. 2010; Ikhlayel 2018; Islam et al. 2020). At both the national and international levels, there is currently substantial study into e-waste management in order to prevent concerns. LCA, MFA, MCA, and EPR are some of the tools that have been created and applied to e-waste management. In industrialized countries, the adoption of a waste electric and electronic equipment (WEEE) regulation (Directive 2002/96/EC) has ushered in a new era in e-waste management. This directive is expected to reduce the disposal of such material and improve environmental quality (EU 2002). Separation of components that can be recycled, as well as the recovery of rare and precious metals, are all part of the research (Wath et al. 2010; Kiddee et al. 2013).

11.1 Life Cycle Assessment (LCA)

A tool for creating eco-friendly electrical devices and reducing e-waste concerns is life cycle assessment. Since the 1990s, there has been a lot of research on the life cycle assessment of electronic devices in terms of environmental impact, eco-design, and product development printers, desktop computers, heating and cooling systems, washing machines, and toys are just a few of the items available are all examples of items that can benefit from LCA. For e-waste management, LCA is commonly employed. Many studies have been done in Europe using LCA to assess the environmental implications of e-waste end-of-life (EoL) treatment (Kiddee et al. 2013). Hischier et al. (2005), for example, evaluated the environmental implications of the Swiss e-waste take-back and recycling systems. In comparison to incineration, the results demonstrated that environmentally, the e-waste recycling system and take-back were clearly helpful.

11.2 Material Flow Analysis (MFA)

MFA is a tool for analyzing the flow of material (e-waste) through recycling facilities, disposal sites, and material stockpiles over time and place. connects the material's origins, paths, and intermediate and final destinations. Material flow analysis is a decision-making tool for waste and environmental management. In nations like China and India, this includes taking into consideration the flow of electronic waste and analyzing it in terms of environmental, economic, and social factors. MFA was employed by Shinkuma and Huong (2009) to look at the flow of e-waste in Asia. They observed that reused Japanese electrical devices are sold in Southeast Asia (e.g., Vietnam). It is a useful tool when data is limited and economic growth is rapid.

11.3 Extended Producer Responsibility (EPR)

EPR is an environmental policy strategy that holds producers accountable for returning products after they have been used and is based on polluter-pays principles. The leader of the program is EU, Japan, Switzerland, and some state provinces of the US and Canada. In terms of e-waste management regulation, Switzerland has been a trailblazer. The ordinance, "the return, the taking back, and the disposal of electric and electronic equipment" were issued by the Swiss Federal Office for the Environment (FOEN). There are four producer responsibility organizations in Switzerland, most of them are non-profitable organizations, SWICO (The Swiss Association for Information, Communication, and Organizational Technology) Recycling Guarantee, SENS (Swiss Foundation for Waste Management), SLRS (Swiss Light Recycling Foundation), and INOBAT (Innovative Waste Management) are the most important waste management organizations in Switzerland (Stakeholder Organization for Battery Disposal). The Specified Home The Electric Household Appliance Recycling Law and the Appliances Recycling (SHAR) Law are the two main rules that govern e-waste in Japan were enacted in 1998 and went into effect in 2001. SHAR was a fictional character. Established to accept e-waste, particularly big household appliances, for recycling (tv, washing machine, etc.) (Kiddee et al. 2013).

11.4 Multi-criteria Analysis (MCA)

MCA is a tool for making decisions that were created to help people make strategic decisions and solve complicated multi-criteria situations that have both qualitative and quantitative features. It is commonly used for solid waste.

But in developing countries, practicing e-waste management is of serious concern owing to informal recycling activities, lack of infrastructure, and lack of awareness and sensitization to handle different types of e-waste's components. While there are no mandatory or effective voluntary take-back programs, legislation, policies, and mechanisms; therefore, it is a very challenging issue.

12 Biotechnological Approach of E-Waste Recycling and Business Opportunities

Biotechnological approaches such as hydrometallurgy and pyrometallurgy are considered as a feasible way for developing a sustainable environment from the e-waste with the help of living micro-organisms. E-waste is the secondary source of metals; therefore, bioleaching is a suitable process for recovering and reusing the metals from waste electrical and electronic equipment. Some cyanogenic bacteria such *Pseudomonas plecoglossicida*, *Pseudomonas fluorescens*, and *Pseudomonas aeruginosa* are being used in bioleaching process for extraction of gold from e-waste. Owing to HCN production, *Chromobacterium violaceum* plays an important role in gold leaching from e-waste under suitable growth conditions (Duran et al. 2010; Yang et al. 2014; Ilyas et al. 2021).

E-waste is emerging as a source of income for the industry and has also opened the door for new jobs. This is because various elements like Au, Pt, Cu, Al, and rare earth metals are present in the e-waste; which is sufficient for recovery (Garlapati 2016; Ilyas et al. 2021). However, it can be said that biotechnological recycling of e-wastes is attracting the interest of scientists; as it recovers efficient energy and valuable metals such as gold, silver, platinum, and palladium from e-waste without destroying the environment; these can easily create new business opportunities; and also potent circular economy (Garlapati 2016). Skilled technology and reduction of toxic chemicals and further research on the treatment of these toxic chemicals are needed, so that an affordable and environmentally friendly process can be created. It will attract innovation and business and also eliminate the incentive to dump e-waste.

13 Conclusion

E-waste production occurs rapidly in developed countries; and now, its production is also occurring at a much faster pace in developing countries. With the advancement of technology, electrical appliances are becoming obsolete, resulting in large quantities of waste. To develop an affordable and environmentally-safe recycling process for e-waste, the categorization and quantification of precious and toxic elements are very crucial. Several management rules, approaches, and legislative policies can finally improve the problems of e-waste. Therefore, the government and other organizations must explore a variety of laws and legislations for the improvement of such goal. But, it is said that the biotechnological approach seems to be an environment-friendly method for the recovery of precise metals.

References

Adam B, Goen T, Scheepers PT, Adliene D, Batinic B, Budnik LT, Duca RC, Ghosh M, Giurgiu DI, Godderis L, Goksel O (2021) From inequitable to sustainable e-waste processing for reduction of impact on human health and the environment. Environ Res 194:110728

Ahirwar R, Tripathi AK (2021) E-waste management: a review of recycling process, environmental and occupational health hazards, and potential solutions. Environ Nanotechnol Monitor Manage 15:100409

Annamalai J (2015) Occupational health hazards related to informal recycling of e-waste in India: an overview. Ind J Occup Environ Med 19(1):61

Bakare AA, Alabi OA, Gbadebo AM, Ogunsuyi OI, Alimba CG (2013) In vivo cytogenotoxicity and oxidative stress induced by electronic waste leachate and contaminated well water. Challenges 4(2):169–187

Baldé CP, Forti V, Gray V, Kuehr R, Stegmann P (2017) The global e-waste monitor 2017: quantities, flows and resources. United Nations University, International Telecommunication Union, and International Solid Waste Association

Bhutta MKS, Omar A, Yang X (2011) Electronic waste: a growing concern in today's environment. Econ Res Int

Brandl H, Bosshard R, Wegmann M (2001) Computer-munching microbes: metal leaching from electronic scrap by bacteria and fungi. Hydrometallurgy 59:319–326

Chen A, Dietrich KN, Huo X, Ho SM (2011) Developmental neurotoxicants in e-waste: an emerging health concern. Environ Health Perspect 119(4):431–438

Chi TD, Lee JC, Pandey BD, Yoo K, Jeong J (2011) Bioleaching of gold and copper from waste mobile phone PCBs by using a cyanogenic bacterium. Miner Eng 24(11):1219–1222

Cucchiella F, D'Adamo I, Lenny Koh SC, Rosa P (2015) Recycling of WEEEs: an economic assessment of present and future e-waste streams. Renew Sustain Energy Rev 51:263–272

Cui JL, Luo CL, Tang CWY, Chan TS, Li XD (2017) Speciation and leaching of trace metal contaminants from e-waste contaminated soils. J Hazard Mater 329:150–158

Das S, Pal D, Sarkar A (2021) Particulate matter pollution and global agricultural productivity. Sustain Agric Rev 50: Emerg Contam Agric 79–107

DH Web desk (2021) Tokyo Olympics 2020 medals are made of your old phones—Here's how. Retrieved 9 September 2021. https://www.deccanherald.com/specials/tokyo-olympics-2020-medals-are-made-of-your-old-phones-heres-how-1014867.html

Duran M, Faljoni-Alario A, Durán N (2010) Chromobacterium violaceum and its important metabolites. Folia Microbiol 55(6):535–547

European Food Safety Authority (2005) Opinion of the scientific panel on contaminants in the food chain [contam] related to the presence of non dioxin-like polychlorinated biphenyls (PCB) in feed and food. EFSA J 3(284):1–137

Fedorenko I (2016) Clean and inclusive? Recycling e-waste in China and India

Ferronato N, Torretta V (2019) Waste mismanagement in developing countries: a review of global issues. Int J Environ Res Public Health 16(6):1060

Forti V, Balde CP, Kuehr R, Bel G (2020) The global e-waste monitor 2020: quantities, flows and the circular economy potential

Fujimori T, Itai T, Goto A, Asante KA, Otsuka M, Takahashi S, Tanabe S (2016) Interplay of metals and bromine with dioxin-related compounds concentrated in e-waste open burning soil from Agbogbloshie in Accra, Ghana. Environ Pollut 209:155–163

Fujimori T, Takigami H (2014) Pollution distribution of heavy metals in surface soil at an informal electronic-waste recycling site. Environ Geochem Health 36(1):159–168

Gamble AV, Givens AK, Sparks DL (2018) Arsenic speciation and availability in orchard soils historically contaminated with lead arsenate. J Environ Qual 47(1):121–128

Gangwar C, Choudhari R, Chauhan A, Kumar A, Singh A, Tripathi A (2019) Assessment of air pollution caused by illegal e-waste burning to evaluate the human health risk. Environ Int 125:191–199

Garlapati VK (2016) E-waste in India and developed countries: management, recycling, business and biotechnological initiatives. Renew Sustain Energy Rev 54:874–881

Great Lakes Electronic Corporation (2021) Tokyo 2020 Olympic medals made from e-waste. Retrieved 9 September 2021. https://www.ewaste1.com/tokyo-2020-olympic-medals/

Gu Z, Feng J, Han W, Wu M, Fu J, Sheng G (2010) Characteristics of organic matter in PM2.5 from an e-waste dismantling area in Taizhou, China. Chemosphere 80(7):800–806

Halim L, Suharyanti YOSEPHINE (2020) E-waste: current research and future perspective on developing countries

Harguinteguy CA, Cirelli AF, Pignata ML (2014) Heavy metal accumulation in leaves of aquatic plant Stuckenia filiformis and its relationship with sediment and water in the Suquía river (Argentina). Microchem J 114:111–118

Heacock M, Kelly CB, Asante KA, Birnbaum LS, Bergman ÅL, Bruné MN, Buka I, Carpenter DO, Chen A, Huo X, Kamel M (2016) E-waste and harm to vulnerable populations: a growing global problem. Environ Health Perspect 124(5):550–555

Hembrom S, Singh B, Gupta SK, Nema AK (2020) A comprehensive evaluation of heavy metal contamination in foodstuff and associated human health risk: a global perspective. In: Contemporary environmental issues and challenges in era of climate change. Springer, Singapore, pp 33–63

Hernandez D (2021) Tokyo's Olympic medals are made entirely from recycled gadgets. Retrieved 9 September 2021. https://www.popularmechanics.com/technology/gear/a28509493/tokyo-2020-olympic-medals-recycled-gadgets/

Hischier R, Wäger P, Gauglhofer J (2005) Does WEEE recycling make sense from an environmental perspective? the environmental impacts of the Swiss take-back and recycling systems for waste electrical and electronic equipment (WEEE). Environ Impact Assess Rev 25(5):525–539

Huang J, Nkrumah PN, Anim DO, Mensah E (2014) Reviews of environmental contamination and toxicology. Rev Environ Contam Toxicol 229:18–34

Ignatuschtschenko E (2017) E-waste management in China: bridging the formal and informal sectors. J Chin Gov 2(4):385–410

Ikhlayel M (2017) Environmental impacts and benefits of state-of-the-art technologies for e-waste management. Waste Manage 68:458–474

Ikhlayel M (2018) An integrated approach to establish e-waste management systems for developing countries. J Clean Prod 170:119–130

Ilankoon IMSK, Ghorbani Y, Chong MN, Herath G, Moyo T, Petersen J (2018) E-waste in the international context—a review of trade flows, regulations, hazards, waste management strategies and technologies for value recovery. Waste Manage 82:258–275

Ilyas S, Srivastava RR, Kim H, Das S, Singh VK (2021) Circular bioeconomy and environmental benignness through microbial recycling of e-waste: a case study on copper and gold restoration. Waste Manage 121:175–185

Islam A, Ahmed T, Awual MR, Rahman A, Sultana M, Abd Aziz A, Monir MU, Teo SH, Hasan M (2020) Advances in sustainable approaches to recover metals from e-waste—a review. J Clean Prod 244:118815

Jiang L, Cheng Z, Zhang D, Song M, Wang Y, Luo C, Yin H, Li J, Zhang G (2017) The influence of e-waste recycling on the molecular ecological network of soil microbial communities in Pakistan and China. Environ Pollut 231:173–181

Kahhat R, Kim J, Xu M, Allenby B, Williams E, Zhang P (2008) Exploring e-waste management systems in the United States. Resour Conserv Recycl 52(7):955–964

Kamczyc A (2021) Medals used in Tokyo Olympics made from recycled materials

Kampa M, Castanas E (2008) Human health effects of air pollution. Environ Pollut 151(2):362–367

Kang HY, Schoenung JM (2005) Electronic waste recycling: a review of US infrastructure and technology options. Resour Conserv Recycl 45(4):368–400

Kiddee P, Naidu R, Wong MH (2013) Electronic waste management approaches: an overview. Waste Manage 33(5):1237–1250

Kit J, Chan Y, Wong MH (2013) A review of environmental fate, body burdens, and human health risk assessment of PCDD/Fs at two typical electronic waste recycling sites in China. Sci Total Environ 463–464:1111–1123

Kumar A, Holuszko M, Espinosa DCR (2017) E-waste: an overview on generation, collection, legislation and recycling practices. Resour Conserv Recycl 122:32–42

Kumar U, Singh DN (2014) Electronic waste: concerns & hazardous threats. Int J Curr Eng Technol 4:802–811

Leader AM, Wang X, Gaustad G (2017) Creating the 2020 tokyo olympic medals from electronic scrap: sustainability analysis. J Metals 69(9):1539–1545

Li W, Achal V (2020) Environmental and health impacts due to e-waste disposal in China—a review. Sci Total Environ 737:139745

Liu Y, Huo X, Xu L, Wei X, Wu W, Wu X, Xu X (2018) Hearing loss in children with e-waste lead and cadmium exposure. Sci Total Environ 624:621–627
Lu C, Zhang L, Zhong Y, Ren W, Tobias M, Mu Z, Ma Z, Geng Y, Xue B (2015) An overview of e-waste management in China. J Mater Cycles Waste Manage 17(1):1–12
Luo C, Liu C, Wang Y, Liu X, Li F, Zhang G, Li X (2011) Heavy metal contamination in soils and vegetables near an e-waste processing site, south China. J Hazard Mater 186(1):481–490
Malkoske T, Tang Y, Xu W, Yu S, Wang H (2016) A review of the environmental distribution, fate, and control of tetrabromobisphenol a released from sources. Sci Total Environ 569:1608–1617
Marshall J (2021) Tokyo 2020: Olympic medals made from old smartphones, laptops. https://www.dw.com/en/tokyo-2020-olympic-medals-made-from-old-smartphones-laptops/a-58325432
Matthews G (1996) PVC: production, properties and uses, vol 587. Publ, The Institute of Materials, London
Mmereki D, Li B, Baldwin A, Hong L (2016) The generation, composition, collection, treatment and disposal system, and impact of e-waste. E-waste Trans Pollut Resour 65–93.
Murthy CVR (2012) Management of e-Waste in the present scenario. Int J Eng Technol 4(5):543
Namias J (2013) The future of electronic waste recycling in the united states: obstacles and domestic solutions. Columbia University, New York, United States, pp 1–66
Nnorom IC, Osibanjo O (2008) Overview of electronic waste (e-waste) management practices and legislations, and their poor applications in the developing countries. Resour Conserv Recycl 52(6):843–858
Osibanjo O, Nnorom IC (2007) The challenge of electronic waste (e-waste) management in developing countries. Waste Manage Res 25(6):489–501
Pariatamby A, Victor D (2013) Policy trends of e-waste management in Asia. J Mater Cycles Waste Manage 15(4):411–419
Peng Y, Wu J, Luo X, Zhang X, Giesy JP, Mai B (2019) Spatial distribution and hazard of halogenated flame retardants and polychlorinated biphenyls to common kingfisher (Alcedo atthis) from a region of South China affected by electronic waste recycling. Environ Int 130:104952
Perkins DN, Drisse MNB, Nxele T, Sly PD (2014) E-waste: a global hazard. Ann Glob Health 80(4):286–295
Pradhan JK, Kumar S (2012) Metals bioleaching from electronic waste by Chromobacterium violaceum and Pseudomonads sp. Waste Manage Res 30(11):1151–1159
Purchase D, Abbasi G, Bisschop L, Chatterjee D, Ekberg C, Ermolin M, Fedotov P, Garelick H, Isimekhai K, Kandile NG, Lundström M (2020) Global occurrence, chemical properties, and ecological impacts of e-wastes (IUPAC Technical Report). Pure Appl Chem 92(11):1733–1767
Rezania S, Taib SM, Din MFM, Dahalan FA, Kamyab H (2016) Comprehensive review on phytotechnology: heavy metals removal by diverse aquatic plants species from wastewater. J Hazard Mater 318:587–599
Rim KT (2016) Effects of rare earth elements on the environment and human health: a literature review. Toxicol Environ Heal Sci 8(3):189–200
Rocchetti L, Veglio F, Kopacek B, Beolchini F (2013) Environmental impact assessment of hydrometallurgical processes for metal recovery from WEEE residues using a portable prototype plant. Environ Sci Technol 47(3):1581–1588
Sankhla MS, Kumari M, Nandan M, Mohril S, Singh GP, Chaturvedi B, Kumar R (2016) Effect of electronic waste on environmental & human health—a review. IOSR J Environ Sci Toxicol Food Technol 10(09):98–104
Sarkar A, Das S, Srivastava V, Singh P, Singh RP (2018) Effect of wastewater irrigation on crop health in the indian agricultural scenario. In: Emerging trends of plant physiology for sustainable crop production. Apple Academic Press, pp 357–371
Sarkar A, Singh A, Agrawal SB (2013) Utilization of fly ash as soil amendments in agricultural fields of North-Eastern Gangetic plains of India: potential benefits and risks assessments. Bull Natl Inst Ecol 23(1–2):9–20
Sepulveda A, Schluep M, Renaud FG, Streicher M, Kuehr R, Hagelüken C, Gerecke AC (2010) A review of the environmental fate and effects of hazardous substances released from electrical and

electronic equipments during recycling: examples from China and India. Environ Impact Assess Rev 30(1):28–41

Shallow T (2021) Tokyo produces Olympic medals throw e-Waste campaign. Retrieved 9 September 2021. https://sustainabilitymag.com/esg/tokyo-produces-olympic-medals-through-e-waste-campaign

Shamim A, Mursheda AK, Rafiq I (2015) E-waste trading impact on public health and ecosystem services in developing countries. Int J Waste Resour 5(4):1–18

Shang H, Wang P, Wang T, Wang Y, Zhang H, Fu J, Ren D, Chen W, Zhang Q, Jiang G (2013) Bioaccumulation of PCDD/Fs, PCBs and PBDEs by earthworms in field soils of an e-waste dismantling area in China. Environ Int 54:50–58

Sharma B, Sarkar A, Singh P, Singh RP (2017) Agricultural utilization of biosolids: a review on potential effects on soil and plant grown. Waste Manage 64:117–132

Shi J, Xiang L, Luan H, Wei Y, Ren H, Chen P (2019) The health concern of polychlorinated biphenyls (PCBs) in a notorious e-waste recycling site. Ecotoxicol Environ Saf 186:109817

Shinkuma T, Huong NTM (2009) The flow of e-waste material in the Asian region and a reconsideration of international trade policies on e-waste. Environ Impact Assess Rev 29(1):25–31

Singh A, Sarkar A, Agrawal SB (2012) Assessing the potential impact of fly ash amendments on Indian paddy field with special emphasis on growth, yield, and grain quality of three rice cultivars. Environ Monit Assess 184(8):4799–4814

Singh N, Li J, Zeng X (2016) Global responses for recycling waste CRTs in e-waste. Waste Manage 57:187–197

Singh RP, Sharma B, Sarkar A, Sengupta C, Singh P, Ibrahim MH (2014) Biological responses of agricultural soils to fly-ash amendment. Rev Environ Contam Toxicol 232:45–60

Sivaramanan S (2013) E-waste management, disposal and its impacts on the environment. Univers J Environ Res Technol 3(5)

Song Q, Li J (2015) A review on human health consequences of metals exposure to e-waste in China. Environ Pollut 196:450–461

Srivastava V, Sarkar A, Singh S, Singh P, de Araujo AS, Singh RP (2017) Agroecological responses of heavy metal pollution with special emphasis on soil health and plant performances. Front Environ Sci 5:64

Srivastava V, Squartini A, Masi A, Sarkar A, Singh RP (2021) Metabarcoding analysis of the bacterial succession during vermicomposting of municipal solid waste employing the earthworm Eisenia fetida. Sci Total Environ 766:144389

Tharappel JC, Lehmler HJ, Srinivasan C, Robertson LW, Spear BT, Glauert HP (2008) Effect of antioxidant phytochemicals on the hepatic tumor promoting activity of 3, 3′, 4, 4′-tetrachlorobiphenyl (PCB-77). Food Chem Toxicol 46(11):3467–3474

Tokyo (2020) Tokyo 2020 medal project: towards an innovative future for all. Retrieved 9 September 2021. https://olympics.com/tokyo-2020/en/games/medals-project/

Wath SB, Vaidya AN, Dutt PS, Chakrabarti T (2010) A roadmap for development of sustainable e-waste management system in India. Sci Total Environ 409(1):19–32

Widmer R, Oswald-Krapf H, Sinha-Khetriwal D, Schnellmann M, Boni H (2005) Global perspectives on e-waste. Environ Impact Assess Rev 25(5):436–458

Wu CC, Bao LJ, Tao S, Zeng EY (2016) Dermal uptake from airborne organics as an important route of human exposure to e-waste combustion fumes

Yang Y, Chen S, Li S, Chen M, Chen H, Liu B (2014) Bioleaching waste printed circuit boards by Acidithiobacillus ferrooxidans and its kinetics aspect. J Biotechnol 173:24–30

Yu J, Williams E, Ju M, Shao C (2010) Managing e-waste in China: policies, pilot projects and alternative approaches. Resour Conserv Recycl 54(11):991–999

Zeng X, Duan H, Wang F, Li J (2017) Examining environmental management of e-waste: China's experience and lessons. Renew Sustain Energy Rev 72:1076–1082

Zeng X, Xu X, Boezen HM, Huo X (2016) Children with health impairments by heavy metals in an e-waste recycling area. Chemosphere 148:408–415

Zhang M (2017) Air pollution and human health risk assessment in e-waste recycling sites and urban indoor environment in South China

Zhang W, Wang H, Zhang R, Yu XZ, Qian PY, Wong MH (2010) Bacterial communities in PAH contaminated soils at an electronic-waste processing center in China. Ecotoxicology 19(1):96–104

Persistent Toxic Substances Released from Uncontrolled E-waste Recycling and Action for the Future

Jaganathan Senthilnathan and Ligy Philip

Abstract Twenty-first century has witnessed an exponential growth in the use of electrical and electronic equipments throughout the world. These e-products, once they are no longer useful, becomes e-waste, and hence gets added to one of the fastest growing waste streams in the world. They contain diverse toxic metals like Pb, Cd, Hg, hexavalent Cr, Ni, As, and Ba and when disposed though the informal recyclers cause pervasive damages to the health and environment. An in-depth discussion on toxicants and pollutants from e-waste produced by the informal sector is presented in this chapter. Though an outcry for eco-friendly e-waste management is rising around the world, the standards of existing protocols for e-waste management in most countries are significantly low due to the intricacies in the formation and implication of policies, lack of proper collection systems, infrastructure, and also low public awareness. The crisis forged by the by-products of informal e-waste disposal; the effects of toxic heavy metals, dioxin and polycyclic aromatic hydrocarbons (PAHs); and their intermediate compounds released during the open combustion of e-waste are explained in this chapter. This highlights the urgency to address the challenges, opportunities, and future prospects of e-waste management at an international level.

1 Introduction

Twenty-first century has witnessed a rapid advancement in the fields of information technology and brisk growth in use of machines in every sector. This has radically changed the lifestyle of people throughout the world. These developments have improved the quality of life of people in every stratum and have also caused the increase in demands for high-end products and appliances which, nowadays, are

J. Senthilnathan (✉) · L. Philip
Environmental & Water Resources Engineering Division, Department of Civil Engineering, Indian Institute of Technology Madras, Chennai 600036, India
e-mail: jsn@iitm.ac.in

L. Philip
e-mail: ligy@iitm.ac.in

M. Jawaid and A. Khan (eds.), *Conversion of Electronic Waste in to Sustainable Products*, Sustainable Materials and Technology,
https://doi.org/10.1007/978-981-19-6541-8_4

exchanged for gadgets with better technology and cheaper market price within a significantly short lifespan of the product. All these have contributed to the increasing amount of electronic equipments which are broken, worn out, dead, or have no more market value. Now, we are facing a serious dilemma in disposal of the electronic equipments, which are no longer fit for use.

1.1 What is E-waste?

E-waste comprises all materials from those electrical or electronic equipment, which are either outdated, unwanted, broken, or are no longer fit for their original intended use and are set apart for recovery, recycling, or disposal (Lundgren 2012; Baldé et al. 2017). The generally acknowledged definition of e-waste is "Electrical or electronic equipment, which is waste, including all components, sub-assemblies, and consumables, which are part of the product at the time of discarding" (EU 2003). It is one of the fastest growing solid waste streams in the planet. The progressively growing volume of e-waste is not surprising given the rising demand and supply of e-products. There are several chemically different constituents present in the e-waste and it is essential to segregate them before sending them for recycling or disposal (MoEF&CC 2015). Different classifications have been proposed to describe e-waste. The classification that is widely accepted is the one put forward by the Waste of Electrical and Electronic Equipment Directive (WEEE), which enlists e-waste into six different categories (EU 2012). As per the classification, the e-waste segregated in 2014 contained 17.4, 13.1, and 10.8 Mt of small equipment, large equipment, and temperature exchange equipment, respectively. Similarly, screens, lamps, and small IT and telecommunication equipment comprised 6.7, 0.9, and 4.7 Mt, respectively (Rotter et al. 2016; Blade et al. 2017; Forti et al. 2020). Since 2014, there are 7,56,000, 36,000, 6,55,000, 6,96,000, and 94,000 tons increase in temperature exchange equipment, lamps, large equipment, small equipment, and small IT and telecommunication equipment, respectively.

1.2 E-waste Challenges: A Global Scenario

The management of e-waste has proven to be unbelievably challenging due to the absence of efficient and inexpensive systems to segregate and manage the toxic and/or valuable ingredients present in it (Needhidasan et al. 2014). Even developed nations with well-established waste management systems are struggling to dispose e-waste properly due to the intricacies in the chemical composition of components of e-waste. Whereas, in developing or underdeveloped countries with little to no policies or infrastructure, e-waste has added challenges to the already existing waste management crisis (Lundgren 2012). It is simply because, several factors like population size, socio-economic development, public awareness, government policies and regulations, etc., which makes significant impact in the successful disposal of e-waste is underdeveloped in those countries. The global volume of electronic and electrical waste produced in year 2016 was 44.7 Mt, and it reached 53.6 Mt and about 7.3 kg/capita in 2019 (Forti et al. 2020; Balde et al. 2017). However, this volume is expected to reach 57.4 Mt in 2021 (Forti et al. 2020; Balde et al. 2017). The forecast of the waste generation per region based on population and GDP growth is given in Fig. 1. The creation and distribution of e-waste is uneven, richer countries produce more when compared to the developing nations. Norway shows a production rate of 28.5 kg per person per year, whereas only less than 2 kg per person is produced in African countries. The top four e-waste-producing countries in Asia are China (10.129 Mt), India (3.23 Mt), Japan (2.569 Mt), and Indonesia (1.618 Mt). It is interesting to note that the number of countries adopting a nationwide e-waste policy has increased from 61 to 78 after 2014 (Forti et al. 2020; Balde et al. 2017).

In India, the disposal and recycling of e-waste has become more and more complicated even when the guidelines for disposal are simple and well established (Vidyadhar 2016; Turaga et al. 2019). The State Pollution Control Boards have not carried out the state-wise inventories since 2012, which is an important input for estimating e-waste production region wise (Vidyadhar 2016; Turaga et al. 2019). Though 407 authorized dismantler/recyclers can treat around ~1,110,103 tons per annum, recycling 20% of the total e-waste generation in India (CPCB 2016), use of these services is low and most of the facilities are operating below their approved capacity level (Turaga et al. 2019). Enforcing the regulations in informal sector is a challenge due to their poor logistics and infrastructure and because of tedious reporting processes involved in complying with the regulations (Sahajwalla and

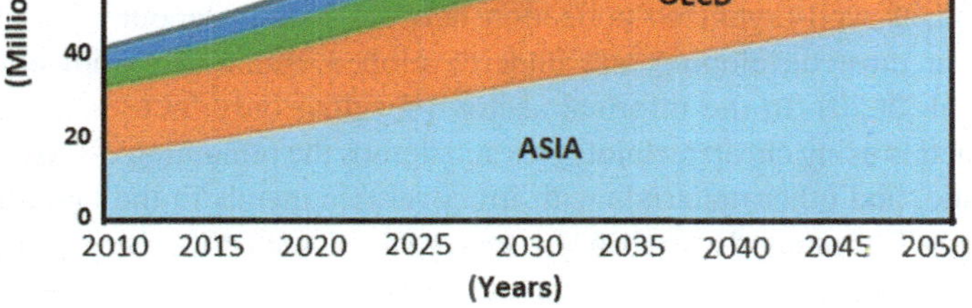

Fig. 1 Forecast of e-waste quantities per region based on population and GDP growth. *Note* REF: Countries from the Reforming Economies of Eastern Europe and the Former Soviet Union; LAM: Countries of Latin America and the Caribbean; MAF: Countries of the Middle East and Africa; OECD: OECD 90 and EU member states and candidates; ASIA: Asian countries with the exception of the Middle East, Japan, and Former Soviet Union states (*Source* Adopted and modified from Parajuly et al. 2019)

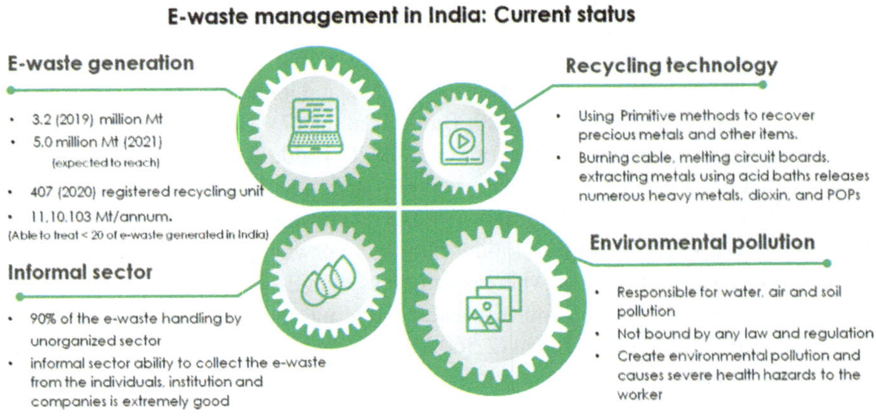

Fig. 2 The status of e-waste management in India and the role of informal sector

Gaikwad 2018). Even though the Government of India developed a scheme that provides up to 50% of the financial support to establish the e-waste management facilities and e-waste businesses (MoEF&CC 2013), it failed to achieve the desired success due to the sloppy performance of the existing recycle centers and substandard e-waste supply chain (MoEF&CC 2013). Also, there is low awareness in public regarding e-waste recycling, environmental hazards, public health, and penalties for improper disposal (Borthakur and Govind 2017; Kwatra et al. 2014). The status of e-waste management in India and the role of informal sector is given in Fig. 2.

1.3 The Role of Informal Sector in E-waste Recycling

In both developing and underdeveloped countries, a proper channel for e-waste management is absent for governing e-waste assemblage and transportation for further processing. This has helped the informal or unorganized recycling units to flourish and its well-organized collection network handles about 90% of e-waste generated in most developing and underdeveloped countries (Wath et al. 2010; Kiddee et al. 2020). In the informal sector, recycling only focuses on recovering precious metals using cheap technologies and dumps the remaining e-waste parts like glass, plastic, and other materials with irrecoverable metals in the landfill (Kiddee et al. 2020). The methods used in the unorganized sector are very primitive and risky and produce large quantities of toxic wastes which adversely affect the living organisms and the environment (Kumar et al. 2017). There is not much awareness among the end-users, households, and small-scale industries regarding the importance of segregation and safe disposal of e-waste (Chaturvedi et al. 2010). They do not possess adequate infrastructure or protective gear. They follow unscientific methods for recycling and extraction of metals and are quite unaware of the risks involved in the process or that it wreaks havoc in nature. They are also ignorant of the health hazards created by the age-old ludicrous approaches in handling e-waste (Borthakur and Singh 2017). They work under dangerous conditions and often from home; thereby their families are also exposed to toxic chemicals (Kumar et al. 2017). The informal sector constitutes an assortment of rag pickers, untrained labor, and small informal businesses that manually perform assemblage, segregation, disassembling, recycling, and riddance of by-products of e-waste (Borthakur 2015). The rag pickers collect, segregate, and sell the e-waste to scrap dealers and the recyclers use old and hazardous techniques and equipments to recycle/treat the e-waste (Gupta and Kumar 2014).

Methods adopted for the recovery of valuable materials in unorganized sector:

I. Dismantling of e-wastes by bare hands, hammers, and screwdrivers.
II. Recovery of Cu and Al by burning of PVC and wires in open air.
III. Recovering valuable metals using the primitive methods like acid baths.
IV. Crude handling of chemicals with no or adequate safety equipment (Borthakur 2020; Chaturvedi et al. 2010).

Informal units are driven by maximization of profits with minimal capital investment, giving low priority to sustainable e-waste recycling and they do not adhere to the e-waste laws (Borthakur 2015; Chaturvedi et al. 2010). Pollutants and toxins produced and hazards caused by unorganized or informal recycle of e-waste are given in Table 1.

2 Hazardous Substances Present in Electrical and Electronic Waste

E-waste consists of different types of toxic components like heavy metals, dioxins, polyaromatic hydrocarbons (PAHs), polyvinyl chloride (PVC), etc. (Perkins et al. 2014; Frazzoli et al. 2010; Kumar and Singh 2014). When practices no more recommended for the recycling of e-waste are used, there is a release of toxic materials which severely damage the quality of air, soil, and groundwater in that area (Li and Achal 2020; Cesaro et al. 2019). A personal computer contains 6% of Pb (wt/wt) in addition to many valuable metals such as Au, Pt, Ag, Cu, Ir, Os, Pd, Rh, and Ru. Further, a large number of thermosetting polymers, C, Pb, Cu, Si, Be, Fe, and Al are obtained during the recycling of e-waste (Wei and Yang 2010). Toxic pollutants present in various e-products and their routes of exposure for humans are given in Table 2. Small-to-trace quantities of metal ions like Cd, Hg, Tl, Am, An, Ba, As, Bi, B, Co, Eu, Ga, Ge, Au, In, Li, Mn, Ni, Nb, Pd, Pt, Rh, Ru, Se, Ag, Ta, Tb, Th, Ti, V, and Y are present in e-waste (Fig. 2) (Li and Achal 2020; Purushothaman et al. 2021; Alabi et al. 2021; Rajput et al. 2021).

E-waste contains polychlorinated and brominated compounds like polychlorinated biphenyls (PCBs), polybrominated diphenyl ethers (PBDEs), polybrominated biphenyls (PBBs), polychlorinated dibenzofurans (PCDFs, released on open combustion), and polychlorinated dibenzodioxins (PCDDs, released on open combustion)

Table 1 Pollutants and toxins produced and hazards caused by or unorganized/informal recycle of e-waste

S. no	E-waste components	Process	Potential occupational hazard	Potential environmental hazard
1	Cathode ray tubes (CRTs)	Breaking and removal of Cu yoke and dumping	Silicosis, Inhalation, or contact with phosphorous, cadmium, or other metals	Lead, barium, and other heavy metals leaching into groundwater, release of toxic phosphorous
2	Printed circuit boards	Disordering and removing computer chips	Tin and lead poisoning. Inhalation of brominated dioxin, beryllium, cadmium, mercury	Air emission of same substances
3	Dismantled printed circuit board processing	Open burning of waste boards to remove inside metals	Toxicity to workers and nearby residents from tin, lead, brominated dioxin, beryllium, cadmium, and mercury inhalation respiratory irritation	Tin and lead contaminate immediate environment and surface and ground waters. Brominated dioxins, beryllium, cadmium and mercury emissions
4	Chips and other gold-plated components	Chemical stripping using nitric and hydrochloric acid along riverbanks	Acid when touches eyes or skin may result in permanent injury. Inhalation of mists, and fumes of acids, Cl_2 and SO_2 gases can cause pulmonary irritation and also serious complications like pulmonary oedema and circulatory failure	Hydrocarbons, heavy metals, brominated substances, etc., discharged directly into river and banks. Acidifies the river destroying fish and flora
5	Plastics from computer and peripherals, e.g., printers, keyboards, etc.	–	–	Emissions of brominated dioxins and heavy metals and hydrocarbons

(continued)

(Chakraborty et al. 2018; Yang et al. 2013; Hites 2004). In addition, burning of e-waste releases different forms of PAHs like acenaphthene, acenaphtylene, anthracene, benz(a)anthracene, benzopyrene derivatives, fluoranthene, chrysene, fluorene, indeno(1,2,3-c,d)pyrene, phenanthrene, and pyrene (Frazzoli et al. 2010;

Table 1 (continued)

S. no	E-waste components	Process	Potential occupational hazard	Potential environmental hazard
6	Shredding and low-temperature melting to be reutilized in poor grade plastics	Probable hydrocarbon, brominated dioxin, and heavy metal exposure	Brominated and chlorinated dioxin, polycyclic aromatic Hydrocarbons (PAHs) are carcinogenic to workers living in the burning area	Hydrocarbon ashes including PAHs discharged to air, water, and soil
7	Miscellaneous computer parts encased in rubber or plastic, e.g., steel rollers	Open burning to recover steel and other metals	Hydrocarbon including PAHs and potential dioxin exposure	Hydrocarbon ashes including PAHs discharged to air, water, and soil
8	Secondary steel or copper and precious metal smelting	Furnace recovers steel or copper from waste including organics	Exposure to dioxins and heavy metals	Emission of dioxins and heavy metals

Source greene.gov.in; Mundada et al. (2004); Al-Anzi et al. (2017); Wath et al. (2010)

Perkins et al. 2014; Kumar and Singh 2014). The various toxic compounds released during the Informal disposal of e-waste and their effect on health are given in Fig. 3.

3 Hazards Caused by Imprudent E-waste Management

In the above section, a variety of pollutants and toxicants released from e-waste were discussed. When methods, which are no longer recommended for e-waste disposal are employed, most of these toxins and pollutants are released into the environment (Needhidasan et al. 2014). These are highly reactive and long-range transport of these pollutants is reported in several studies. When animals and humans are exposed to these chemicals, they are readily affected. Due to bioaccumulation these chemicals are transferred through the food chain (Akintokun et al. 2017). In human, the bioaccumulation of these chemicals causes acute and chronic health hazards according to their potency. Following are the major class of chemicals and pollutants released from e-waste and the repercussions they create in nature and human health (Needhidasan et al. 2014; Akintokun et al. 2017).

Table 2 Toxic pollutants present in a various e-products, and their routes of exposure for humans

Sl. no	Compounds	Component of electrical and electronic equipment	Route of exposure
Persistent organic pollutants (POPs)			
1	Brominated flame retardants: polybrominated biphenyls	Fire retardants for electronic equipment	Ingestion, inhalation, dermal absorption, and transplacental
2	Polychlorinated biphenyls	Dielectric fluids, lubricants and coolants in generators, capacitors and transformers, fluorescent lighting, ceiling fans, dishwashers, and electric motors	Ingestion, inhalation, dermal contact, and transplacental
Dioxins			
3	Polychlorinated dibenzodioxins and dibenzofurans	Released as a combustion by-product	Ingestion, inhalation, dermal contact, and transplacental
4	Dioxin-like polychlorinated biphenyls	Transformers, capacitors, softening agents for paint, glue, plastic, dielectric fluids, coolants in generators, capacitors and transformers, fluorescent lighting, lubricants, ceiling fans, dishwashers, and electric motors	Ingesting animal fats, inhalation, and dermal contact
5	Perfluroalkyls	Perfluroalkyls fluoropolymers in electronics	Ingestion, dermal contact, inhalation, and transplacental
Polyaromatic hydrocarbons			
6	Acenaphtene, acenaphtylene, anthracene, benz(a)anthracene, benzopyrene derivatives, chrysene, fluoranthene, fluorene, indeno(1,2,3-c,d)pyrene, phenanthrene, pyrene, etc.	Released as a combustion by-product	Ingestion, inhalation, and dermal contact
Elements			
7	Arsenic	Semiconductors, diodes, microwaves, LEDs (light-emitting diodes), solar cells	Inhalation, ingestion, and via dermal absorption
8	Barium	Electron tubes, filler for plastic and rubber, lubricant additives, and fluorescent lamps	Ingestion, inhalation, and dermal contact

(continued)

Table 2 (continued)

Sl. no	Compounds	Component of electrical and electronic equipment	Route of exposure
9	Beryllium	Power supply boxes, computers, X-ray machines, ceramic components of electronics, and motherboard	Inhalation, ingestion, and transplacental
10	Brominated flame- roofing agent (PBBs)	Casing, circuit boards (plastic), cables, and PVC cables	Inhalation, ingestion, and via dermal absorption
11	Cadmium	Batteries, pigments, solder, alloys, circuit boards, computer batteries, monitor cathode ray tubes (CRTs) switches, springs, connectors, PCBs, batteries, infrared detectors, semi-conductor chips, ink or toner photocopying machines, cathode ray tubes, and mobile phones	Ingestion and inhalation
12	Chromium	Dyes/pigments, switches, solar	Inhalation, ingestion, and via dermal absorption
13	Cobalt	Insulators	Inhalation and dermal absorption
14	Copper	Conducted in cables, copper ribbons, coils, circuitry, pigments	Inhalation, ingestion, and via dermal absorption
15	Hexavalent chromium	Anticorrosion coatings, data tapes, and floppy disks	Inhalation and ingestion
16	Lead	Lead rechargeable batteries, solar, transistors, PVC (polyvinyl chloride) stabilizers, lasers, LEDs, thermoelectric elements, printed circuit boards (PCBs), cathode ray tubes, light bulbs, televisions (1.5–2.0 kg per monitor), and lithium batteries	Inhalation, ingestion, and dermal contact
17	Liquid crystal	Displays	Dermal or ingestion
18	Lithium	Mobile phones, equipments for photography, and video recording (batteries)	Inhalation, ingestion, and dermal contact

(continued)

Table 2 (continued)

Sl. no	Compounds	Component of electrical and electronic equipment	Route of exposure
19	Mercury	Components in copper machines and steam irons; batteries in clocks and pocket calculators, switches, LCDs thermostats, sensors, monitors, cells, PCBs, and cold cathode fluorescent lamps (1–2 g per device)	Inhalation, ingestion, and dermal contact
20	Nickel	Alloys, batteries, relays, semiconductors, and pigments	Inhalation, ingestion, dermal contact, and trans placental
21	PCBs (polychlorinated biphenyls)	Transformers, capacitors, softening agents for paint, glue, plastic, dielectric fluids, coolants, capacitors and transformers, fluorescent lighting, ceiling fans, dishwashers, and electric motors	Inhalation, ingestion, and dermal contact
22	Rare-earth metals like La, Ce, Pr, Nd, Pm, Sm, Eu, Gd, Tb, Dy, Ho, and Er	The central processing unit; motherboard; and different parts of the electronics items, PCBs, and cathode ray tubes	Inhalation, ingestion, dermal absorption, and trans placental
23	Selenium	Photoelectric cells, pigments, photocopiers, fax machines	Inhalation, ingestion, and dermal contact
24	Silver	Capacitors, switches (contacts), batteries, resistors	Inhalation and dermal absorption
25	Zinc	Steel, brass, alloys, disposable and rechargeable batteries, luminous substances, cathode ray tubes, and metal coatings	Ingestion and inhalation

Source Alabi et al. (2021); Frazzoliet al. (2010); Perkins et al. (2014); Kumar and Singh (2014)

3.1 Heavy Metals: Impact on Health and Environment

Heavy metals are present in nearly all aspects of modern life. Heavy metals or their compounds can be found in electronic components, electrodes, wiring, and solar panels where they may be used as conductors, semiconductors, or insulators. The heavy metals which are used in the manufacturing of electronic equipment are As, Ba, Cd, Cr, Co, Cu, Pb, Hg, Ni, and Zn (Herawati et al. 2000). Similarly, rare earth metals like La, Ce, Pr, Nd, Pm, Sm, Eu, Gd, Tb, Dy, Ho, and Er are present within various components of electronic equipment (He et al. 2005). When they become e-waste, recycling them using informal methods causes heavy metals to leach into the

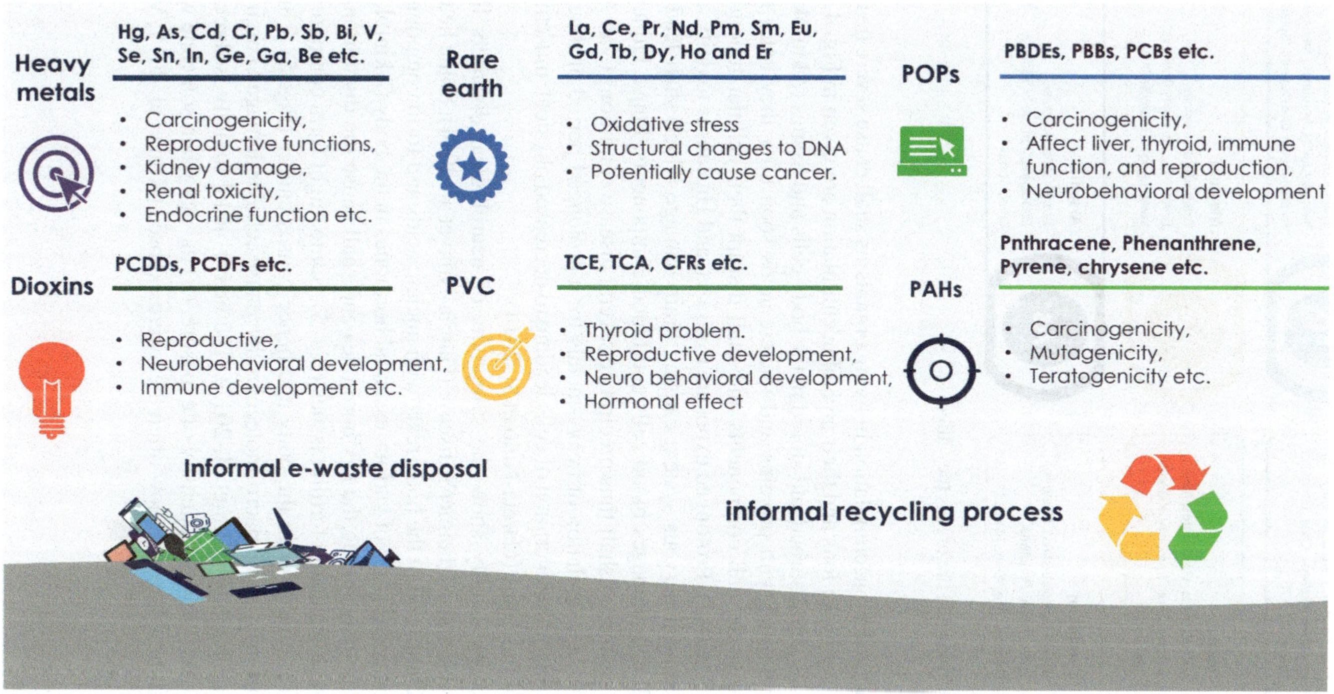

Fig. 3 The various toxic compounds released during the Informal disposal of e-waste and their effect on health. *Note* PBDEs: polybrominated diphenyl ethers; PBBs: polybrominated biphenyls; PCBs: polychlorinated biphenyls; TCE: trichloroethylene; TCA: trichloroethane; CFRs: chlorinated flame retardants; PAHs: polyaromatic hydrocarbons (Ref: Rim et al. 2013; Perkins et al. 2014; Frazzoli et al. 2010; Kumar and Singh 2014; Li and Achal 2020; Hashmi and Varma 2019)

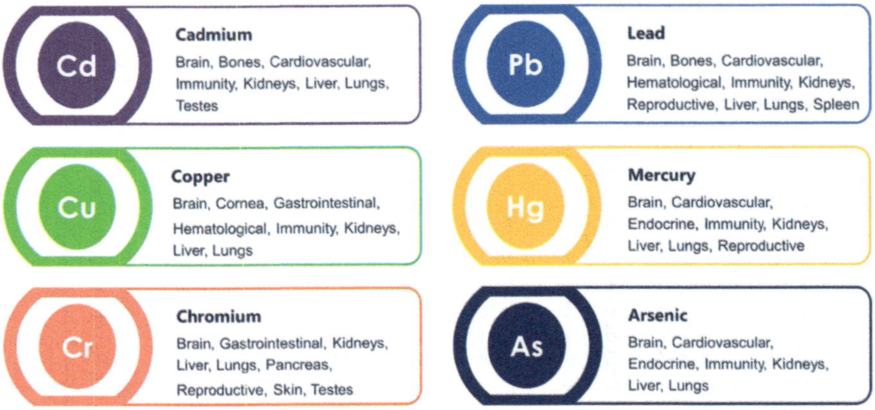

Fig. 4 Heavy metal ions and the organs they affect

environment and their bioaccumulation in living organisms shall cause severe health issues. Toxic heavy metal ions and their area of accumulation are given in Fig. 4. In humans, the toxins get absorbed and stored in the body cells and tissues (Akintokun et al. 2017). They bind with proteins and nucleic acids soon after they enter the cells. This destroys the cell components and shall disrupt their cellular functions (García-Niño et al. 2014). Prolonged exposure to Cr, As, and Hg causes toxicity and shall affect the central nervous system causing brain damage which shall lead to mental illness. It also damages blood cells and blood components, lungs, kidneys, and other vital organs and shall thus precipitate several disease conditions (Van Ael et al. 2012). Also, accumulation of heavy metals for a prolonged period of time in body may result in the degeneration of cells leading to decreased physical, muscular, and neurological functions (García-Niño et al. 2014).

These conditions usually show the symptoms of neurological disorders like Parkinson's disease and Alzheimer's disease (García-Niño et al. 2014). Some heavy metals when accumulated in the body react with nucleic acids and form compounds that mimic the characteristics of various essential hormones in the body (Akintokun et al. 2017). Some may mimic the sex hormones and shall cause the disruption of reproductive function causing fertility issues (Fig. 5). Some might form compounds that shall act as carcinogens in the body and hence causes cancer (Fig. 5). Thus, they are notorious in creating macro-molecules that cause cellular damages, carcinogenesis, and neurotoxicity (Kim et al. 2015). The heavy metal toxicity is directly proportional to the (i) level of intensity of exposure, (ii) the age when someone was exposed, and (iii) how long and how often they were exposed (Kim et al. 2015; Van Ael et al. 2012).

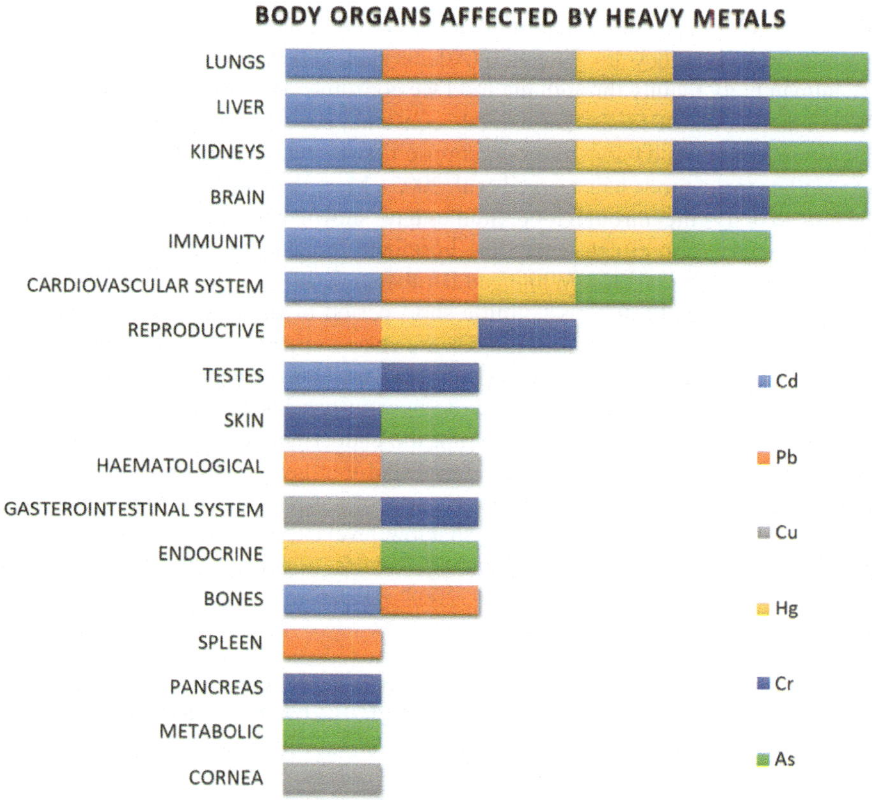

Fig. 5 Human organs affected by toxic heavy metals like Cd, Pb, CU, Hg, Cr, and As

3.2 *Hazards Caused by Dioxins on Human and Nature*

Dioxins are a group of pollutants like PCDDs and PCDFs, which are produced when e-waste undergoes informal process of recycling (Addink and Olie 1995). They are found to be having a wide spectrum of potential toxicity (WHO 2017). About 419 types of dioxin-related compounds have been discovered so far and among these only about 30 are considered to be toxic (WHO 2017). However, assessment of its affliction on plant and animal life is challenging due to their complex composition (Needhidasan et al. 2014). Chlorinated dioxins including PCDDs and PCDFs are produced when copper is extracted by burning wire; the coating of polyvinyl chloride (PVC) used in cables produces chlorinated dioxins when burned in open air (Addink and Olie 1995; Zhang et al. 2015). Lesser known brominated dioxins are thermal degradation products of brominated flame retardants (BFRs). These are plastic additives designed for prevention of accidental fires (Zhang et al. 2015). Mixed brominated/chlorinated dioxins are also generated during e-waste burning (Aracil et al. 2005). But their features have not been well characterized because of the difficulty in analyzing their individual roles in producing a particular effect as they are always produced in large varieties whenever they are generated (Adams et al. 2000). Nowadays, dioxins enter the human body mainly through the food we consume. The animal products, due to the contamination of foods that animals consume, are stained by these chemicals. Dioxins when ingested are stored in adipose tissues, which store fat, and therefore accumulate in the food chain (Aracil et al. 2005). More than 90% of dioxins that enter the human body enter through food. Studies have shown that exposure to dioxins may cause adverse health effects, including hormonal imbalances, reproductive system issues, cancer, and diabetes (White and Birnbaum 2009).

Heavy exposure to dioxins within a short period of time can lead to chloracne (Steenland et al. 1999). This is a skin disease characterized by acne-like lesions that usually manifest on the face and upper body (Fig. 6). This can happen if there has been a significant contamination event. Long-term exposure damages the nervous system at developing stage and shall also disrupt the endocrine and reproductive systems (Steenland et al. 1999). Studies have suggested that heavy exposure to dioxins in workplace over years may boost the prospect of cancer (Fig. 6) (Steenland et al. 1999). Studies on animals also suggest that low-level exposure to dioxins over a long time, or a high-level exposure at early stages of life, might result in reproductive or neurological problems (Mocarelli et al. 2008).

3.3 Noxious Polycyclic Aromatic Hydrocarbons and Their Health Impact

PAHs like acenaphthene, acenaphtylene, anthracene, benz(a)anthracene, benzopyrene derivatives, chrysene, fluoranthene, fluorene, indeno(1,2,3-c,d)pyrene, phenanthrene, pyrene, etc. are present all over the world due to the long-term uncontrolled production from informal recycling processes practiced throughout the country (Rengarajan et al. 2015; Czub et al. 2008). Many PAHs are mutagenic, carcinogenic, teratogenic, and immunotoxic to anything alive (Jin et al. 2020a). The PAHs' toxic effects are directly proportional to the route, duration, and the heaviness of exposure. This determines the severity of toxicity produced (Rengarajan et al. 2015; Jin et al. 2020b).

PAHs are produced from several sources. Each of them carries different levels of risk to health (Fu et al. 1999). The soil-bound PAHs were studied and were classified as follows, according to their Incremental Lifetime Cancer Risk (ILCR) in humans. The divisions are (i) ingestion of PAHs through food carried the highest risk of cancer, i.e., 98.1–99.3%; (ii) absorption of PAHs through skin, i.e., via dermal contact was found to be causing 0.66–1.83% of cancer; and (iii) inhalation of PAH was found to be having the lowest risk of cancer, i.e., 0.03–0.04% (Rengarajan et al. 2015; Petit et al. 2019; Liu et al. 2020; Patel et al. 2020). The toxicity of PAHs is also dependent on the health and age of the affected person (Huo et al. 2019). The sudden exposure to high amount of PAHs shall cause eye irritation, regurgitation, diarrhoea,

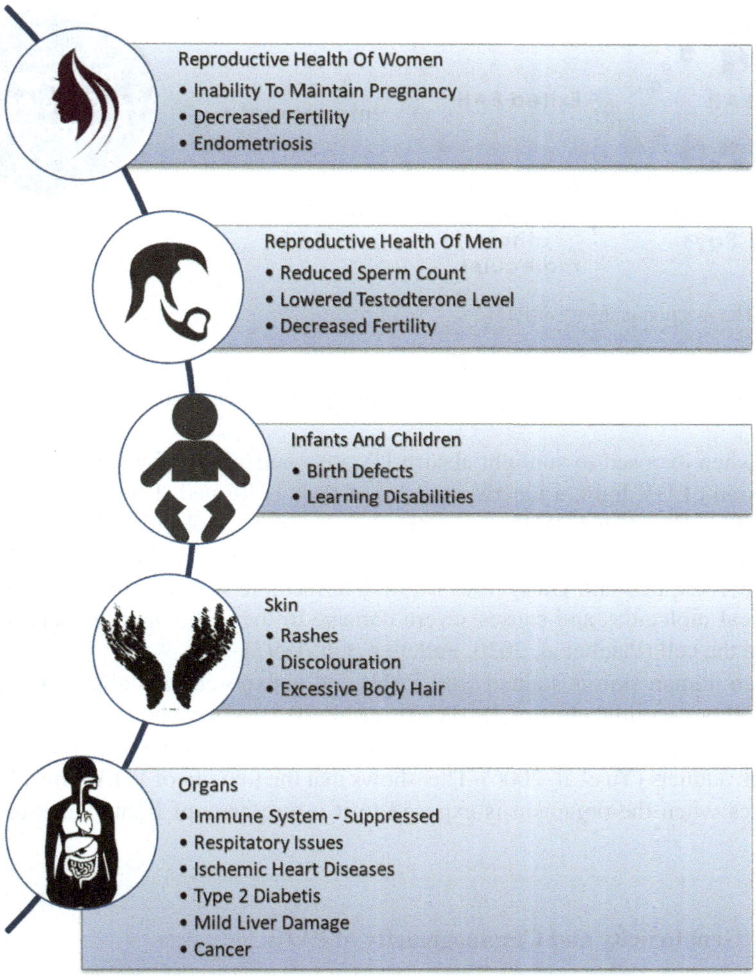

Fig. 6 Dioxins and their effects on human

reduced brain functions, skin irritation, and inflammation (Lu et al. 2016; Wu et al. 2020). PAHs like Naphthalene, anthracene, and benzo(a)pyrene cause irritation on skin of humans and animals on contact (Colmsjö et al. 1984). Prolonged exposure to PAHs will cause chronic health issues like cataracts, renal diseases, hepatic damages, dyspnoea, compromised immunity, and pulmonary issues like asthma (Fu et al. 1999; Wu et al. 2020). Naphthalene, which is still used as a pesticide, if reaches inside the body in heavy doses, through breathing or by consumption, shall induce the destruction of RBCs in blood stream (Huo et al. 2019; Lu et al. 2016; Wu et al. 2020).

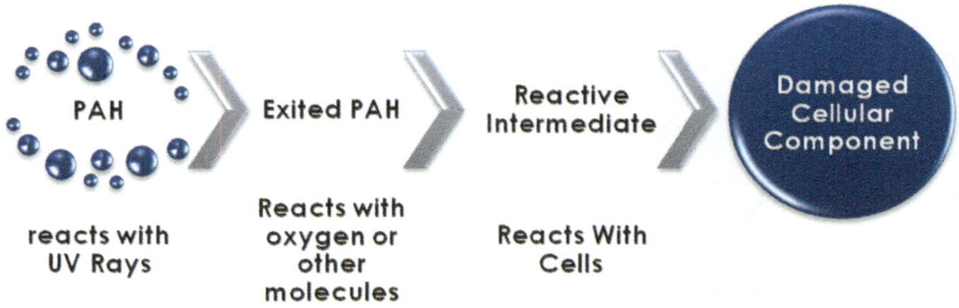

Fig. 7 The reaction pathway of PAH

3.3.1 Photo-Toxicity of Polycyclic Aromatic Hydrocarbons

PAHs when exposed to sunlight absorb UV rays and visible light (Fig. 7). On the absorption of UV light, a reactive species of PAH is formed (Patel et al. 2020; Yu et al. 2006). This exited PAH reacts with oxygen or other molecules. Thus, a reactive intermediate is generated due to the transfer of electron by the excited PAHs in the cells (Newsted and Giesy 1987). The intermediate of PAHs interacts with the biological molecules and causes severe damage to the cell membrane and nucleic acids in the cell (Patel et al. 2020; Pelletier et al. 1997).

When human skin is stained with PAHs and is exposed to sunlight, irradiation causes photo-modification of PAHs (Yu et al. 2006). This photo-modified PHAs causes DNA single-strand cleavage, oxidation of DNA bases, and formation of DNA covalent adducts (Yu et al. 2006). This shows that the toxicity of PHAs considerably increases when the organism is exposed to it in presence of light (Pelletier et al. 1997).

3.3.2 Genotoxicity and Carcinogenicity of PAHs

Detoxification of blood in mammals occurs in the hepatic system through a series of catalytic reactions involving cytochrome P450 and many other oxidase enzymes by the generation of water-soluble epoxide glutathione conjugates (Patel et al. 2020; Binková and Srám 2004). However, when some PAHs are metabolized, reactive intermediates like diolepoxides, quinones, and hydroxyalkyl derivatives are formed which the liver cannot extract from the body (Rengarajan et al. 2015; Abdel-Shafy and Mansour 2016). These reactive intermediates form covalent adducts with nucleic acids thus changing the nature of the particular strand. This causes genotoxicity. Based on the PAHs toxicity, it was classified into four different groups by "The International Agency for Research on Cancer (IARC)". The divisions are (i) as carcinogenic to humans, (ii) as probably carcinogenic to humans, (iii) as possibly carcinogenic to humans, and (iv) as not classifiable as carcinogenic to humans (Patel et al. 2020; Rengarajan et al. 2015). Benzo(a)pyrene has been discovered as being the most potent carcinogen among PAHs and issued as an exposure marker for risk assessments (Delgado-Saborit et al. 2011). The accepted threshold limit for PAHs exposure is 10^{-6}, and there is always a chance (45%) of getting carcinogenic risk, if exposed beyond this limit (Petit et al. 2019). Further, prolonged exposure of PAHs may cause, tumor and cancer in lung, skin, oesophagus, colon, pancreas, bladder, and women's breast (Delgado-Saborit et al. 2011; Petit et al. 2019).

3.3.3 Teratogenicity, Immunotoxicity, and Reproductive Toxicity

On studies conducted in animals, exposure to PAHs like naphthalene, benzo(a)anthracene, and benzo(a)pyrene has shown severe damage or death to the embryos in utero (Patel et al. 2020). These compounds disrupt the normal function of sex hormones as they resemble antioestrogens and/or antiandrogens and hence binds with oestrogen and androgen receptors (Fig. 8) (Patel et al. 2020). In pregnant women, it produces adverse effects like low birth weight, premature delivery, and congenital heart defects in baby (Abdel-Shafy and Mansour 2016; Patel et al. 2020). They also cause non-cancer reproductive system-related health issues in humans (Abdel-Shafy and Mansour 2016). Prolonged exposure of PAHs affects the quality of sperm, proper functioning of male gonads, and viability of egg. It also damages the DNA strands in oocytes (Fig. 8). Impairment to ovarian functions due to structural damage, polycystic ovary syndrome, infertility, spontaneous abortion, and premature birth are also few other complications (Delgado-Saborit et al. 2011; Petit et al. 2019; Patel et al. 2020). The exposed children were found to be prone to behavioral disorders and low IQ (Petit et al. 2019).

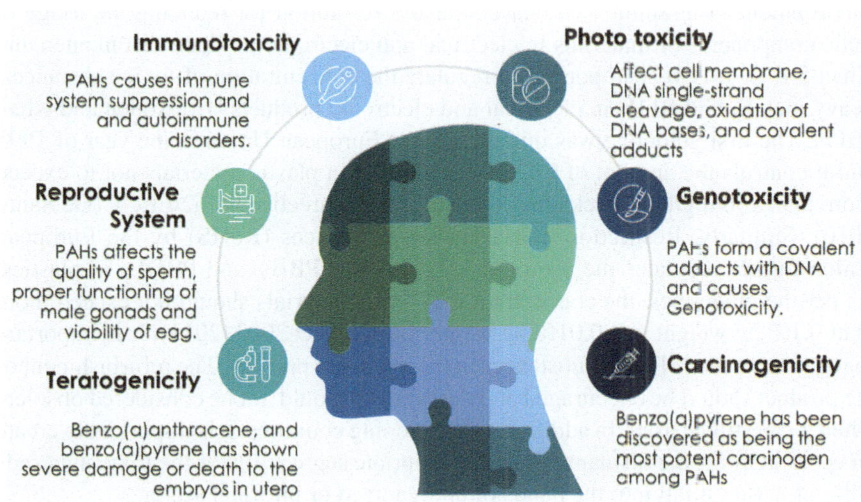

Fig. 8 Noxious polycyclic aromatic hydrocarbon and their impact on health

PAHs cause immune system suppression and may cause the development of tumor, hypersensitivity (allergy), and autoimmune disorders (Burchiel and Luster 2001; Burchiel and Gao 2014). When exposed to PAHs for a significant period of time, degenerative changes in the structure and functions of bone marrow were observed and this compromises the viability of blood cells and weakens the immune system (Hrudkova et al. 2004). PAHs inhibit the pre-B, pre-T, and myeloid cell development. In addition, it disrupts the cytokine production by macrophages and monocytes (Burchiel and Gao 2014). PAHs decrease the immunity of body by binding with aryl hydrocarbon receptors (AhRs). This causes an increase in the level of cytochrome P450 enzyme and shall produce toxic oxidative and electrophilic metabolites (Marris et al. 2020). The amount of PAH required to produce immunotoxicity is much higher than the amount needed to produce cancer (Patel et al. 2020).

3.4 Challenges and Opportunities in Substituting Toxicants in E-products

In the recent past, a multitude of programs were executed, to create awareness among the various stakeholders regarding the presence of toxicants in e-products (Abdel-Shafy and Mansour 2016). Also, "the International Electronics Manufacturing Initiative" is working to reduce or eliminate the use of poisonous chemicals in electrical and electronic products and replace them with safer alternatives. There were more than 100 governmental, non-governmental, and industrial representatives gathered at the International Conference on Chemicals Management in Geneva, hosted by UN Environment Programme, who have signed a resolution for reducing the usage of toxic components or materials in electrical and electronic products (Honkonen and Khan 2017). It is very important to regulate the concentration of toxic substances, heavy metals, and BFRs in electrical and electronic products (Honkonen and Khan 2017). The first guideline was initiated by the European Union in the year of 1995 and it controls the amount of Cd, Cr, Hg, and Pb in plastic materials not to exceed more than 100 mg/kg ("Packaging Directive"-EC Directive 94/62/EEC), (De Santo 2010). Similarly, Restriction of Hazardous Substances (RoHS) by the European Union further regulates the use of Pb, Hg, Cr(VI), PBBs, and PBDEs in plastics. As per the guidelines, the concentration of these materials should not exceed more than 0.1% by weight and 0.01% by weight for Cd (EU 2003, 2011). It is important to regulate the use of toxic substances in the upstream process. The refurbishment of the product should be encouraged and the product should not be considered obsolete when it can still be used. In addition, in developing countries, it is important to create an ecosystem for the dismantling and appropriate segregation of the toxic materials without letting it fall into the hands of unorganized or informal sector.

3.5 Action to Replace PVC and PBDE

PVC is the most widely used synthetic plastic polymer and more than 40 million tons of PVC is produced each year. When PVC comes in contact with chlorinated compounds, undesirable toxic by-products like PCBs, PCDDs, and PCDFs are generated (Patel et al. 2020). Various studies have shown that Cu present in the electric cables and wires when extracted by burning in open air, the PCDDs and PCDFs are created (Chan and Wong 2013; Wang et al. 2009). In the recent times, this has considerably reduced as there is a control on this burning by the implementation of appropriate laws by Govt. It is important to reduce the use of PVC, which will release PCDDs and PCDFs to the environment. There is an increasing demand to identify PVC-free alternatives for gadgets, hospital equipment, and in building construction (Wang et al. 2009). Similarly, there are different commercial PBDEs flame-retardant products like Penta-BDE (used in polyurethane foams), octa-BDE (acrylonitrile/butadiene/styrene resin), and deca-BDE used in HD-polystyrene (Schecter et al. 2003). Among different PBDEs' flame retardant, Deca-BDE is considered as a stable retardant compared to penta-BDE and octa-BDE (Guardia et al. 2006). However, recent studies have shown that debromination occurred during the photolytic and anaerobic degradation and leads to production of toxic intermediate by-products (Sharkey et al. 2020). Hence, the use of PBDEs as flame retardants is banned in several countries.

4 Future Prospective

The challenges in controlling and reducing the activities that lead to the release of persistent toxic compound present in plastic, glass, and other materials of the e-waste have been extensively discussed. It is very important and essential to develop alternatives to the currently used components of e-products, which are free from toxic materials and are also eco-friendly. In addition, it is important to find cost-effective recycling technology, which wouldn't emit toxic compounds to air, land, and water. The eco-friendly technologies used widely for the recovery of metals are hydrometallurgy and pyrometallurgy. This incurs high operating cost, high consumption of raw materials, and slow rate of metal recovery. On the other hand, greener technology like chelation technique is novel but not widely used as a substitute for the above-mentioned techniques. However, recovery and recycling of the chelating agents still needs to be addressed for it to the most efficient metal recovery process.

Waste mobile phones contain 15–20 different plastic parts, which contain different concentrations of Pb, Cd, Be, Cr, Sb, As, Hg, and BFRs in it. Thus, the burning of plastic waste from mobile phone and other e-waste by unorganized sector leads to the release of various toxic compounds produced by the oxidation of the above-mentioned metals. This poses a serious danger to the environment and human health. In developing countries, drafting and implementing a stringent policy is an important and foremost challenging step in the implementation of a robust e-waste management. In addition, policymakers have to find more innovative methods to engage with the public and recycling agents and improve their awareness regarding alternative solutions to the existing problems. It is important to minimize the presence of toxic materials in the electronic products and refurbishment of the electronic products should be encouraged. In developing countries, adequate infrastructure for the recycling of e-waste materials should be planned and initiated. Recently, few sustainable and eco-friendly technologies for the recovery of precious metals from PCBs and separation of plastic from e-waste have been developed by the Centre for Materials for Electronics Technology (C-MET), India and Central Institute of Plastic Engineering and Technology (CIPET), India. Developing sustainable and eco-friendly technologies in developing countries are the first step in the right direction to achieve robust e-waste management. The government of developing countries should create awareness among all stakeholders to channelize the e-waste to authorized recycling centers.

Acknowledgements The authors thank Professor B.S Murty, Department of Civil Engineering, IIT Madras, and Professor V.R. Muraleedharan, Department of Humanities and Science, IIT Madras, India, for the discussions and support.

References

Abdel-Shafy HI, Mansour MSA (2016) Review on polycyclic aromatic hydrocarbons: source, environmental impact, effect on human health and remediation. Egypt J Pet 25: 107–123

Adams B, Buekens AJ, Joannes EW (2000) Dioxin-emissions from a MSWI related to memory effects in a 2-stage wet scrubber. Organohalogen Compd 46: 178–181

Addink R, Olie K (1995) Mechanisms of formation and destruction of polychlorinated dibenzo-p-dioxins and dibenzofurans in heterogeneous systems. Environ Sci Tech 29:1425–1435

Akintokun AK, Onatunde OO, Shittu OB, Okeyode IC, Taiwo MO (2017) Bioaccumulation of heavy metals using selected organisms isolated from electronic waste dumpsite of two south-western states in Nigeria. Appl Environ Res 39(2):29–40

Alabi OA, Adeoluwa YM, Huo X, Xu X, Bakare AA (2021) Environmental contamination and public health effects of electronic waste: an overview. J Environ Health Sci Eng 19:1209–1227

Al-Anzi1 BS, Al-Burait AA, Thomas A, Ong CS (2017) Assessment and modelling of E-waste generation based on growth rate from different telecom companies in the State of Kuwait. Environ Sci Pollut Res 24: 27160–27174

Aracil I, Font R, Conesa JA (2005) Semivolatile and volatile compounds from the pyrolysis and combustion of polyvinyl chloride. J Anal Appl Pyrol 74:465–478

Baldé CP, Forti V, Gray V, Kuehr R, Stegmann P (2017) The global e-waste monitor, United Nations University (UNU), International Telecommunication Union (ITU) and International Solid Waste Association (ISWA), Bonn/Geneva/Vienna. ISBN electronic version, pp 978–992

Binková B, Srám RJ (2004) The genotoxic effect of carcinogenic PAHs, their artificial and environmental mixtures (EOM) on human diploid lung fibroblasts. Mutat Res 22(1–2):109–121

Borthakur A (2015) Generation and management of electronic waste in India: an assessment from stakeholders' perspective. Int J Dev Soc 31:220–248

Borthakur A (2020) Policy approaches on e-waste in the emerging economies: a review of the existing governance with special reference to India and South Africa. J Clean Prod 252:119885

Borthakur A, Govind M (2017) Emerging trends in consumers' E-waste disposal behaviour and awareness: A worldwide overview with special focus on India. Resour Conserv Recycl 117:102–113

Borthakur A, Singh P (2017) Researches on informal e-waste recycling sector: it's time for a 'Lab to Land' approach. J Haz Mater 323:730–732

Burchiel SW, Gao J (2014) Polycyclic aromatic hydrocarbons and the immune system. In: Vohr H-W (ed) Encyclopedia immunotoxicology. Springer, Berlin, pp 1192–1194. https://doi.org/10.1007/978-3-642-27786-3

Burchiel SW, Luster MI (2001) Signalling by environmental polycyclic aromatic hydrocarbons in human lymphocytes. Clin Immunol 98:2–10

Central Pollution Control Board (2016) List of registered e-waste dismantlers/recyclers in the country. http://cpcb.nic.in/cpcbold/List_of_E-waste_Recycler_as_on_29.12.2016.pdf

Cesaro A, Belgiorno V, Gorrasi G, Viscusi G, Vaccari M, Vinti G, Salhofer S (2019) A relative risk assessment of the open burning of WEEE. Environ Sci Pollut Res 26(11):11042–11052

Chakraborty P, Selvaraj S, Nakamura M, Prithiviraj B, Cincinelli A, Bang JJ (2018) PCBs and PCDD/Fs in soil from informal e-waste recycling sites and open dumpsites in India: levels, congener profiles and health risk assessment. Sci Total Env 621:930–938

Chan JKY, Wong MH (2013) A review of environmental fate, body burdens and human health risk assessment of PCDD/Fs at two typical electronic waste recycling sites in China. Sci Total Environ 463–464:1111–1123

Chaturvedi A, Arora R, Ahmed S (2010) Mainstreaming the informal sector in e-waste management. In: Urban, industrial and hospital waste management conference, Ahmedabad, Gujarat

Colmsjö A, Rannug A, Rannug U (1984) Some chloro derivatives of polynuclear aromatic hydrocarbons are potent mutagens in salmonella typhimurium. Mutat Res 135:21–29

Czub G, Wania F, McLachlan MS (2008) Combining long-range transport and bioaccumulation considerations to identify potential arctic contaminants. Environ Sci Technol 42:3704–3709

De Santo EM (2010) 'Whose science?' precaution and power-play in European marine environmental decision-making. Mar Policy 34:414–420

Delgado-Saborit JM, Stark C, Harrison RM (2011) Carcinogenic potential, levels and sources of polycyclic aromatic hydrocarbon mixtures in indoor and outdoor environments and their implications for air quality standards. Environ Int 37(2):383–392

EU (2003) Directive 2002/96/EC of the European parliament and the council of 27 January (2003) on waste electrical and electronic equipment (WEEE)-joint declaration of the European parliament, the council and the commission relating to article 9

EU (2011) Directive 2011/65/EU of the European Parliament and of the Council of 8 June 2011, on the restriction of the use of certain hazardous substances in electrical and electronic equipment (recast)

EU (2012) Directive 2012/19/EU of the European parliament and the council of 4 July (2012) on waste electrical and electronic equipment (WEEE), https://eur-lex.europa.eu/legal-content/EN/TXT/PDF/?uri=CELEX:32012L0019&from=EN

Forti V, Baldé CP, Kuehr R, Bel G (2020) The global E-waste monitor 2020: quantities, flows and the circular economy potential. United Nations University (UNU)/United Nations Institute for Training and Research (UNITAR)—co-hosted SCYCLE Programme, International Telecommunication Union (ITU) & International Solid Waste Association (ISWA), Bonn/Geneva/Rotterdam

Frazzoli C, Orisakwe OE, Dragone R, Mantovani A (2010) Diagnostic health risk assessment of electronic waste on the general population in developing countries' scenarios. Environ Impact Assess Rev 30:388–399

Fu PP, Von-Tungeln LS, Chiu LH, Own ZY (1999) Halogenated-polycyclic aromatic hydrocarbons: a class of genotoxic environmental pollutants. J Environ Sci Health C 17:71–109

García-Niño WR, Pedraza-Chaverrí J (2014) Protective effect of curcumin against heavy metals induced liver damage. Food Chem Toxicol 69:182–201

Guardia MJL, Hale RC, Harvey E (2006) Detailed polybrominateddiphenyl ether (PBDE) congener composition of the widely used penta-, octa-, and deca-PBDE technical flame-retardant mixtures. Environ Sci Technol 40(20):6247–6254

Gupta V, Kumar A (2014) E-waste Status and Management in India. J. Inform. Engg. Applications. 4:9

Hashmi MZ, Varma A (2019) Electronic waste pollution, environmental occurrence and treatment technologies. In: Hashmi MZ, Varma A (eds). Springer Science and Business Media LLC

He ZL, Yang XE, Stoffella PJ (2005) Trace elements in agroecosystems and impacts on the environment. J Trace Elem Med Bio 19(2–3):125–140

Herawati N, Suzuki S, Hayashi K, Rivai IF, Koyoma H (2000) Cadmium, copper and zinc levels in rice and soil of Japan, Indonesia and China by soil type. Bull Environ Contam Toxicol 64: 33–39

Hites RA (2004) Polybrominateddiphenyl ethers in the environment and in people: a meta-analysis of concentrations. Environ Sci Technol 38:945–956

Honkonen T, Khan SA (2017) Chemicals and waste governance beyond 2020. Exploring pathways for a coherent global regime. Nordic Council of Minister

Hrudkova M, Fiala Z, Borska L, Novosad J, Smolej L (2004) The effect of polycyclic aromatic hydrocarbons to bone marrow. Actamedica (Hradec Kralove). Supplementum 47: 75–81

Huo X, Wu Y, Xu L, Zeng X, Qin Q, Xu X (2019) Maternal urinary metabolites of PAHs and its association with adverse birth outcomes in an intensive e-waste recycling area. Environ Pollut 245:453–461

Jin R, Bu D, Liu G, Zheng M, Lammel G, Fu J, Yang L, Li C, Habib A, Yang Y, Liu X (2020a) New classes of organic pollutants in the remote continental environment—Chlorinated and brominated polycyclic aromatic hydrocarbons on the Tibetan Plateau. Environ Int 137:105574

Jin R, Zheng MH, Lammel G, Bandowe BAM, Liu GR (2020b) Chlorinated and brominated polycyclic aromatic hydrocarbons: sources, formation mechanisms, and occurrence in the environment. Prog Energ Combust Sci 76:100803

Kiddee P, Pradhan JK, Mandal S, Biswas JK, Sarkar B (2020) Handbook of electronic waste management. Elsevier, pp 1–18

Kim HS, Kim YJ, Seo YR (2015) An overview of carcinogenic heavy metal: molecular toxicity mechanism and prevention. J Cancer Prev 20(4):232–240

Kumar U, Singh D (2014) Electronic waste: concerns & hazardous threats. Int J Curr Eng Technol 4:802–811

Kumar R, Verma A, Rajput N (2017) E-waste management in India: Formal vsinformal sectors. Int J Recent Adv Multidiscip Res 4(8):2720–2726

Kwatra S, Pandey S, Sharma S (2014) Understanding public knowledge and awareness on e-waste in an urban setting in India: a case study for Delhi. Manag Environ Qual Int J 25(6):752–765

Li W, Achal V (2020) Environmental and health impacts due to e-waste disposal in China—a review. Sci Total Environ 737:139745

Liu R, Shengtao M, Yu Y, Li G, Yu Y, An T (2020) Field study of PAHs with their derivatives emitted from e-waste dismantling processes and their comprehensive human exposure implications. Environ Int 144:106059

Lu SY, Li YX, Zhang JQ, Zhang T, Liu GH, Huang MZ, Li X, Ruan JJ, Kannan K, Qiu RL (2016) Associations between polycyclic aromatic hydrocarbon (PAH) exposure and oxidative stress in people living near e-waste recycling facilities in China. Environ Int 94:161–169

Lundgren K (2012) The global impact of e-waste: addressing the challenge. International Labour Organization, Geneva

Marris C, Kompella SN, Miller M, Incardona JP, Brette F, Hancox JC, Sorhus E, Shiels HA (2020) Polyaromatic hydrocarbons in pollution: a heart-breaking matter. J Physiol 598:227–247

Mocarelli P, Gerthoux PM, Patterson DG, Milani S, Limonta G, Bertona M, Signorini S, Tramacere P, Colombo L, Crespi C, Brambilla P, Sarto C, Carreri V, Sampson EJ, Turner WE, Needham LL (2008) Dioxin exposure, from infancy through puberty, produces endocrine disruption and affects human semen quality. Environ Health Perspect 116(1):70–77

Ministry of Environment, Forests and Climate Change (MoEF&CC) (2015) Implementation of e-waste (management and handling) rules, 2011 by Central Pollution Control Board. Report No.30 of 2015

MoEF&CC (2013) Invitation of proposals for establishment of common treatment and disposal facilities for hazardous, bio-medical and e-wastes viz., Treatment, Storage and Disposal Facilities (TSDFs), Secured Land Fills (SLFs), Integrated E-Waste Recycling and Treatment, New Delhi. http://moef.gov.in/wp-content/uploads/2018/03/RFD-2013-14.pdf

Mundada MN, Kumar S, Shekdar AV (2004) E-waste: a new challenge for waste management in India. Int J Environ Stud 61(3):265–279

Needhidasan S, Samuel M, Chidambaram R (2014) Electronic waste—an emerging threat to the environment of urban India. J Environ Health Sci Eng 12–36

Newsted JL, Giesy JP (1987) Predictive models for photo-induced acute toxicity of polycyclic aromatic hydrocarbons to daphnia magna, strauss (cladocera, crustacea). Environ Toxicol Chem 6:445–461

Parajuly K, Kuehr R, Awasthi AK, Fitzpatrick C, Lepawsky J, Smith E, Widmer R, Zeng X Future e-waste scenarios (2019) StEP (Bonn), UNU ViE-SCYCLE (Bonn) & UNEP IETC (Osaka)

Patel AB, Shaikh S, Jain KR, Desai C, Madamwar D (2020) Polycyclic aromatic hydrocarbons: sources, toxicity, and remediation approaches. Front Microbiol 11:562813

Pelletier MC, Burgess RM, Ho KT, Kuhn A, McKinney RA, Ryba SA (1997) Phototoxicity of individual polycyclic aromatic hydrocarbons and petroleum to marine invertebrate larvae and juveniles. Environ Toxicol Chem 16(10):2190–2199

Perkins DN, Drisse MB, Nxele T, Sly PD (2014) E-waste: a global hazard. Ann Glob Health 80(4):286–295

Petit P, Maitre A, Persoons R, Bicout DJ (2019) Lung cancer risk assessment for workers exposed to polycyclic aromatic hydrocarbons in various industries. Environ Int 124:109–120

Purushothaman M, Inamdar MG, Muthunarayanan V (2021) Socio-economic impact of the e-waste pollution in India. Mater Today: Proc 37(2):280–283

Rajput RR, Ahuja Nigam N (2021) An overview of E-waste, its management practices and legislation in present Indian context. J Appl Nat Sci 13(1): 34–41

Rengarajan T, Rajendran P, Nandakumar N, Lokeshkumar B, Rajendran P, Nishigaki I (2015) Exposure to polycyclic aromatic hydrocarbons with special focus on cancer Asian Pac J Trop Biomed 5:182–189

Rim KT, Koo KH, Park JS (2013) Toxicological evaluations of rare earths and their health impacts to workers: a literature review. Saf Health Work 4:12–26

Rotter VS, Maehlitz P, Korf N, Chancerel P, Huisman J, Habib H, Herreras L, Ljunggren SM, Hallberg A (2016) ProSUM deliverable 4.1—waste flow studies. 1–100

Sahajwalla V, Gaikwad V (2018) The present and future of e-waste plastics recycling. Curr Opin Green Sustain Chem 13:102–107

Schecter A, Pavuk M, Päpke O, Ryan JJ, Birnbaum L, Rosen R (2003) Polybrominateddiphenyl ethers (PBDEs) in U.S. mothers' milk. Environ Health Perspect 111(14): 1723–1729

Sharkey M, Harrad S, Abdallah MA, Drage DS, Berresheim H (2020) Phasing-out of legacy brominated flame retardants: the UNEP stockholmconvention and other legislative action worldwide. Environ Int 144:106041

Steenland K, Piacitelli L, Deddens J, Fingerhut M, Chang LI (1999) Cancer, heart disease, and diabetes in workers exposed to 2,3,7,8-tetrachlorodibenzo-p-dioxin. J Natl Cancer Inst 91(9): 779–786

Turaga RMR, Bhaskar K, Sinha S, Hinchliffe D, Hemkhaus M, Arora R, Chatterjee S, Khetriwal DS, Radulovic V, Singhal P, Sharma H (2019) E-Waste Management in India: issues and strategies. J Decis Makers 44(3):127–162

Van Ael E, Covaci A, Blust R, Bervoets L (2012) Persistent organic pollutants in the Scheldt estuary: environmental distribution and bioaccumulation. Environ Int 48:17–27

WHO/UNEP global surveys of PCDDs, PCDFs, PCBs and DDTs in human milk and benefit-risk evaluation of breast-feeding. Arch Toxicol 91: 83–96

Vidyadhar A (2016) E-waste in transition—from pollution to resource. Edited by Florin-Constant in Mihai, pp 121–158

Wang H-M, Yu Y-J, Han M, Yang S-W, Yang Y (2009) Estimated PBDE and PBB congeners in soil from an electronics waste disposal site. Bull Environ Contam Toxico 83: 789–793

Wath SB, Vaidya AN, Dutt PS, Chakrabarti T (2010) A roadmap for development of sustainable E-waste management system in India. 409(1): 19–32

Wei B, Yang L (2010) A review of heavy metal contaminations in urban soils, urban road dusts and agricultural soils from China. Microchem J 94(2):99–107

White SS, Birnbaum LS (2009) An overview of the effects of Dioxins and Dioxinlike compounds on vertebrates, as documented in human and ecological epidemiology. J Environ Sci Health C: Environ Carcinog Ecotoxicol Rev 27(4): 197–211

Wu H, Lu L, Chen J, Zhang C, Liu W, Zhuang S (2020) Inhibited nitric oxide production of human endothelial nitric oxide synthase by nitrated and oxygenated polycyclic aromatic hydrocarbons. Environ Sci Technol 54:2922–2930

Yang Q, Qiu X, Li R, Liu S, Li K, Wang F, Zhu P, Li G, Zhu T (2013) Exposure to typical persistent organic pollutants from an electronic waste recycling site in Northern China. Chemosphere 91(2):205–211

Yu H, Xia Q, Yan J, Herreno-Saenz D, Wu Y-S, Tang I-W, Fu PP (2006) Photoirradiation of polycyclic aromatic hydrocarbons with UVA light—a pathway leading to the generation of reactive oxygen species, lipid peroxidation, and DNA damage. Int J Environ Res Public Health 3(4): 348–354

Zhang M, Buekens A, Jiang X, Li X (2015) Dioxins and polyvinylchloride in combustion and fires. J Sustain Circular Econ 33(7):630–643

Overview of E-Waste Reverse Logistics: How to Promote the Return of Electronic Waste to the Production Chain

Andressa Soares Barrozo, Emanuele Caroline Araujo dos Santos, Julia de Vargas Biehl, Carline Fabiane Stalter, Joice Pinho Maciel, Daiane Calheiro Evaldt, José Luís Cardoso da Silva, Regina Célia Espinosa Modolo, Feliciane Andrade Brehm, and Carlos Alberto Mendes Moraes

Abstract Waste Electrical Electronic Equipment (WEEE) is the type of waste that grows the most in generation in Brazil and the world. This is due to planned obsolescence, the development of technology, and high consumption rates. If not well managed, this waste can cause negative environmental impacts and bring risks to human health, due to the diversity of toxic substances that may be present in it. This chapter analyzes the relationship of various countries with WEEE management. It can be noted that many countries are in the process of developing laws on this subject, while others are still far behind. Unfortunately, a large portion of this waste is not properly destined, having illegal and/or uncertain outcomes. In Brazil, the Sectoral Agreement for the Reverse Logistics of Electronics was signed in 2019, and is currently in its initial structuring process. However, Brazil has a great socio-environmental and economic responsibility regarding the implementation of a reverse logistics system. Cooperatives represent a good possibility to facilitate this process in municipalities as providers of environmental services.

Keywords Small-sized E-waste · Urban mining · E-waste · Reverse logistics

E. C. A. dos Santos · C. F. Stalter · D. C. Evaldt · R. C. E. Modolo · C. A. M. Moraes (✉)
Graduate Program of Mechanical Engineering, University of Vale do Rio dos Sinos - UNISINOS, São Leopoldo, RS, Brasil
e-mail: cmoraes@unisinos.br

A. S. Barrozo · J. P. Maciel · J. L. C. da Silva · R. C. E. Modolo · F. A. Brehm · C. A. M. Moraes
Graduate Program of Civil Engineering, University of Vale do Rio d'os Sinos - UNISINOS, São Leopoldo, RS, Brasil

J. de Vargas Biehl
Graduated Program in Environmental Engineering, University of Vale do Rio dos Sinos - UNISINOS, São Leopoldo, RS, Brasil

M. Jawaid and A. Khan (eds.), *Conversion of Electronic Waste in to Sustainable Products*, Sustainable Materials and Technology,
https://doi.org/10.1007/978-981-19-6541-8_5

1 Introduction

In 2019, it was found that, every year, an average of 2.5 million metric tons more of electronic equipment (EE) are consumed (Forti et al. 2020). According to Forti et al. (2020), the increase in the consumption of this equipment is directly linked to global economic development, that is, the increase in income, urbanization, and greater mobility, allied to industrialization.

Regarding the waste from these post-consumer products, the world generated 53.6 million metric tons (Mt) in 2019, meaning a growth of 9.2Mt since 2014. In addition, as the amount of Waste Electrical and Electronic Equipment (WEEE) has been increasing at a rate of 2 Mt/year, the total global generation by 2030 is expected to reach 74.7 Mt (Forti et al. 2020).

Of the 53.6 Mt of WEEE that are discarded, about 17.4 Mt (32.4%) were small equipment, and 13.1 Mt (24.4%) were large equipment; in addition, there are the categories of temperature exchange equipment (10.8 Mt), screens and monitors (6.7 Mt), small IT and telecommunications equipment (4.7 Mt), and lamps (0.9 Mt) (Forti et al. 2020).

According to Nicolai (2016), due to planned obsolescence, technology growth, and market consumption, WEEE is the waste that grows the most in generation in Brazil, as well as in the world. In addition, they have several toxic substances, and can reach up to 100 different contaminants, which can cause contamination of the environment and risks to human health, if they are not well managed (Pradiian 2013; Kiddee et al. 2013; Nicolai 2016; Silveira et al. 2020).

This waste presents a great complexity, due to the variety of constituent materials in them, such as metals, polymers, ceramics, and composite materials, of which many of these present a high added value, such as gold (Xavier and Lins 2018; Silveira et al. 2020). Furthermore, EEE may contain critical materials, i.e. materials of great economic importance and presenting a supply risk, such as the elements neodymium, praseodymium, and dysprosium (European Commission 2017).

In many cases, the concentration of some metals in WEEE can be higher than in natural ores; with this, this type of waste has been considered a secondary ore, also called "urban mines", and the practice of recovering metals contained in this waste is called urban mining (Pradiian 2013; Cossu and Williams 2015; Nicolai 2016; Havlik 2015; Willersinn and Bart 2017; Giese et al. 2018; Xavier et al. 2019; Santos 2021; Santos et al. 2021). According to Nancharaiah et al. (2016), both solid and liquid waste that contain critical metals are increasingly being seen as secondary sources for obtaining metals.

Urban mining can be considered one of the main tools of circular economy, which is nothing more than a new way of thinking about the cycles of materials, in which it is expected to increase their service life, as well as seek closed cycles, promote reuse, repair, and remanufacturing, and, when these possibilities are exhausted, recycling (Ellen Macarthur Foundation 2011, 2012, 2016, 2017, 2019; Duthie and Lins 2017; Xavier and Lins 2018; Giese et al. 2018; Ottoni et al. 2020).

The collection and recycling of WEEEs, formally documented worldwide, was 9.3 Mt in 2019, meaning only 17.4% of the total generation in the same year, that is, recycling processes are not keeping up with the global growth of waste generated, as there has been a growth of only 0.4 Mt per year since 2014 (Forti et al 2020). Meanwhile, 44.3 Mt (82.6%) of WEEEs generated in 2019 have not been documented, so both their fate and the environmental impact caused by them are uncertain, but vary by region (Forti et al. 2020).

Higher income countries have a more developed recycling infrastructure, thus.

- Only about 8% of the electronic waste generated is disposed of in specific bins and ends up going to landfills or incinerated;
- In some cases, when equipment can still be repaired, they are generally shipped, as second-hand products, to low- or middle-income countries. There is still the practice of illegally exporting this waste, under the pretext of being second-hand or as scrap. In any case, about 7–20% of WEEEs may be shipped to other countries;
- Finally, much of the undocumented waste must end up entering flows of other waste, such as polymers and metals (Forti et al. 2020).

Low- or middle-income countries, on the other hand, do not yet have a well-developed management infrastructure or, in some cases, they do not even exist yet. With this, the fate of a significant part of the WEEEs becomes a responsibility of the informal sector (Forti et al. 2020).

2 Reverse Logistics System in the World

In 2019, the annual global generation of WEEE was 53.6 million metric tons, about 7.3 kg per capita. Only 17.4% of the waste generated is documented, being properly collected and recycled. Out of a total of 193 countries, there are 78 that have current laws and regulations regarding the generation of this type of waste (Forti et al. 2020).

Asia was the continent that presented the highest rates of electronic waste generation, with 24.9 Mt, followed by America (13.1 Mt), Europe (12 Mt), Africa (2.9 Mt), and Oceania (0.7 Mt), also in 2019. However, of all the continents, Europe has the best recycling rates (42.5%), followed by Asia (11.7%), the Americas (9.4%), and Oceania (8.8%), while the African continent has the lowest rate (0.9%) (Forti et al. 2020).

In 2016, Brazil was the second largest generator of this type of waste in the Americas, behind only the USA, and generated about 1.5 Mt, meaning a per capita generation of 7.4 kg/hab. In the ranking of generated volumes, the country was in the seventh position in that year, corresponding to 3.4% of the world total (Baldé et al. 2017b; Xavier and Lins 2018). In 2019, its generation of WEEEs reached 2.143 kt (remaining in the second position of the Americas), and the per capita generation passed the 10 kg/hab/year (Forti et al. 2020).

Europe is the continent with the highest rate of WEEE collected and recycled, with 42.5%, followed by Asia with 11.7%, the Americas with 9.4%, Oceania with 8.8%, and finally the African continent with 0.9% (Forti et al. 2020).

An international regulatory framework, signed by 165 countries and been in force since 1992, is the so-called Basel Convention, addressing the Control of Transboundary Movements of Hazardous Wastes and Their Disposal (Sant'anna et al. 2014).

Below, we will see what the laws of some countries are like, such as Switzerland, the United States of America, Canada, and China.

- **Europe: Switzerland**

Since July 1998, Switzerland has had an ordinance in place for the return, removal, and disposal of the EEE, with this function being the responsibility of the manufacturer and importer (Khetriwal et al. 2007). However, it was not necessary to create this ordinance so that the reverse logistics of electronic waste began to be practiced. Through the *Swiss Foundation for Waste Management*, a voluntary project of electronics manufacturers, the collection and management of waste generated began to be carried out, leaving to the state the function of only regularizing this action (Sant'anna et al. 2014).

One of the components of the Swiss electronic waste management system is the Extended Producer Responsibility (EPR). According to Khetriwal et al. (2007), the definition of this term is "an environmental protection strategy that makes the product manufacturer responsible for the life cycle and, especially, for the reverse logistics, recycling, and final disposal of the product".

In Switzerland, those responsible for the WEEE management system are the government, manufacturers, and members of the Producer Responsibility Organisation, distributors and resellers, consumers, and recyclers. To fulfill the obligations of producer responsibility (EPR), there is the so-called Producer Responsibility Organisation (PRO), which ensures the proper management of all types of electronic waste, from financing, collection, transportation, and the control system (Khetriwal et al. 2007).

When purchasing electronic products, consumers pay the advanced recycling fee (ARF), ensuring all the logistics are carried out after use. Producers can then choose to participate in a PRO or have their own system for returns (Khetriwal and Jain 2021).

Despite the high WEEE generation in the country—around 23.4 kg per capita in 2019, according to Forti et al. (2020)—the management model, when thinking of e-waste, seems to be a world reference (Sant'anna et al. 2014).

- **America: United States and Canada**

Both the United States of America as well as Canada do not have federal legislation on the management of WEEEs, however, some states/provinces have some form of legislation. In the case of the US, 25 states and the District of Columbia have some laws regarding this waste, covering around 75–80% of the population. However, the scopes and impacts vary greatly in each state, making it difficult for a proper collection approach (Forti et al. 2020).

As for Canada, there is no such legislation at the national level, since the federal agency does not have such authority. However, 12 provinces have some regulations in place, where programs are managed by the industry (Forti et al. 2020).

However, regarding WEEE management, the US presents some general measures, at the federal level, in order to prevent impacts and adverse effects from the incorrect disposal of this waste. With this, there is the Resource Conservation and Recovery Act (RCRA), which some equipment falls under when meeting certain requirements. In addition, certifications have been adopted for recyclers (Responsible Recycling (R2) and E-Stewards), which federal agencies are oriented to use (Forti et al. 2020).

With respect to other countries in America, only a few have been able to establish laws regarding WEEEs. The main forces related to the management of this waste in the region are Mexico, Costa Rica, Colombia, and Peru, where work has been done to improve the systems already established. Meanwhile, Brazil and Chile have been laying the groundwork to begin implementing formal regulations specifically for this waste (Forti et al. 2020).

In general, according to Forti et al. (2020), there is a problem in the region, which is the heterogeneity of regulations; that is, there are differences in approaches such as Extended Producer Responsibility (EPR) versus shared responsibility versus public sector programs. Furthermore, the jurisdictions vary between federal, state, and municipal, as well as divergences in definitions of responsibilities, categories, and principles, among others (Forti et al. 2020).

- **Asia: China**

China has legislation in force for the collection and treatment of fourteen types of electronic waste; among them are the most common, such as televisions, air conditioners, washing machines, computers, telephones, and others. In 2019, the country had a generation of 10.1 Mt of e-waste, being one of the largest producers of this type of waste in the world (Forti et al. 2020).

Because it is a very populous country, China has a great demand for electronic equipment, as well as large manufacturing industries for this equipment. The ability to recycle this type of waste is enormous, as, with formal regulations and facilities, more than 70 million equipment is dismantled annually (Forti et al. 2020).

We can name four legislations that have contributed to the management of waste from electrical and electronic equipment in China. They are the Technical, Policy on Pollution Prevention, and Control of the Waste Electrical and Electronic Products, in 2006; the Ordinance on the Management of the Prevention and Control of Pollution from Electronic and Information Products, in 2007; the Administrative Measures on Pollution Prevention of WEEE, in 2008; and the Regulation on the Management of the Recycling and Disposal of Waste Electrical and Electronic Equipment, in the year 2011 (Wang et al. 2013).

The responsibility for the management of this type of waste, according to the legislation, belongs to the producers, including importers, retailers, and collection, recycling, and treatment companies (Wang et al. 2013). Producers must pay a fee for the treatment of each equipment produced, except for exported products. Producers are still responsible for prioritizing materials that do not harm the environment, such as hazardous ones, and should use materials conducive to recycling and reuse after use (Khetriwal and Jain 2021).

In 2017, stimulating the recycling of electronic waste to be developed, new policies such as the Plan for the Implementation of the Extended Producer Responsibility System were defined. The Chinese State Council has proposed goals to recycle 50% of electronic waste by 2025, as well as increase the use of recycled materials in the production of new equipment to 20% (Wen 2018).

3 Reverse Logistics System in Brazil

Federal Law No. 12,305/2010—National Solid Waste Policy[1] (PNRS) in Brazil establishes important principles to allow the country to make the necessary progress in facing the main environmental, social, and economic problems arising from the inadequate management of solid waste. PNRS addresses sectoral Reverse Logistics agreements, which are acts of a contractual nature, signed between the government and manufacturers, importers, distributors or sellers, and consumers, aiming at the implementation of shared responsibility for the life cycle of products.

In the second half of 2019, the Sectoral Agreement for the Implementation of a Reverse Logistics System of Electronic Products and their Components was signed in Brazil. However, it was only in February 2020, when Decree No. 10,240 was published, which replicates the content of the agreement of the implementation of a reverse logistics system for this waste, that implementation began (Barrozo et al. 2021).

The officially signed sectoral agreement proposal is connected to the National Solid Waste Policy, and shares responsibilities between manufacturers, importers, distributors, and sellers, regardless of public authorities. In addition, the agreement establishes an accurate classification of which objects should be directed by the system, which facilitates the understanding of their environmental and occupational safety obligations (Santos et al. 2021).

Making WEEE return to the recycling production chain is directly related to the principles of industrial ecology and circular economy, in which it is proposed to preserve and improve natural capital by controlling finite stocks and balancing the flow of renewable resources (Demajorovic et al. 2016).

This signed agreement establishes the general conditions for the implementation of a reverse logistics system in a shared management model that includes manufacturers, importers, distributors, sellers, consumers, and public authorities. Decree No. 10,240 of 2020 establishes the role of each actor in reverse logistics, according to Table 1.

[1] Política Nacional de Resíduos Sólidos.

Table 1 Obligations of the actors of the EEE production, distribution, and commercialization chain in the reverse logistics system

Actors	Obligations
Manufacturers and importers	I—give an environmentally appropriate final destination, preferably for recycling, to one hundred percent of the electronic products that are received by the system; II—inform the Performance Monitoring Group of the objective criteria for the calculation of the mass balance of electronic products, observing the parameters established in art. 48, especially (a) the estimate of the average unit weight of each of the electronic products included in this Decree that are commercialized on the domestic market in the base year 2018; and (b) the periodic updating of the estimates referred to in point "a", according to the evolution of the weight of the products in different base years; III—participate in the execution of communication plans and non-formal environmental education; and IV—make available to Sisnama[2] members, when requested, the report to verify compliance with the actions of their responsibility provided for in this Decree, protecting the confidentiality of information, upon request and justification
Distributors	I—encourage, through its representative entities or through agreements or contracts, the adherence to the management entities or individual participation in the reverse logistics system of retail establishments that are part of its commercial chain; II—inform retail establishments that are part of their commercial chain about the process of operationalization of the reverse logistics system; III—make available or cost the physical spaces for the consolidation points to be used in the reverse logistics system, in compliance with the requirements of the basic operating manual; and IV—make available to Sisnama members, when requested, the report to verify compliance with the actions of their responsibility provided for in this Decree, protecting the confidentiality of information, upon request and justification

(continued)

[2] Sistema Nacional do Meio Ambiente (in free translation: National System of the Environment).

Table 1 (continued)

Actors	Obligations
Sellers	I—inform consumers, at the points of reception, about their responsibilities; II—receive, pack, and temporarily store the electronic products discarded by consumers at the points of reception, and return these products to manufacturers and importers, observing the requirements of the basic operating manual and the formal instrument signed with the managing entity or with the company; III—participate in the execution of communication plans and non-formal environmental education; and IV—make available to Sisnama members, when requested, the report to verify compliance with the actions of their responsibility provided for in this Decree, protecting the confidentiality of information, upon request and justification

Source Brasil (2010) and Santos et al. (2021)

Consumers, on the other hand, need to segregate and store WEEE in order to dispose of them appropriately at the specific receiving points of the reverse logistics system, observing the procedures and guidelines regarding disposals that are contained in the product manuals, or other means of communication provided for in the decree (Brazil 2020).

According to Santos and Jacobi (2021), Brazilian states have only a regulatory character, acting through their ministries and environmental agencies. Among some attributions of the public power, we can mention.

- Registering the manufacturer and other actors in the EEE chain as importers, distributors, and retail and wholesale sellers;
- Approving the schemes of the recycling plans established by the actors;
- Monitoring the validity of the information reported by the actors (with regard to the amount of EEE placed on the market and the amount of WEEE collected and recycled);
- Facilitating the placement of voluntary collection points in public areas;
- Facilitating communication on optimal WEEE disposal practices by the population.

Therefore, the public power, together with the others involved in this chain, is responsible for the effectiveness of the actions aimed at ensuring compliance with the National Solid Waste Policy, as well as the guidelines and other determinations established in the construction of this law (Brasil 2010).

- **Goals set by the sector agreement**

The agreement provides for two phases, the first being dedicated to the structuring of the system and the second related to its implementation and operationalization, with annual goals. This agreement defines the role of each one within reverse logistics, and offers security in case any environmental institution comes to make questions or demands (Barrozo et al. 2021).

The goals of the Sectoral Agreement predict that, in 2021, the generating companies must correctly collect and allocate 1%, by weight, of what they put on the market in 2018, the year defined as the base. In the coming years, the targets rise to 3, 6, and 12%, and reach 17% by 2025 (Brazil 2021).

Another goal is the increase of collection points, in which all cities with more than 80 thousand people, approximately 400 municipalities, should have voluntary delivery points installed in places easily accessible to consumers, such as retail stores, educational institutions, and municipal squares, among others, by 2025 (Barrozo et al. 2021).

There will be more than 5 thousand collectors, who will serve approximately 65% of the Brazilian population. By the end of 2021, the goal is to reach 60 cities in 13 states, with 600 collection points. For this to happen, retail must hurry to form partnerships with electronic reverse logistics management systems, and install the collectors in strategic stores. Communication actions and awareness-raising campaigns regarding adequate disposal are also planned (Barrozo et al. 2021).

The obstacles to this management are exactly at the stage of waste collection, given that, so far, there are few voluntary collection points in the country. In addition, there is a need to establish environmental awareness campaigns to educate the population about the importance of disposing of WEEE in an appropriate way and its appropriate place of disposal, in order to have a return and continuity in this process.

However, with the aim of promoting the implementation of the reverse logistics system and qualifying opinion-makers, institutions leaders, associations, and the city management, it is necessary to support the implementation of the system, and follow the agreement, which predicts communication and awareness-raising campaigns for the public through the printed press, television, radio, outdoor; busdoor, bus, train, and subway station panels, social media campaigns, caravans, lectures, and other events (Brazil 2021).

The fulfillment of the goals of the reverse logistics system will depend on the effective participation of the actors in the life cycle of the electronic products, according to the individual and shared limits and assignments.

4 Recycling and Processing of WEEE—Private Companies and Cooperatives

The surge in the consumption of electronics, according to the Brazilian Association of the Electrical and Electronic Industry[3] (ABINEE), and the turnover of this industry in Brazil during the global COVID-19 pandemic, reached the value of R$ 173.2 billion in 2020, indicating a nominal growth of 13% compared to 2019 (R$ 153.0 billion) (ABINEE 2021).

[3] Associação Brasileira da Indústria Elétrica Eletrônica.

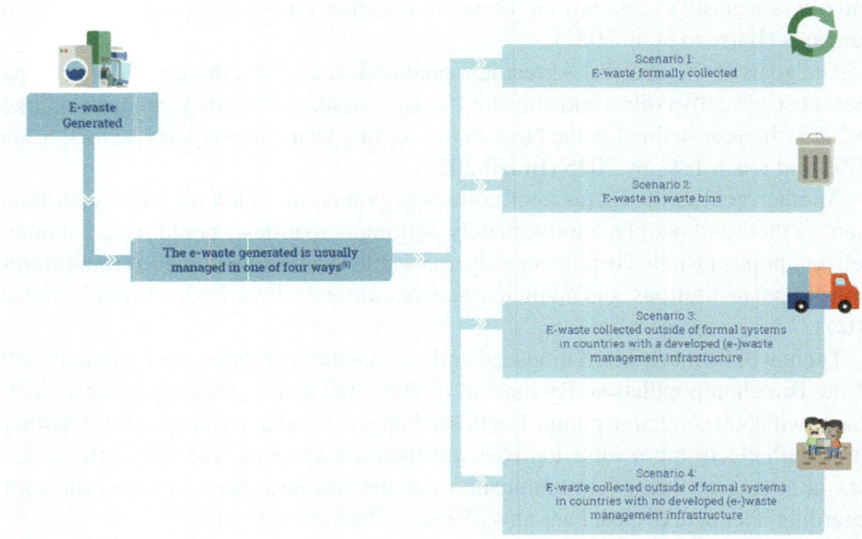

Fig. 1 Four possible scenarios to WEEE management (*Source* Forti et al. 2020)

In 2019, the largest generation of WEEE in Latin America is the responsibility of Brazil, having generated 2.4 million tons, a per capita production of 10.2 kg. On a global scale, Brazil is the fifth largest generator of this type of waste in the world, only behind China (10.1 million tons), the USA (6.9 million tons), India (3.2 million tons), and Japan (2.5 million tons) (Forti et al. 2020).

It is estimated that only 20% of WEEE in the world is properly processed (collected and recycled) by companies specializing in its recycling (Baldé et al. 2017b). According to the Brazilian Agency for Industrial Development[4] (ABDI), in 2013, approximately 94 recycling companies specialized in WEEE that were duly regularized (formal) operated in the country.

Since the entry of WEEE is a fundamental step, as this waste is the main raw material for the recycling process, Forti et al. (2020) state the existence of 4 distinct WEEE management scenarios, illustrated in Fig. 1, where.

- **Scenario 1—Formal collection**: activity usually supported by federal legislation, in which waste is collected in general by the companies in charge, producers, and/or the government. They can be discarded at retailers, municipal collection points, or by collection services. With this, the waste is sent to appropriate treatment, where the valuable materials are recovered in a technical and environmentally correct way.
- **Scenario 2—Disposal in bins**: the post-consumer equipment is disposed of in domestic waste bins, mixed with other types of waste; with this, the WEEE ends

[4] Agência Brasileira de Desenvolvimento Industrial.

up being treated together with the other waste, in which they are incinerated or disposed of in landfills.

- **Scenario 3—Collection outside formal systems in countries with developed WEEE management**: in this situation, the waste is collected by autonomous waste companies/sellers, and ends up being commercialized in different lines, the main ones being the recycling of polymer and metal. Here, in general, the waste is not treated in suitable facilities, and the waste can be exported.
- **Scenario 4—Collection outside formal systems in countries without developed WEEE management**: door-to-door collection by informal freelancers takes place; they may collect or buy such waste both from residences, as well as from companies and public institutions. The equipment is sold and can be repaired, or disassembled; in this second case, the recyclers perform burning, leaching, and melting of the waste for recovery of metals, in an amateur way, causing environmental and public health impacts.

Barrozo et al. (2021) conducted a survey of the internal management of WEEE cooperatives in the Metropolitan Region of the state of Rio Grande do Sul/Brazil, in which it was possible to find similarities between cooperatives and private companies in relation to the process of entry of this waste; for cooperatives, there are also collection campaigns, which are mobilization actions so that, on a given day and time, people can dispose of their WEEE to cooperatives.

Despite the existence of these three different ways of obtaining raw material, the collection of WEEE represents one of the biggest challenges for the recycling activity, due to the irregularity of the input flow. Barrozo et al. (2021) claim that there is a difficulty in the management of reception in relation to the minimum volume that must arrive in cooperatives and companies to make the process economically viable, since there is no frequency in the collection.

After collection, WEEE recycling by the specialized unit begins. According to the United Nations Environmental Programme management manual (UNEP 2007), this process occurs through three levels of processing:

- Level 1 consists of a set of procedures, such as the destruction of data and the classification and disassembly of the collected WEEE. After the waste is classified according to the presence of components for reuse or the presence of elements with hazardous substances (which will require specific safety procedures), disassembly may begin, which usually occurs manually. At this level, disassembled plastics, cables, printed circuit boards, and other WEEE are transported to the second level. Oils, CFCs, mercury switches, batteries, and capacitors are taken for suitable disposal. The materials used in this procedure are simple, such as a hammer, screwdriver, and drill, among others. The sorted and disassembled WEEE proceeds to level 2 of the recycling process. Companies that perform only level 1 sell the disassembled WEEE to other recyclers, as shown in Fig. 1.
- At level 2 of the recycling process, the technological density increases, so that the WEEE that was disassembled in the previous step passes through machines that will continue with the process of physical fragmentation of the waste, precisely

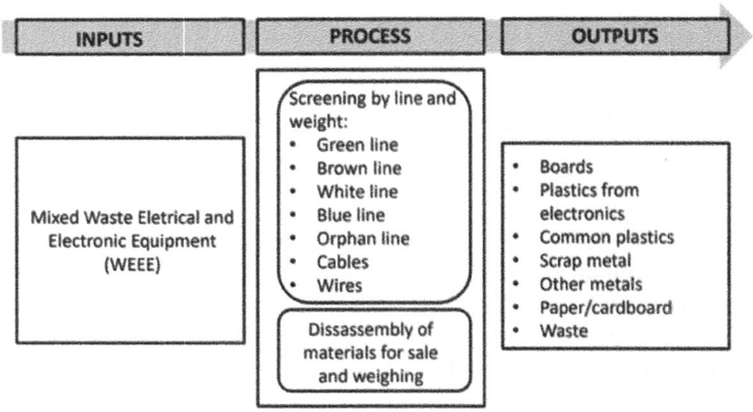

Fig. 2 Representation of Level 1 of WEEE recycling, inputs, and outputs Coopertec (WEEE Cooperative from the Municipality of Canoas/RS) (*Source* Barrozo et al. 2021)

through its crushing. These machines vary according to the material processed (plastic, ferrous metals, non-ferrous metals, and precious metals).

- Level 3 represents the process of processing the treated material. For reasons of capitalization, recycling at level 3 is often not carried out in Brazil, so the fractions obtained in the previous step can be sold to recycling companies that have appropriate techniques for completion, such as pyrometallurgy, hydrometallurgy, and electrometallurgical. Steel mills are able to reuse recycled ferrous and non-ferrous metals.

Figure 2 shows the process of level 1 with its inputs and outputs.

The management entities of the sector are ABREE—Brazilian Association of Recycling of Electronics and Household Appliances,[5] and GREEN ELETRON—National Waste Management of Electronic and Electrical Equipment[6] (Brazil 2021). They reported that, in 2019, 384.5 tons of WEEE were collected and 258 collection points were installed in Brazil (SINIR 2019).

It is hypothesized that the agreement with the municipalities would provide, in addition to the changes in the processes, more adequate infrastructure in relation to the place of installation of cooperatives, fleet of vehicles to carry out the collection, control tools for storage, expanded disclosure of the routes carried out in the neighborhoods, and facilitated disposal for the consumer (Barrozo et al. 2021).

[5] Associação Brasileira de Reciclagem de Eletroeletrônicos e Eletrodomésticos.

[6] Gestora para Resíduos de Equipamentos Eletroeletrônicos Nacional.

5 Final Remarks

Regarding the management of WEEE in the world, it is clear that there is no consensus among countries regarding legislation, even if there is a practice of exporting this waste. In addition, many are still in the process of developing laws regarding this issue, while others are still far behind.

A large portion of this waste is not properly destined, with illegal and/or uncertain outcomes. This is a delay with regard to sustainable development, as well as tools such as industrial ecology and circular economy, because materials of great added value are lost, which tends to cause serious damage to the environment and people.

There are several techniques of recovery of these materials that are being developed, and with the consumption and generation values of these recoveries already published. The low recycling rates of this type of waste prove the urgent need for the development of large-scale, sustainable, and viable techniques combined with the valorization of this source of raw material, in the field of urban mining.

As the fifth largest generator of WEEE on the planet, Brazil has a great socio-environmental and economic responsibility with regard to the implementation of a reverse logistics system. Cooperatives represent a possibility to facilitate this process in municipalities as providers of environmental services. The relationship of cooperatives with municipal governments is foreseen in the PNRS, which encourages these partnerships to promote productive inclusion and job generation, with a consequent strengthening of these institutions.

Regarding the WEEE Reverse Logistics Sector Agreement, there are no official data regarding the fulfillment of goals that measure the progress of the agreement from the beginning of its implementation in 2019 until the current year. The agreement, although late, seeks to solve the previously mentioned challenges, especially focusing on the expansion of voluntary collection points through environmental education strategies, in order to communicate with the population about the appropriate ways to dispose of waste, given that, so far, there are few voluntary collection points in the country.

As a consequence of this awareness, when the increase in the regular flow of WEEE in cooperatives and recyclers becomes truly effective, they will be a key process to making the activity economically viable. Thus, it provides opportunities for the capitalization of companies to structure the Level 3 of recycling within the national territory, facilitating logistics and sales, as well as adding value to the WEEE recycling process.

Finally, it is important to highlight that, in order to comply with the principles of industrial ecology, circular economy, and sustainable development goals, it is essential that society pressure EEE manufacturers rethink the sustainability of the EEE, as well as its design, to ensure a decrease in the generation of non-recyclable waste, and greater ease for reverse logistics of obsolete EEE.

References

Decreto Nº 10.240, de 12 de fevereiro de 2020, Regulamenta o inciso VI do caput do art. 33 e o art. 56 da Lei nº 12.305, de 2 de agosto de 2010, e complementa o Decreto nº 9.177, de 23 de outubro de 2017, quanto à implementação de sistema de logística reversa de produtos eletroeletrônicos e seus componentes de uso doméstico. Disponível em: http://www.planalto.gov.br/ccivil_03/_ato 2019-2022/2020/decreto/D10240.htm. Acesso em: 29 nov. 2021

ABDI – AGÊNCIA BRASILEIRA DE DESENVOLVIMENTO INDUSTRIAL. Lo-gística reversa de equipamentos eletroeletrônicos: análise de viabilidade técnica e econômica. Brasília: ABDI, 2013. Disponível em: http://www.abdi.com.br/Estudo/Logistica%20reversa%20de%20r esiduos_.pdf. Searched in: 29 nov. 2021

ABINEE - ASSOCIAÇÃO BRASILEIRA DA INDÚSTRIA ELÉTRICA ELETRÔNICA. Desempenho do Setor - Dados atualizados em março de 2021. Available in: http://www.abinee.org.br/abinee/decon/decon15.htm. Searched in: 29 nov. 2021

art. 56 da Lei nº 12.305, de 2 de agosto de 2010, e complementa o Decreto nº 9.177, de 23 de outubro de 2017, quanto à implementação de sistema de logística reversa de produtos eletroeletrônicos e seus componentes de uso doméstico. Presidência da República, Secretaria Geral, Brasília – DF, 2020. Available in: http://www.planalto.gov.br/ccivil_03/_ato2019-022/2020/decreto/D10 240.htm. Searched in: 29 de jul. 2021

Baldé CP et al (2017a) The global E-waste monitor—2016. Bonn/Geneva/Vienna: United Nations University (UNU). International Telecommunication Union (ITU) & International Solid Waste Association (ISWA). Available in: https://www.itu.int/en/ITU-D/Climate-Change/Documents/GEM%202017/Global-E-waste%20Monitor%202017%20.pdf Searched in: 10 nov. 2021

Baldé CP, Forti V, Gray V, Kuehr R, Stegmann P (2017b) The global E-waste monitor—2017, United Nations University (UNU). International Telecommunication Union (ITU) & International Solid Waste Association (ISWA), Bonn/Geneva/Vienna. Available in: https://www.itu.int/en/ITU-D/Climate-Change/Documents/GEM%202017/Global-E-waste%20Monitor%202017%20.pdf. Searched in: 6 jun 2021

Barrozo AS (2021) ANÁLISE DE RECICLABILIDADE DE REEE PÓS CONSUMO NA REGIÃO METROPOLITANA DE PORTO ALEGRE – RS; IX ENSUS – Encontro de Sustentabilidade em Projeto – UFSC – Florianópolis – 19 a 21 de maio de 2021. Available in: https://repositorio.ufsc.br/handle/123456789/228804. Searched in: 10 nov. 2021

Brasil (2019) Acordo Setorial para a Implementação de Sistema de Logística Reversa de Produtos Eletroeletrônicos de uso Domésticos e seus Componentes. 10 de outubro de 2019. Available in: http://consultaspublicas.mma.gov.br/eletroeletronicos/wp-content/uploads/2019/07/Con sulta_PublicaEEE2030.07.2019.pdf. Searched in: 28 jul. 2021

Brasil (2010) Lei federal n. 12.305, de 2 de agosto de 2010. Institui a Política Nacional de Resíduos Sólidos; altera a Lei no 9.605, de 12 de fevereiro de 1998; e dá outras providências. Presidência da República, Casa Civil, Brasília – DF. Disponível em: http://www.planalto.gov.br/ccivil_03/_ato2007-2010/2010/lei/l12305.htm. Acesso em: 28 jul. 2021

Cossu R, Williams ID (2015) Urban mining: concepts, terminology, challenges. Waste Manag 45:1–3. Available in: https://doi.org/10.1016/j.wasman.2015.09.040. Searched in: 01 nov. 2021

Demajorovic J et al (2016) Logística reversa de REEE em países em desenvolvimento: desafios e perspectivas para o modelo brasileiro. Ambiente e Sociedade, 19, abr./jun. Available in: https://www.redalyc.org/articulo.oa?id=31746369007. Searched in: 10 nov. 2021

Duthie ACR, Lins FF (2017) A economia circular e sua relação com a mineração. Brasil Mineral, n 374, pp 66–70. Available in: http://mineralis.cetem.gov.br/bitstream/cetem/2074/1/Ana%20C ristina%20Ribeiro%20Duthie.pdf. Searched in: 10 nov. 2021

Ellen Macarthur Foundation (2011) Explaining the circular economy: rethink progress. Available in: https://ellenmacarthurfoundation.org/videos/explaining-the-circular-economy-rethink-progress. Searched in: 15 ago. 2021

Ellen Macarthur Foundation (2012) Recycling and the circular economy: what's the difference? Available in: https://ellenmacarthurfoundation.org/topics/circular-economy-introduction/overview.. Searched in:10 jun. 2021
Ellen Macarthur Foundation (2016) Circular economy diagram. Available in: https://ellenmacarthurfoundation.org/circular-economy-diagram. Searched in: 15 ago. 2021
Ellen Macarthur Foundation (2017) Circular consumer electronics: an initial exploration. Available in: https://ellenmacarthurfoundation.org/circular-consumer-electronics-an-initial-exploration. Searched in:15 ago. 2021
Ellen Macarthur Foundation (2019) What is a circular economy? Available in: https://ellenmacarthurfoundation.org/topics/circular-economy-introduction/overview. Searched in: 15 ago. 2021
European Commission (2017) Study on the review of the list of critical raw materials—publications Office of the EU [WWW Document]. Publ. Off. EU. Available in: https://publications.europa.eu/en/publication-detail/-/publication/08fdab5f-9766-11e7-b92d-01aa75ed71a1/language-en. Accessed in: 24 oct. 2021
Forti V, Baldé CP, Kuehr R, Bel G (2020) The global E-waste monitor 2020—quantities, flows, and the circular economy potential. United Nations University (UNU)/United Nations Institute for Training and Research (UNITAR)—co-hosted SCYCLE Programme. International Telecommunication Union (ITU) & International Solid Waste Association (ISWA), Bonn/Geneva/Rotterdam. Available in: http://ewastemonitor.info/. Searched in: 15 aug. 2021
Giese EC, Xavier LH, Lins FAF (2018) Biomineração Urbana: O futuro da reciclagem de resíduos eletroeletrônicos. Brasil Mineral, n. 385, pp 36–39. Available in: https://www.cetem.gov.br/antigo/images/periodicos/2018/biomineracao-urbana.pdf. Searched in: 01 nov. 2021
Havlik T (2015) Foreword. In: Veit HM, Bernardes AM (eds) Eletronic waste—recycling techniques, 1 edn. Springer, Košice, Slovakia. E-book, pp vii–viii. Available in: https://doi.org/10.1007/978-3-319-15714-6. Searched in: 01 nov. 2021
Khetriwal DS, Jain G (2021) International e-waste management practice-country factsheets from twelve jurisdictions. Sustainable Recycling Industries (RSI). Available in: https://www.sustainable-recycling.org/reports/international-e-waste-management-practice-country-factsheets-from-twelve-jurisdictions/. Searched in: 19 nov. 2021
Khetriwal DS, Kraeuchi P, Widmer R (2007) Producer responsibility for e-waste management: key issues for consideration e learning from the Swiss experience. J Environ Manag. Disponívl em: https://pubmed.ncbi.nlm.nih.gov/18162284/. Acesso em: 1 nov. 2021
Kiddee P, Naidu R, Wong MH (2013) Electronic waste management approaches: an overview. Waste Manag 33:1237–1250. Available in: https://doi.org/10.1016/j.wasman.2013.01.006. Searched in: 01 nov. 2021
Nancharaiah YV, Mohan SV, Lens PNL (2016) Biological and bioelectrochemical recovery of critical and scarce metals. Trends Biotechnol 34(2):137–155. Available in: https://doi.org/10.1016/j.tibtech.2015.11.003. Searched in: 01 nov. 2021
Nicolai FNP (2016) Mineração urbana: avaliação da economicidade da recuperação de componentes ricos em Au a partir de resíduo eletrônico (e-waste). 242f. Tese (doutorado em em Engenharia de Materiais) – Universidade Federal de Ouro Preto, Belo Horizonte. Available in: https://www.repositorio.ufop.br/handle/123456789/7251. Searched in: 19 nov. 2021
Ottoni M, Dias P, Xavier LH (2020) A circular approach to the e-waste valorization through urban mining in Rio de Janeiro, Brazil. J Clean Prod 261:1–13. Available in: https://doi.org/10.1016/j.jclepro.2020.120990. Searched in: 17 nov. 2021
Pradiian JK (2013) Environmental impact assessment and bioleaching of metals from electronic waste (E-Waste). Tese (Doutorado em Filosofia e Biotecnologia). Jaypee University of Information Technology, Waknaghat, Índia. Available in: https://www.mdpi.com/2313-4321/5/3/22. Searched in: 19 nov. 2021
Sant'anna LT, Machado RTM, Brito MJ (2014) Os Resíduos Eletroeletrônicos no Brasil e no Exterior: Diferenças Legais e a Premência de uma Normatização Mundial. Revista de Gestão Social e Ambiental - RGSA, São Paulo, vol 8, n 1, pp 37–53, jan./abr., 2014. Available in: https://rgsa.emnuvens.com.br/rgsa/article/view/822. Searched in: 10 oct. 2021

Santos ECA, AVALIAÇÃO TÉCNICA, ECONÔMICA E AMBIENTAL PARA RECUPERAÇÃO DE MATERIAIS A PARTIR DE LÂMPADAS LED PÓS-CONSUMO (2021) 177f. Dissertação (Mestrado em Engenharia Civil) – Universidade do Vale do Rio dos Sinos, São Leopoldo – RS – Brasil, 2021. Available in: https://www.scielo.br/j/esa/a/gZgg9y4kV5RrgK8Mv6J9YNh/%3Fformat%3Dpdf&lang%3Dpt. Searched in: 10 oct. 2021

Santos ECA, Silva JLC, Calheiro D, Moraes CAM (2021) BENEFICIAMENTO DE LÂMPADAS LED PÓS-CONSUMO EM UMA COOPERATIVA DE RECICLAGEM DE RESÍDUOS ELETROELETRÔNICOS. Mix Sustentável, Florianópolis – SC, 2021b. Available in: https://ojs.sites.ufsc.br/index.php/mixsustentavel/article/view/4880. Searched in: 10 oct. 2021

Santos KL (2017) Uma nova pobreza urbana: financeirização do consumo e novos espaços periféricos em São Paulo. São Paulo: Alameda. Available in: https://www.redalyc.org/pdf/5139/513951681010.pdf. Searched in: 01 nov. 2021

Santos KL, Jacobi PR (2021) O acordo setorial para gestão de Resíduos Eletroeletrônicos: Desafios e possibilidades; 10 anos da Política de Resíduos Sólidos: caminhos e agendas para um futuro sustentável; São Paulo: IEE-USP: OPNRS. Available in: http://iee.usp.br/?q=pt-br/publicacao-iee/10-anos-da-pol%C3%ADtica-de-res%C3%ADduos-s%C3%B3lidos-caminhos-e-agendas-para-um-futuro-sustent%C3%A1vel.. Searched in: 10 oct. 2021

Silveira TA, Santos ECA, Colling AV, Moraes CAM, Brehm FA (2020) E-waste management and the conservation of geochemical scarce resources. In: Khan A, Inamuddin, Asiri AM (eds) E-waste re-cycling and management—present scenarios and environmental issues, 1 edn. Springer Nature Switzerland AG, pp 179–200.

SINIR (2019) Sistema Nacional de Informações sobre a Gestão dos Resíduos Sólidos. Available in: https://sinir.gov.br/component/content/article/2-sem-categoria/474-acordo-setorial-de-eletroeletronicos. Searched in: 28 jul. 2021

Slam A, Ahmed T, Awual MR, Rahman A, Sultana M, Aziz AA, Monir MU, Teo SH, Hasan M (2020) Advances in sustainable approaches to recover metals from e-waste—a review. J Clean Prod 244. Available in: https://doi.org/10.1016/j.jclepro.2019.118815. Searched in: 15 oct. 2021

UNEP – United Nations Environment Programme (2007) E-waste management manual, vol 2. Available in: http://www.unep.or.jp/ietc/Publications/spc/EWasteManual_Vol2.pdf. Searched in: 29 nov. 2021

Wang F et al (2013) E-waste in China: a country report. United Nations University (UNU)

Wen Z et al (2018) Recovery of key metals in the electronics industry in the People's Republic of China: an opportunity in circularity. In: World economic forum

Willersinn S, Bart H (2017) Reactive extraction and critical raw materials: industrial recovery of tungsten. Chem Ing Tech 89(1–2):82–91. Available in: https://onlinelibrary.wiley.com/doi/abs/10.1002/cite.201600079#:~:text=https%3A//doi.org/10.1002/cite.201600079. Searched in: 01 nov. 2021

Xavier LH, Giese EC, Duthie ACR, Lins FAF (2019) Sustainability and the circular economy: a theoretical approach focused on e-waste urban mining. Resour Policy. Available in: https://doi.org/10.1016/j.resourpol.2019.101467. Searched in: 15 oct. 2021

Xavier LH, Lins FA (2018) Mineração Urbana de resíduos eletroeletrônicos: uma nova fronteira a explorar no Brasil. Brasil Mineral. n 379, pp 22–26. Available in: https://www.cetem.gov.br/antigo/images/periodicos/2018/mineracao-urbana.pdf. Searched in: 15 oct. 2021

E-Plastic Waste Use as Coarse-Aggregate in Concrete

Swaib Semiyaga, Pamela Opio Acheng, Racheal Wesonga, Moses J. Matovu, and Musa Manga

Abstract Natural resources are rapidly deteriorating due to factors such as rapid population growth, technological developments, and changes in consumption habits, with increasing harm from electronic wastes (e-waste). E-plastics alone make up to 21% of the e-waste collected globally, and their management or recycling is a growing concern as they pose detrimental effects on the environment and public health. Construction industry is continuously adopting e-plastic waste for use as aggregate or fiber in concrete. Several studies have been conducted to test the applicability and properties of the resultant concrete when e-plastic is used as a partial substitute material for both coarse and fine aggregate in concrete of various grades. The results have yielded positive outcomes as well as criticisms. Therefore, this chapter presents an overview of published research regarding the physical properties of e-plastic aggregate and its potential use as replacement for coarse aggregate in concrete, clearly highlighting the effects of e-plastic waste on the physical, mechanical, durability, and thermal properties of concrete. The effects were found to be dependent on the type and characteristics of the e-plastic aggregate, the replacement rates, concrete grade, and the use of additives like superplasticizers and fly ash. Therefore, it is recommended that for use as an aggregate in concrete, the e-plastic waste ought to have a high specific gravity, low water absorption levels, and

P. O. Acheng · R. Wesonga · M. Manga
Department of Construction Economics and Management, Makerere University, P.O Box 7062, Kampala, Uganda

S. Semiyaga (✉) · M. J. Matovu
Department of Civil and Environmental Engineering, Makerere University, P.O Box 7062, Kampala, Uganda
e-mail: semiyaga@gmail.com

M. Manga
Faculty of Engineering, Department of Civil Engineering, Ndejje University, P.O Box 7088, Kampala, Uganda

The Water Institute at UNC, Department of Environmental Sciences and Engineering, Gillings School of Global Public Health, University of North Carolina at Chapel Hill, Chapel Hill, NC, USA

M. Jawaid and A. Khan (eds.), *Conversion of Electronic Waste in to Sustainable Products*, Sustainable Materials and Technology,
https://doi.org/10.1007/978-981-19-6541-8_6

favorable and comparable abrasion resistance, crushing, and impact value to natural coarse aggregates.

1 Introduction to E-Plastic Wastes

Electronic wastes (e-wastes) grow fast, globally at an annual rate of 3–4% (Sahajwalla and Gaikwad 2018). E-wastes are described as electrical or electronic devices that have been discarded or used as electronics destined for reuse, resale, salvage, recycling, or disposal (Gull and Subramanian 2014; Manjunath 2016). In many instances, e-waste is disposed of by incineration, and a smaller share is recycled or disposed of in landfills. E-wastes from various sources, whether domestic or commercial, contain more than 1000 types of substances that could be valuable, non-hazardous, or hazardous (Lakshmi and Nagan 2011; Luhar and Luhar 2019). If not well managed, the exposure e-waste to people and the environment is reported to have detrimental influences (Luhar and Luhar 2019; Masuduzzaman et al. 2018; Singh and Pandey 2017).

Plastics in e-wastes are approximately 21%, making it the second largest proportion after iron and steel (Kumar and Malik 2016; Lakshmi and Nagan 2011). The percentage of plastics in the e-waste is dependent on the categories of electrical and electronic equipment, with toys and monitoring and control instruments containing more than 50% plastic weight, while lighting and medical equipment containing less than 5% of plastic (Stenvall 2013). E-plastic waste presents a great challenge in collection and treatment, and given the ever-growing use of electronic devices especially mobile phones and computers, its management and recycling is a growing concern (Makri et al. 2019; Manjunath 2016). As such, the construction industry has started to adopt e-plastic waste for use as aggregate or fiber in concrete. Recycled e-plastic waste has been researched as a partial replacement for coarse aggregate (Manjunath 2016; Sahajwalla and Gaikwad 2018). However, there is limited understanding of the composition and potential of recycling of plastics (Pivnenko and Astrup 2018). This chapter discusses the composition of common e-plastics used in concrete, their properties, and behavior of concrete made with partial replacement of coarse aggregates with e-plastic waste aggregates.

2 E-Plastic Types, Composition, and Potential Reuses

The type and composition of e-plastic highly dictates the type and potential use, especially with regard to its application in concrete. This is due to the difference in characteristics of the different types of e-plastics as well as their recycle-ability.

2.1 Types and Composition of E-Plastics

E-plastic is mainly made of resins (30–60% of plastic content), which determine the features and applications (Usuki 2018). Based on heat resistance, these e-plastics can be classified under two broad categories, thermoplastic and thermosetting resins (Inamuddin and Lichtfouse 2021; Usuki 2018). Thermoplastics are reported to have low heat resistance and stiffness properties, but with good impact toughness (Usuki 2018). They soften when heated and harden on cooling without changing the properties and appearance of the materials, and thus can easily be processed and molded. This consequently increases the bonding capacity of concrete when thermoplastics are used as aggregate. Thermosetting e-plastics, on the other hand, can only be molded, set, and hardened due to their resistance to heat (Inamuddin and Lichtfouse 2021). However, different e-plastic waste types can be blended to improve the properties of the resulting concrete. The most common e-plastic waste types that fall under the two broad categories of thermoplastic and thermosetting include the following:

- ABS, acrylonitrile–butadiene–styrene.
- EPS, expanded polystyrene.
- HIPS, high-impact polystyrene.
- PA, polyamide.
- PC, polycarbonate.
- PC-ABS, acrylonitrile–butadiene–styrene.
- PP, polypropylene.
- PS, polystyrene.
- PVC, polyvinyl chloride.

The biggest percentage of e-plastic wastes including PS, ABS, HIPS, PC-ABS, and PP are thermoplastic in nature. These are reported to be the most predominantly collected wastes, with ABS mostly applied and collected from computers, computer monitors, and printer, and HIPS from mostly television housing (Grigorescu et al. 2019). A large percentage of these e-plastic wastes have also been found to contain hazardous substances including brominated flame retardants (BFR), chlorine (Cl), cadmium (Cd), and antimony (Sb), which potentially complicates their recycling process given their hazardous effects (Forti et al. 2020). Brominated Flame Retardants (BFR) are majorly used to reduce the flammability of products and appliances and are commonly used for the outer casing for computers, printed wiring boards, connectors, wires, and cables.

2.2 E-Plastic Waste Application in Construction Industry

E-plastic waste has potential for various applications in construction industry depending on the type and properties. For example, polypropylene (PP) wastes which are hard, flexible, and resistant to chemical attack can be used as aggregates in asphalt mixtures and polystyrene (PS) wastes, which are stiff and brittle, are used in mechanically low stress parts such as insulation materials. On the other hand, polyvinyl chloride (PVC) which is hard, rigid, and with excellent chemical properties has a potential use as aggregate in cement-based materials (da Silva et al. 2021; Grigorescu et al. 2019; Kamaruddin et al. 2017). In concrete, e-plastic waste can be used in two forms, such as plastic aggregate (PA) or plastic fiber (PF). Whereas PAs are crushed or shredded pieces of plastic used to replace coarse and fine aggregate in the production of lightweight concrete, PFs are enhanced through the incorporation of fiber materials such as glass and carbon, and are used as reinforcement to replace the common steel fiber to provide improved mechanical strength and stability (Kamaruddin et al. 2017). However, when using PAs in concrete, several factors including the type, level of substitution, shape, and size of the e-plastic aggregate, as well as the water-cement ratio, need to be considered as they have varying effects on the physical, mechanical, and durability characteristics of the resultant concrete.

2.3 Preparation of E-Plastic Waste for Use in Concrete

E-plastic waste aggregates have been used with different shapes and sizes. E-plastic wastes are prepared by chopping, grinding, or shredding into varying sizes (Mills and Tatara 2016) as presented in Fig. 1. In practice, e-plastic waste aggregate of sizes between 9.5 and 20 mm has been used to produce concrete of varying grades. In general, aggregates of size 20 mm were associated with stronger concretes.

3 Material Properties of E-Plastic Aggregates

The mechanical properties of e-plastic concrete depend on e-plastic type, replacement proportion, and mix design. Aggregates are estimated to occupy between 70 and 80% of the volume of concrete (Alexander and Mindess 2010). Therefore, on a microscopic level, concrete can be considered as a two-phase material, consisting of aggregate particles dispersed in a matrix of the cement paste (Mehta and Monteiro 2001). The selection of aggregates therefore has a strong influence on the mix proportions, strength, structural and durability properties of fresh and hardened concrete, and economy of concrete (Neville and Brooks 2010). In addition, the aggregate/particle shape and surface texture have impact on the properties of fresh and hardened concrete (Neville and Brooks 2010). To be used as an aggregate in concrete, the plastic ought to have low water absorption levels so as to minimize competition with the cementitious matrix for water, have a favorable and comparable abrasion resistance, crushing, and impact value to natural coarse aggregates. Table 1 summarizes the properties of e-plastic waste aggregates used in production of concrete by different researchers.

Fig. 1 Varying particle sizes of e-plastic waste aggregates; **a** chopped/grounded/shredded e-plastic aggregate (Mills and Tatara 2016); **b** E-plastic of size 0–2 mm (Bulut and Şahin 2017); **c** E-plastic of size of 2–4 mm (Bulut and Şahin 2017); **d** E-plastic of size 4–8 mm (Bulut and Şahin 2017); and **e** E-plastic of size 20 mm (Shinu and Needhidasan 2020)

Table 1 Properties of e-plastic waste aggregates

Source	E-plastic waste properties									
	Preparation method	E-plastic waste composition	Shape	Max. size (mm)	Specific gravity	Fineness modulus (mm)	Bulk density (kg/m^3)	Crushing value (%)	Impact value (%)	Water absorption (%)
Santhanam and Anbuarasu (2020)	–	Printed circuit boards	Chipped	20	0.8	–	–	–	–	0.04
Shinu and Needhidasan (2020)	–	–	–	20	0.8	7	–	–	–	0
Needhidasan et al. (2020b)	–	–	–	20	1.3	7.36	–	–	–	0.01
Ahirwar et al. (2016)	–	–	–	–	1.2	2.5	–	2.35	1.95	0.04
Prashant and Kumar (2019)	*Manually broken*	Keyboards	–	20	–	–	–	–	–	–
Suleman and Needhidasan (2020)	–	Printed circuit boards	Chipped	20	1.28	–	–	–	–	0
Manjunath (2016)	*Manually crushed*	–	–	20	1.1	7.59–9.18	–	–	–	0
Gavhane et al. (2016)	*Manually crushed*	Monitors, keyboards, mouse, and C.P.U. of computers	Angular	–	0.84	–	–	2.83	–	0

(continued)

Table 1 (continued)

Source	E-plastic waste properties									
	Preparation method	E-plastic waste composition	Shape	Max. size (mm)	Specific gravity	Fineness modulus (mm)	Bulk density (kg/m^3)	Crushing value (%)	Impact value (%)	Water absorption (%)
Sabău and Vargas (2018)	*Manually broken*	Computer housings	–	9.5	0.93	5.23	490	–	–	0.2
Lakshmi and Nagan (2011)	*Manually crushed*	–	Angular	–	1.01	1.86–2.78	–	<2	–	<0.2
Santhanam et al. (2020)	*Manually chipped*	Computer equipment and accessories	Chipped	20	1.3	7.36–8.13	–	–	–	0
Rathore and Rawat (2019)	–	–	Angular	10–20	1.1	–	–	–	–	–
Parsons and Nwaubani (2020)	–	Polycarbonate and acrylonitrile-butadiene-styrene	–	–	1.19	6	580	–	–	–
Parsons and Nwaubani (2020)	–	High-impact polystyrene	–	–	1.39	6	580	–	–	–
Parsons and Nwaubani (2020)	–	Acrylonitrile-butadiene-styrene	–	–	0.94	6	520	–	–	–

(continued)

Table 1 (continued)

Source	E-plastic waste properties									
	Preparation method	E-plastic waste composition	Shape	Max. size (mm)	Specific gravity	Fineness modulus (mm)	Bulk density (kg/m^3)	Crushing value (%)	Impact value (%)	Water absorption (%)
Needhidasan et al. (2020a)	*Manually cut*	Printed circuit boards	*Angular*	20	0.8	–	–	–	–	0.04
Dhanabal et al. (2021)	*Manually crushed*	CRT television, computer parts (motherboards, keyboards)	–	–	1.13	–	–	0.9	1.2	–
Arivalagan (2020)	–	–	Irregular	–	1.32	3.1	600	–	–	–
Singh and Pandey (2017)	–	–	–	4.75–20	1.2	2.5	–	2.35	1.95	0.04
Kumar and Baskar (2014)	–	High–impact polystyrene	Flaky	12	1.29	–	515.3	–	–	–
Luhar and Luhar (2019)	–	–	–	–	0.84–2.65	–	–	2.0–3.26	1.95–4.33	0.04–0.5
General ranges				4.75–20	0.8–2.65	1.86–9.18	490–600	0.9–3.26	1.2–4.33	0–0.5

3.1 Aggregate Crushing Value and Impact Value

The aggregate impact value for the e-plastic aggregates was in the range of 1.2–4.33%, which is an indicator of a strong material, since the values are less than 10% (Ahirwar et al. 2016; Dhanabal et al. 2021; Luhar and Luhar 2019; Singh and Pandey 2017). This can be explained by the fact that the e-plastics just break into individual particles with minimal crumbling or formation of finer particles. Similarly, the aggregate crushing value generally ranges between 0.9 and 3.26 (Ahirwar et al. 2016; Dhanabal et al. 2021; Gavhane et al. 2016; Lakshmi and Nagan 2011; Luhar and Luhar 2019; Singh and Pandey 2017). This may be due to limited amount of fine material produced when crushed since the crushing strength is greatly influenced by the characteristic of the parent material (Mehta and Monteiro 2001). Typically, the e-plastics are crushed into independent particles with minimal crumbling.

3.2 Specific Gravity

E-plastic waste materials have specific gravity ranging between 0.8 and 2.65. The specific gravity is used to classify aggregates into lightweight, normal weight, and heavy. The weight eventually has an impact on the overall fresh and hardened concrete, where e-plastic ranges qualify them to be lightweight (Dhir et al. 2017). The specific gravity for e-plastic wastes compares with the most commonly used rock and natural aggregates, which have values between 2.60 and 2.70 (Chandra and Berntsson 2002; Mehta and Monteiro 2001).

In addition, the specific gravity can be loosely defined as the measure of strength of the aggregate. In reference to published data, there is generally a significant relationship between the specific gravity of the e-plastic waste aggregates and its compressive strength as demonstrated in Fig. 2 (Dhanabal et al. 2021; Gavhane et al. 2016; Lakshmi and Nagan 2011; Manjunath 2016; Parsons and Nwaubani, 2020; Rathore and Rawat 2019; Sabău and Vargas 2018; Suleman and Needhidasan 2020). The analysis of results from multiple studies demonstrates that e-plastic waste aggregates of higher specific gravities were associated with concretes of higher compressive strengths. This suggests that the e-plastic aggregates with higher specific gravity may present high compressive strength, and hence have minimal planes of weakness and fractures (Amuda et al. 2014). Therefore, where improved compressive strength is required, the use of e-plastic waste with higher specific gravities ought to be used.

3.3 Bulk Density

E-plastic wastes have bulk density ranging between 490 and 600 kg/m^3 while that of most natural aggregates range between 1300 and 1750 kg/m^3 (Arivalagan 2020;

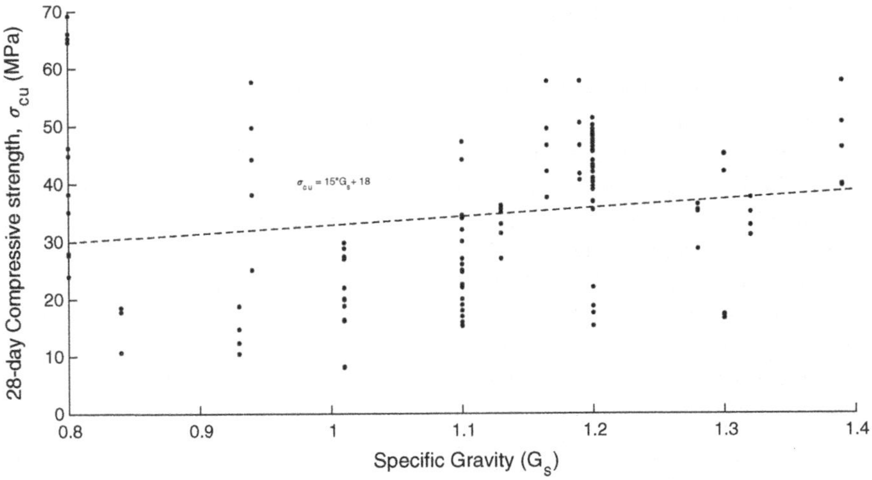

Fig. 2 Influence of e-plastic waste aggregate specific gravity on the compressive strength

S. Kumar & Baskar 2014). This implies that the e-plastic wastes can be used in optimum proportions to provide lightweight concretes.

4 Properties of Concrete Containing E-Plastic Waste Aggregates

4.1 Physical and Mechanical Properties

The physical and mechanical properties of concrete in both fresh and hardened state are vital especially when the concrete is to be used in loaded reinforced concrete structures. These properties depend on the type/composition of e-plastic, preparation method, and the mix design. Although there is paucity of studies on the physical properties of e-plastic concrete, this section addresses two of the common concrete properties of workability and density. In addition, the impact of e-plastic waste on the mechanical properties of concrete, including compressive strength, flexural strength, and split tensile strength, has been widely studied, with different researchers obtaining varying results as presented in Table 2.

4.1.1 Slump/Workability

The slump cone test is the most common method used to determine the workability or consistency of the prepared concrete mix. Workability is the ease with which concrete

Table 2 Summary of the physical–mechanical properties of various e-plastic waste concrete

Source	E-plastic waste concrete preparation				E-plastic %	Concrete density (kg/m^3)	Compressive strength (MPa)			Flexural strength (MPa)			Split tensile Strength (MPa)			Slump (mm)
	w/c ratio	Additives	E-plastic shape/type	E-plastic max. size (mm)			7	14	28	7	14	28	7	14	28	
Santhanam and Anbuarasu (2020)	0.28	Super plasticizer	Chipped	20	0	–	20.96	29.78	65.25	2.20	3.10	4.10	1.31	1.96	3.30	270
					8	–	21.71	30.94	64.61	2.40	3.40	4.30	1.40	2.10	3.53	250
					12	–	22.5	31.48	66.09	2.70	3.60	4.70	1.70	2.60	3.90	240
					16	–	22.74	31.82	69.22	2.90	3.80	4.90	2.43	3.43	4.90	225
Shinu and Needhidasan (2020)	0.40	N/A	–	20	0	2548	32.58	37.70	46.25	2.18	4.48	4.54	2.72	3.77	4.63	
					12	2497	31.16	33.21	44.85	2.11	4.02	4.20	2.59	3.32	4.09	
					17	2411	29.50	31.20	38.24	2.01	3.96	4.01	2.45	3.12	3.82	
					22	2329	26.40	28.16	35.15	1.94	3.72	3.84	2.20	2.81	3.01	
Needhidasan et al. (2020b)	0.28	Super plasticizer	–	20	0	–	33.98	40.33	45.05	2.20	3.10	4.20	2.80	3.10	4.90	270
					12	–	28.83	35.24	41.95	2.40	3.30	4.30	1.80	2.10	3.50	250
					17	–	26.76	37.20	44.93	2.90	3.80	4.80	2.40	3.40	4.90	240
					22	–	25.14	37.03	41.95	3.70	4.80	5.40	4.40	5.40	6.70	225
Ahirwar et al. (2016)			–	–	0	–	31.46	37.16	42.85	–	–	–	–	–	–	27
					5	–	30.91	35.71	40.51	–	–	–	–	–	–	31
					10	–	30.26	34.84	39.42	–	–	–	–	–	–	38
					15	–	29.32	33.09	36.85		–	–	–	–	–	47
					20	–	27.12	33.00	38.87	–	–	–	–	–	–	61
					25	–	26.78	31.78	36.77	–	–	–	–	–	–	70

(continued)

Table 2 (continued)

Source	E-plastic waste concrete preparation				E-plastic %	Concrete density (kg/m^3)	Compressive strength (MPa)			Flexural strength (MPa)			Split tensile Strength (MPa)			Slump (mm)
	w/c ratio	Additives	E-plastic shape/type	E-plastic max. size (mm)			7	14	28	7	14	28	7	14	28	
					30	–	25.92	30.67	35.41	–	–	–	–	–	–	73
Ahirwar et al. (2016)		Fly ash 10%			0	–	38.91	47.46	49.98	–	–	–	–	–	–	36
					5	–	38.78	47.05	48.02	–	–	–	–	–	–	43
					10	–	38.67	46.78	47.46	–	–	–	–	–	–	47
					15	–	37.76	44.36	47.17	–	–	–	–	–	–	54
					20	–	35.72	42.74	45.64	–	–	–	–	–	–	67
					25	–	32.24	41.27	42.61	–	–	–	–	–	–	79
					30	–	31.56	40.42	41.89	–	–	–	–	–	–	84
Ahirwar et al. (2016)		Fly ash 20%			0	–	37.02	45.90	48.23	–	–	–	–	–	–	46
					5	–	36.89	45.49	46.27	–	–	–	–	–	–	53
					10	–	36.78	45.22	45.71	–	–	–	–	–	–	60
					15	–	35.87	42.80	45.42	–	–	–	–	–	–	68
					20	–	33.83	41.18	43.89	–	–	–	–	–	–	78
					25	–	30.35	39.71	40.86	–	–	–	–	–	–	87
					30	–	29.67	38.86	40.14	–	–	–	–	–	–	95
Ahirwar et al. (2016)		Fly ash 30%			0	–	40.03	48.45	51.21	–	–	–	–	–	–	55
					5	–	39.90	48.04	49.25	–	–	–	–	–	–	62

(continued)

Table 2 (continued)

Source	E-plastic waste concrete preparation				E-plastic %	Concrete density (kg/m³)	Compressive strength (MPa)			Flexural strength (MPa)			Split tensile Strength (MPa)			Slump (mm)
	w/c ratio	Additives	E-plastic shape/type	E-plastic max. size (mm)			7	14	28	7	14	28	7	14	28	
					10	–	39.79	47.77	48.69	–	–	–	–	–	–	73
					15	–	38.88	45.35	48.40	–	–	–	–	–	–	81
					20	–	36.84	43.73	46.87	–	–	–	–	–	–	89
					25	–	33.36	42.26	43.84	–	–	–	–	–	–	95
					30	–	32.68	41.41	43.12	–	–	–	–	–	–	106
Prashant and Kumar (2019)	0.40	N/A		20	0	–	24.26	–	34.92	4.00	–	4.53	–	–	–	–
					10	–	24.97	–	29.58	4.33	–	5.66	–	–	–	–
					15	–	20.80	–	26.22	4.27	–	4.87	–	–	–	–
					20	–	17.58	–	23.76	3.00	–	4.20	–	–	–	–
Suleman and Needhidasan (2020)			Chipped	20	0	–	22.15		35.40	3.32		5.24	2.06		3.16	96
					5.5	–	20.31		36.33	3.66		5.30	2.32		3.42	81
					11	–	19.28		35.10	4.10		5.37	2.39		3.53	65
					16.5	–	16.98		28.62	2.93		3.92	1.92		2.62	44
Manjunath (2016)	0.50	N/A		20	0	–	36.00	44.81	47.18	1.50	3.00	4.35	4.30	4.66	4.90	128
					10	–	33.18	41.25	44.07	1.20	2.90	4.40	4.30	4.40	4.80	114
					20	–	19.90	17.95	24.69	1.00	2.00	4.30	3.15	5.00	5.40	90
					30	–	16.39	19.03	22.15	0.75	1.25	2.50	2.40	3.10	3.80	75

(continued)

Table 2 (continued)

Source	E-plastic waste concrete preparation				E-plastic %	Concrete density (kg/m^3)	Compressive strength (MPa)			Flexural strength (MPa)			Split tensile Strength (MPa)			Slump (mm)
	w/c ratio	Additives	E-plastic shape/type	E-plastic max. size (mm)			7	14	28	7	14	28	7	14	28	
Gavhane et al. (2016)	0.50	N/A	Angular	–	0	2223	14.10	17.08	18.55	–	–	–	–	–	–	–
					10	2176	14.07	16.59	17.80	–	–	–	–	–	–	–
					20	–	8.67	10.74	10.72	–	–	–	–	–	–	–
Sabău and Vargas (2018)	0.53	N/A		9.5	0	2426	13.20	16.20	18.80	–	–	–	–	–	–	62.5
					40	2119	11.80	12.90	14.80	–	–	–	–	–	–	75
					50	1991	8.93	10.70	12.40	–	–	–	–	–	–	75
					60	1897	8.78	10.30	10.50	–	–	–	–	–	–	100
Lakshmi and Nagan (2011)		N/A	Angular		0	–	16.81	27.23	28.79	–	–	–	–	–	–	–
					4	–	–	–	19.89	–	–	–	–	–	–	–
					8	–	–	–	18.80	–	–	–	–	–	–	–
					12	–	–	–	16.40	–	–	–	–	–	–	–
					16	–	–	–	16.23	–	–	–	–	–	–	–
					20	–	–	–	8.25	–	–	–	–	–	–	–
					24	–	–	–	8.10	–	–	–	–	–	–	–
Lakshmi and Nagan (2011)		10% Fly ash	Angular	–	0	–	16.81	27.23	28.79	–	–	–	–	–	–	–
					4	–	–	–	27.03	–	–	–	–	–	–	–
					8	–	–	–	27.38	–	–	–	–	–	–	–

(continued)

Table 2 (continued)

Source	E-plastic waste concrete preparation				E-plastic %	Concrete density (kg/m^3)	Compressive strength (MPa)			Flexural strength (MPa)			Split tensile Strength (MPa)			Slump (mm)
	w/c ratio	Additives	E-plastic shape/type	E-plastic max. size (mm)			7	14	28	7	14	28	7	14	28	
					12	–	17.12	27.73	29.79	–	–	–	–	–	–	–
					16	–	–	–	26.98	–	–	–	–	–	–	–
					20	–	–	–	21.97	–	–	–	–	–	–	–
					24	–	–	–	20.10	–	–	–	–	–	–	–
Santhanam et al. (2020)	0.50		Chipped	20	0	–	15.55	15.24	17.31	1.30	2.10	2.60	1.25	1.31	1.15	120
					10	–	16.44	17.43	16.58	1.60	2.40	1.66	1.25	1.25	1.26	98
					20	–	15.98	15.81	17.06	1.56	2.46	2.90	1.68	1.38	1.56	95
Rathore and Rawat (2019)	0.50		Angular	20	0	–	–	–	27.00	–	–	3.50	–	–	–	–
					5	–	–	–	22.50	–	–	4.00	–	–	–	–
					10	–	–	–	20.00	–	–	4.00	–	–	–	–
					15	–	–	–	19.00	–	–	4.10	–	–	–	–
					20	–	–	–	17.00	–	–	3.50	–	–	–	–
					25	–	–	–	16.00	–	–	3.20	–	–	–	–
					30	–	–	–	15.50	–	–	3.00	–	–	–	–
Rathore and Rawat (2019)	0.50		Angular	15	0	–	–	–	27.00	–	–	3.50	–	–	–	–
					5	–	–	–	30.00	–	–	3.90	–	–	–	–
					10	–	–	–	32.00	–	–	3.91	–	–	–	–

(continued)

Table 2 (continued)

Source	E-plastic waste concrete preparation				E-plastic %	Concrete density (kg/m^3)	Compressive strength (MPa)			Flexural strength (MPa)			Split tensile Strength (MPa)			Slump (mm)
	w/c ratio	Additives	E-plastic shape/type	E-plastic max. size (mm)			7	14	28	7	14	28	7	14	28	
					15	–	–	–	32.00	–	–	3.92	–	–	–	–
					20	–	–	–	25.00	–	–	3.40	–	–	–	–
					25	–	–	–	22.00	–	–	3.10	–	–	–	–
					30	–	–	–	15.30	–	–	2.90	–	–	–	–
Rathore and Rawat (2019)	0.50		Angular	10	0	–	–	–	27.00	–	–	3.50	–	–	–	–
					5	–	–	–	34.00	–	–	3.20	–	–	–	–
					10	–	–	–	34.10	–	–	3.10	–	–	–	–
					15	–	–	–	34.50	–	–	3.00	–	–	–	–
					20	–	–	–	26.00	–	–	2.90	–	–	–	–
					25	–	–	–	22.50	–	–	2.80	–	–	–	–
					30	–	–	–	18.00	–	–	2.70	–	–	–	–
Parsons and Nwaubani (2020)	0.52	Super plasticizer	(PC/ABS)	–	0	–	46.40	–	57.60	–	–	–	–	–	–	–
					5	–	43.50	–	50.40	–	–	–	–	–	–	–
					10	–	40.70	–	46.50	–	–	–	–	–	–	–
					20	–	37.50	–	41.60	–	–	–	–	–	–	–
					30	–	36.40	–	40.50	–	–	–	–	–	–	–
Parsons and Nwaubani (2020)	0.52	Super plasticizer	(HIPS)	–	0	–	46.40	–	57.60	–	–	–	–	–	–	–

(continued)

Table 2 (continued)

Source	E-plastic waste concrete preparation				E-plastic %	Concrete density (kg/m^3)	Compressive strength (MPa)			Flexural strength (MPa)			Split tensile Strength (MPa)			Slump (mm)
	w/c ratio	Additives	E-plastic shape/type	E-plastic max. size (mm)			7	14	28	7	14	28	7	14	28	
					5	–	42.20	–	50.50	–	–	–	–	–	–	–
					10	–	39.50	–	46.10	–	–	–	–	–	–	–
					20	–	35.50	–	39.80	–	–	–	–	–	–	–
					30	–	38.50	–	39.50	–	–	–	–	–	–	–
Parsons and Nwaubani (2020)	0.52	Super plasticizer	(ABS)	–	0	–	46.40	–	57.60	–	–	–	–	–	–	–
					5	–	42.20	–	49.70	–	–	–	–	–	–	–
					10	–	37.80	–	44.20	–	–	–	–	–	–	–
					20	–	34.00	–	38.10	–	–	–	–	–	–	–
					30	–	22.10	–	25.10	–	–	–	–	–	–	–
Parsons and Nwaubani (2020)	0.52	Super plasticizer	(PC/ABS, HIPS)	–	0	–	46.40	–	57.60	–	–	–	–	–	–	–
					5	–	42.90	–	49.40	–	–	–	–	–	–	–
					10	–	41.40	–	46.50	–	–	–	–	–	–	–
					20	–	36.80	–	42.00	–	–	–	–	–	–	–
					30	–	34.90	–	37.50	–	–	–	–	–	–	–
Needhidasan et al. (2020a)			Angular	20	0	–	16.11	25.00	28.00	7.00	10.00	15.00	2.21	2.00	1.85	120
					10	–	16.36	23.50	27.70	4.50	5.00	6.30	2.58	2.15	1.93	105
					12.5	–	16.90	21.30	24.00	4.00	4.80	5.00	4.50	3.58	2.99	95

(continued)

Table 2 (continued)

Source	E-plastic waste concrete preparation				E-plastic %	Concrete density (kg/m^3)	Compressive strength (MPa)			Flexural strength (MPa)			Split tensile Strength (MPa)			Slump (mm)
	w/c ratio	Additives	E-plastic shape/type	E-plastic max. size (mm)			7	14	28	7	14	28	7	14	28	
Dhanabal et al. (2021)	0.47	N/A	–	–	0	–	21.07	–	36.18	–	–	–	–	–	–	–
					5	–	21.49	–	35.80	–	–	–	–	–	–	–
					10	–	23.39	–	35.00	–	–	–	–	–	–	–
					15	–	24.02	–	33.00	–	–	–	–	–	–	–
					20	–	25.07	–	27.00	–	–	–	–	–	–	–
Arivalagan (2020)	0.45		Irregular	–	0	–	16.00	19.52	31.00	1.50	3.00	4.40	4.10	4.67	4.90	90
					10	–	17.50	21.35	32.73	1.20	2.85	4.40	4.15	4.40	4.40	110
					20	–	19.00	23.18	37.50	1.00	2.23	4.50	3.10	4.90	5.50	100
					30	–	18.00	21.96	35.00	0.85	1.30	2.90	2.50	3.10	3.75	93
Singh and Pandey (2017)	0.5	N/A	–	–	0	–	17.67	–	22.00	2.94		3.22				35
					5	–	16.56	–	18.72	2.84		3.02	–	–	–	43
					10	–	15.23	–	17.53	2.73		2.93	–	–	–	47
					15	–	13.17	–	15.31	2.54		2.73	–	–	–	54

can be mixed, transported, placed, and finished without segregation (Siddique et al. 2008). Workability is affected by factors such as the mix proportion of e-plastic wastes, particle size and shape, and water-cement ratio. Various researchers have investigated the effect of replacing varying proportions of e-plastic waste as fine and coarse aggregate in concrete. The slump values obtained depicted medium and acceptable workability for e-plastic concrete (Table 2); however, no form of consistency was noted with reference to a particular water-cement ratio, shape, or cement grade. Whereas a few studies found a constantly increasing trend in workability with an increase in e-plastic content (Ahirwar et al. 2016; Singh and Pandey 2017; Sabău and Vargas 2018), the majority observed a decreasing trend synchronizing with an increase in e-plastic content (Siddique et al. 2008; Kumarl and Baskar 2014; Manjunath 2016; Prashant and Kumar 2019; Santhanam et al. 2020; Brahmacharimayum et al. 2019; Arivalagan 2020; Suleman and Needhidasan 2020). In some cases, the increase in workability was limited to 10% levels of substitution of course aggregate with e-plastic aggregate beyond which a decrease in workability was observed (Arivalagan 2020; Brahmacharimayum et al. 2019; Kumar and Baskar 2014).

The increase in slump values is generally attributed to the non-absorptive nature of e-plastic, which neither absorbs nor adds water to the concrete mix, thus having more free water and consequently increasing the workability. However, even with the smooth and slippery surface that increases the flow and ease of placement of the e-plastic concrete, a finishing problem was reported especially after a 10% replacement of the coarse aggregate. Thus, the overall reduction in workability was mainly attributed to the poor shape, size, and texture of the e-plastic aggregate which greatly affects the consistency of the concrete.

4.1.2 Concrete Density

The density of concrete usually depends on the type of aggregate used (Jóźwiak-Niedźwiedzka and Lessing 2019). Due to the lightweight nature of e-plastic, its incorporation in concrete as aggregate is bound to decrease the fresh and dry densities of the concrete. Several studies observed a reduction in density values of e-plastic concrete for different concrete grades and mixes, when compared to conventional concrete (Gavhane et al. 2016; Kumar and Baskar 2014; Sabău and Vargas 2018; Shinu and Needhidasan 2019).

As depicted in Fig. 3, a high correlation exists between the replacement rate and concrete density as presented using the regression equation $\sigma_{\overline{20}} = -472\alpha + 2223$ for grade 20 concrete and $\sigma_{30} = -873\alpha + 2436$ for grade 30 concrete with almost the same water-cement ratio. This generally showed that a 1% replacement rate leads to a 21% and 11% reduction in the density of grade 20 and grade 30 concrete, respectively.

Therefore, concrete produced with more than 40% e-plastic waste content is more likely to fall under the lightweight category with a density of not more than 2200 kg/m^3 (Fig. 3). The overall decrease in density with e-plastic waste addition is attributed to the difference in density of the e-plastic which is reported to be about

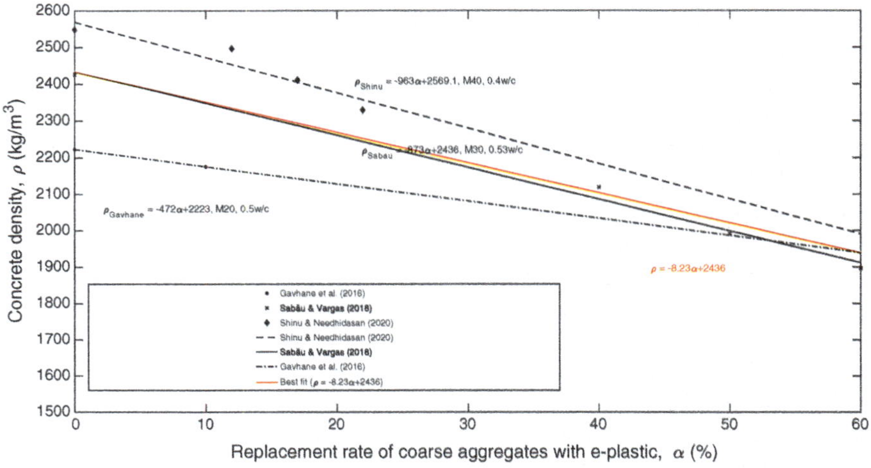

Fig. 3 Comparison of density results of e-plastic waste concrete from different researchers

70% lower than that of conventional coarse aggregate (Kumar and Baskar 2014). This indicates the potential of using e-plastic concrete for non-load bearing lightweight concrete elements, such as concrete panels for facades (Manjunath 2016).

4.1.3 Compressive Strength

The compressive strength is a measure of structural performance. In practice, most concrete is designed primarily to resist compressive stresses that arise from structural loads. Studies on ordinary and standard concrete of grades (M5–M20) and (M25–M50), respectively, demonstrate that increasing the proportion of e-plastic waste as a partial replacement for coarse aggregates generally results in a reduction in the 7-, 14-, and 28-day compressive strengths of ordinary concrete. This decrease may be attributed to (i) a decrease in the adhesive strength between the materials (Manjunath 2016) or a weak bond between e-plastic waste and cement paste (Shinu and Needhidasan 2020), (ii) the fact that e-plastic waste is hydrophobic materials hence limiting cement hydration (Manjunath 2016); and (iii) the possible increase in matrix porosity, which decreases the density and so the compressive strength (Shinu and Needhidasan 2020).

A regression analysis as demonstrated in Fig. 4 shows that as the replacement rate (α) of the coarse aggregates with e-plastic waste increases, the compressive strength reduces by Eq. 2 where β is the percentage negative influence of the e-plastic waste on the compressive strength and α is the percentage replacement rate of coarse aggregates with e-plastic waste. Equation 2 demonstrates that for 10% replacement of coarse aggregates with e-plastic waste has a 6.9% negative impact on the compressive strength of concrete.

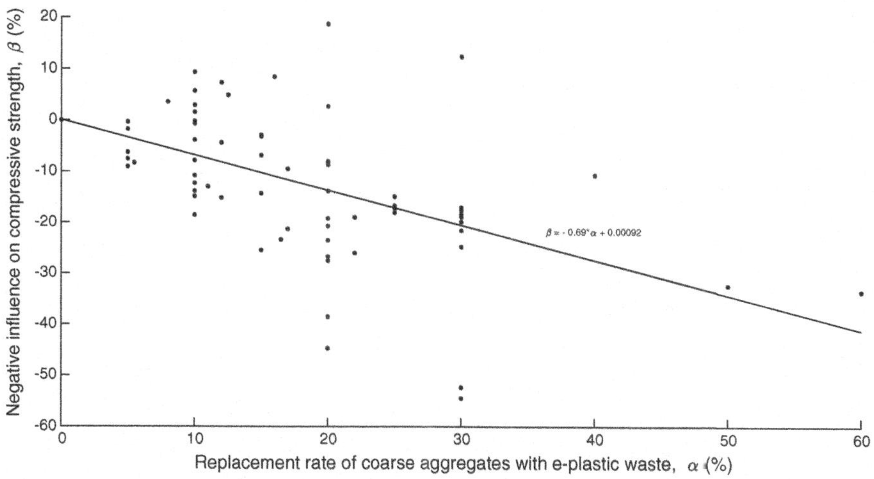

Fig. 4 Relationship between the negative influence of e-plastic waste and the replacement rate

$$\beta = -0.69\alpha + 0.00092 \tag{1}$$

This behavior has been presented in studies where an M20 e-plastic waste concrete produced from crushed e-plastic waste (Gavhane et al. 2016) lost 4% of its design strength and a similar e-plastic concrete from computer equipment and accessories (Santhanam et al. 2020) lost 4.22% of its design strength at a 10% replacement rate. Alike, a replacement rate up to 60% from computer housings of a maximum size of 9.5 mm resulted in reductions in the compressive strength up to 44.15% (Sabău and Vargas 2018). The modified concrete in the study by Sabău and Vargas (2018) achieved at least 74%, 62%, and 53% of the design strength, for 40%, 50%, and 60% replacement rate, respectively.

However, a further analysis of the influence of the replacement rate of e-plastic waste on the compressive strength shows varying results for different concrete grades as demonstrated in Fig. 5. While concrete grades M20 to M40 showed reduction in the compressive strength with increasing replacement rates, M60 concrete realized improvements. This positive was observed where additives like a superplasticizer or fly ash are used (Needhidasan et al. 2020b; Santhanam and Anbuarasu 2020). The strength improvement is attributed to the presence of silica in the fly ash hence enhancing the binding of the cementitious matrix. In addition, the superplasticizers are known to promote cement hydration and inhibit early setting. Therefore, it is recommended from Fig. 5 that a replacement rate up to 30% for M20 concrete, up to 15% for M30, and 38% for M40 will achieve the characteristic strength for the respective concrete grades.

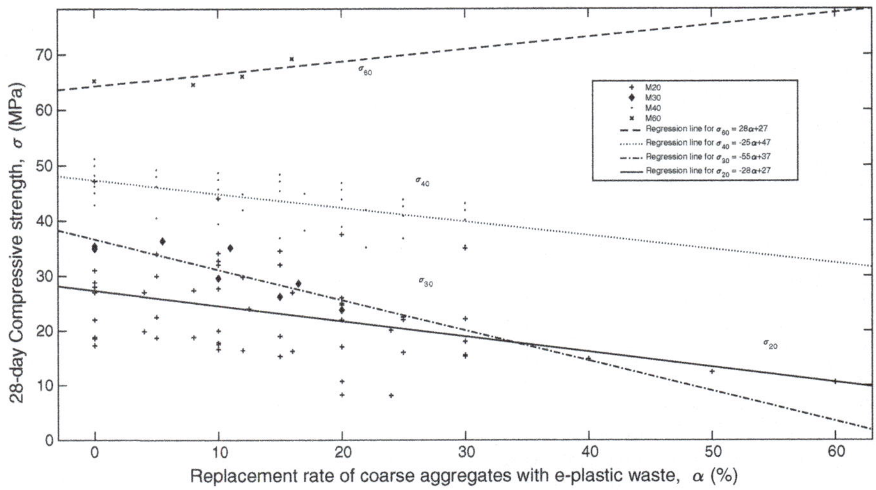

Fig. 5 Influence of e-plastic waste replacement rate for varying concrete grades

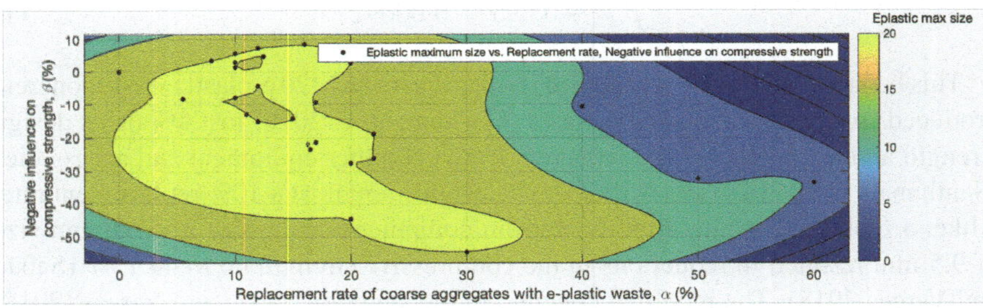

Fig. 6 Relationship between compressive strength and e-plastic waste maximum size

Effect of E-Plastic Waste Aggregate Size on the Compressive Strength

The size of coarse aggregates is known to have an impact on the compressive strength of concrete. An analysis on the effect of e-plastic maximum size on the compressive strength shows a positive correlation as the e-plastic waste aggregates with greater sizes are seen to have a lower impact on the reduction of the compressive strength as presented in Fig. 6.

This can be explained by the fact that the amount of water needed to achieve a similar workability in larger sized aggregates as compared to the smaller sized aggregates is reduced leading to a lower water-cement ratio and hence higher compressive strength. This phenomenon is explained in Fig. 7 where the e-plastic waste aggregate of larger sizes is associated with lower water-cement ratios and hence higher compressive strengths. However, the loss in compressive strength of the concrete with every 10% replacement rate was found to be consistent for all the water-cement ratios at an average of 9.5 to 11.5% strength loss. This implies that the e-plastic

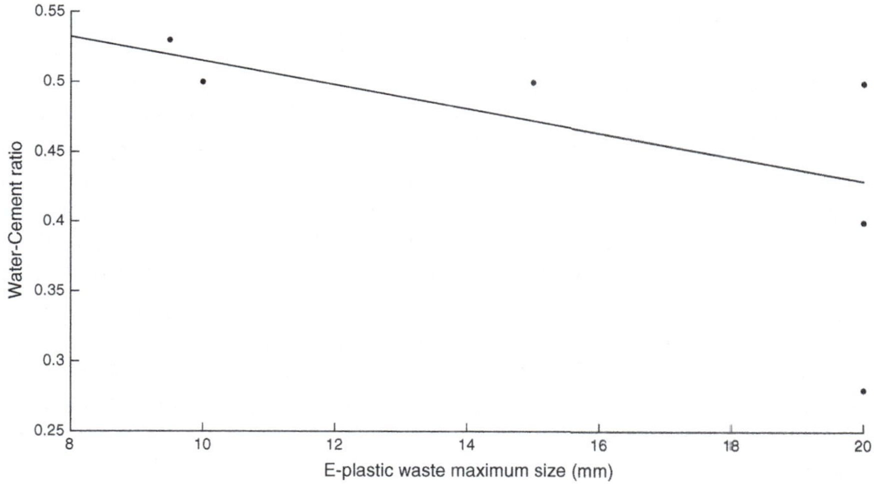

Fig. 7 Influence of e-plastic waste size on the water-cement ratio

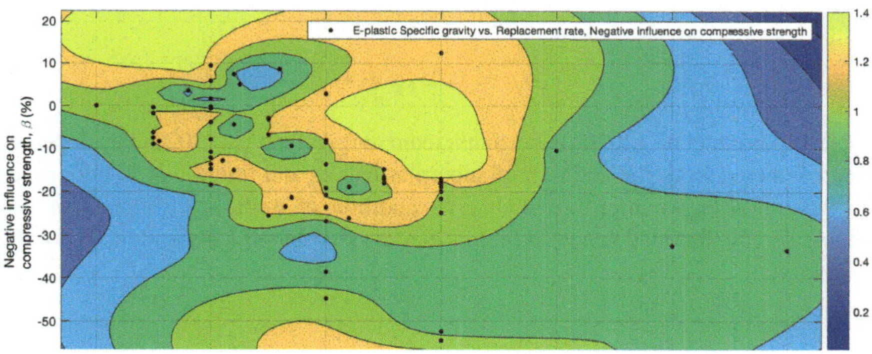

Fig. 8 Effect of e-plastic specific gravity on compressive strength

waste aggregate composition has little effect on the water demand for concrete and the influence only arises from the size of the aggregate.

A further analysis as demonstrated in Fig. 8 shows that the e-plastic waste aggregates with higher specific gravities generally produced concretes of higher strength. This is because the specific gravity can be loosely termed as a measure of the strength of the aggregates. This implies that the aggregates with higher specific gravities contributed greatly to the strength of the concrete. This phenomenon is no different with e-plastic aggregates from Fig. 8 where at the different replacement rates (α), e-plastics waste aggregates with higher specific gravities (G_s) were associated with stronger concretes.

4.1.4 Flexural Strength

Flexural strength is a measure of an unreinforced structural member's ability to resist failure in bending (M. Shinu and Needhidasan 2020). The partial replacement of coarse aggregates with e-plastic waste has been found to reduce flexural strength as demonstrated in Fig. 9. The relationship has been presented using the regression equation $\sigma_{fr} = -5.5\alpha + 4.8$; where a 10% replacement rate leads to an 11% reduction in the flexural strength. This has been demonstrated by Shinu and Needhidasan (2020) that reported a 7% decline for a 12% replacement and with Singh and Pandey (2017) where a 5%, 10%, and 15% replacement led to a reduction in the 28-day flexural strength by 6.22%, 9.01%, and 15.22%, respectively.

A further review as demonstrated in Fig. 10 on the different concrete grades demonstrated that the flexural strength of standard concrete grades of M40 and M60 showed improvements in the flexural strength by up to 0.2% and 1.3%, respectively, for a 1% replacement rate.

However, this improvement was observed only where superplasticizers were used as demonstrated in Fig. 11. The e-plastic concrete made using superplasticizers behaved according to Eq. 2 where the flexural strength improved by 12.75% for every 10% replacement of coarse aggregates with e-plastic.

$$\sigma_{fr} = 0.051\alpha + 4 \quad (2)$$

Needhidasan et al. (2020b) and Santhanam and Anbuarasu (2020) developed e-plastic concrete with the addition of superplasticizers and found that in both cases the flexural strength improved with each addition of e-plastics. Santhanam and Anbuarasu (2020) found that an 8% replacement led to an improvement of 5% for the

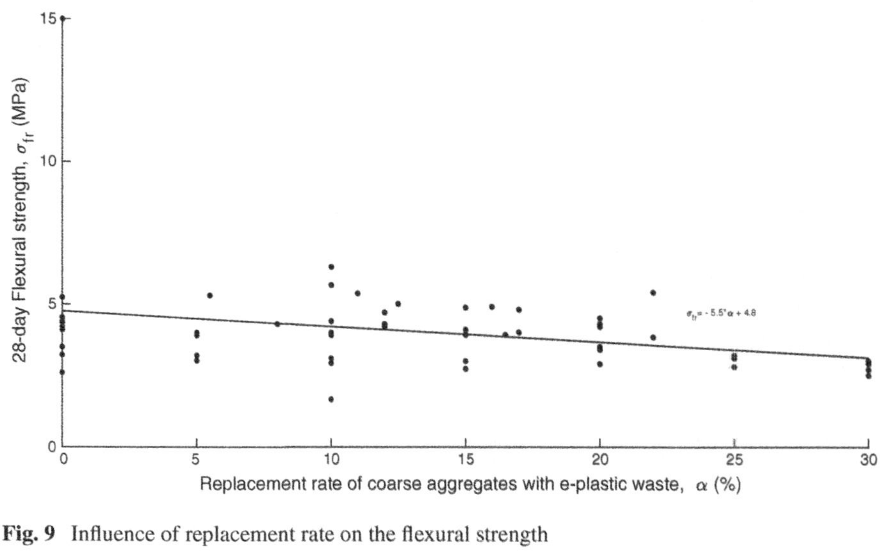

Fig. 9 Influence of replacement rate on the flexural strength

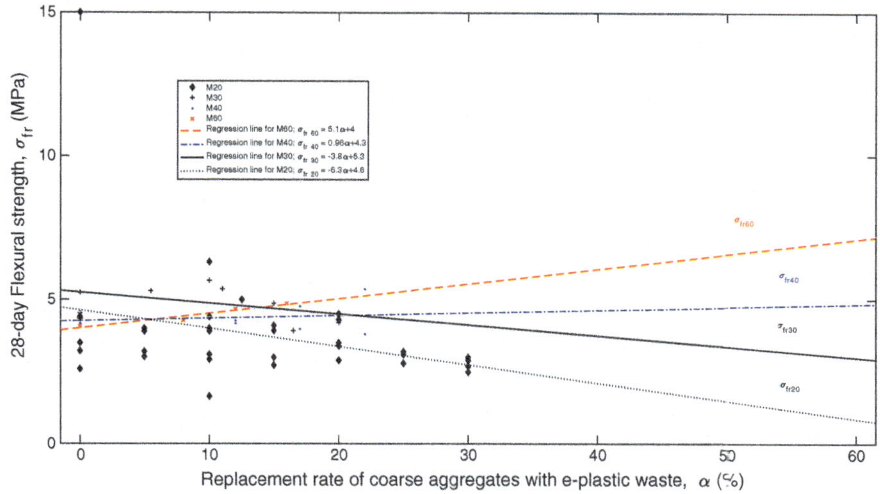

Fig. 10 Influence of replacement rate on the flexural strength of different concrete grades

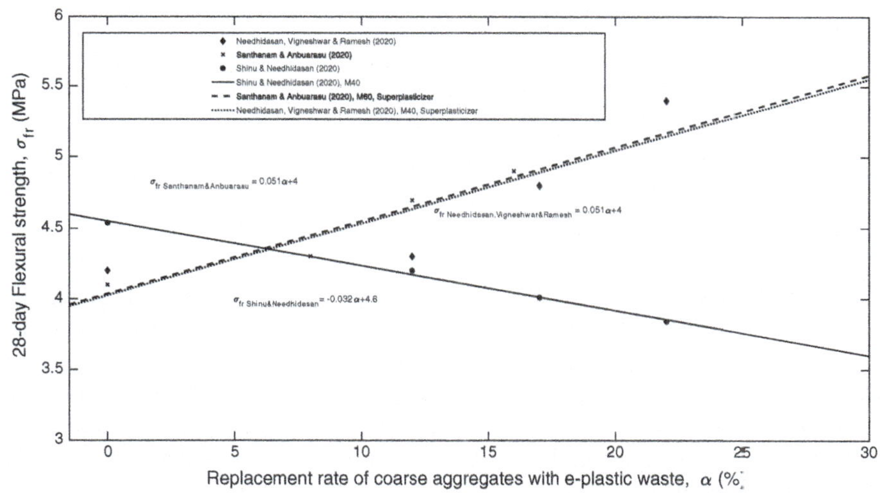

Fig. 11 Variation of flexural strength for standard concrete (M40-M60)

flexural strength, and with each additional 4% replacement realized a 5% incremental improvement. Relatedly, Needhidasan et al. (2020b) study found an improvement in the flexural strength up to 29% in comparison to the conventional concrete for a 22% replacement.

The correlation between the flexural strength and compressive strength has been described using Eq. 3 For the e-plastic concrete, the statistical analysis as demonstrated in Fig. 12 established a relationship between the flexural strength and compressive strength as described by Eq. 4.

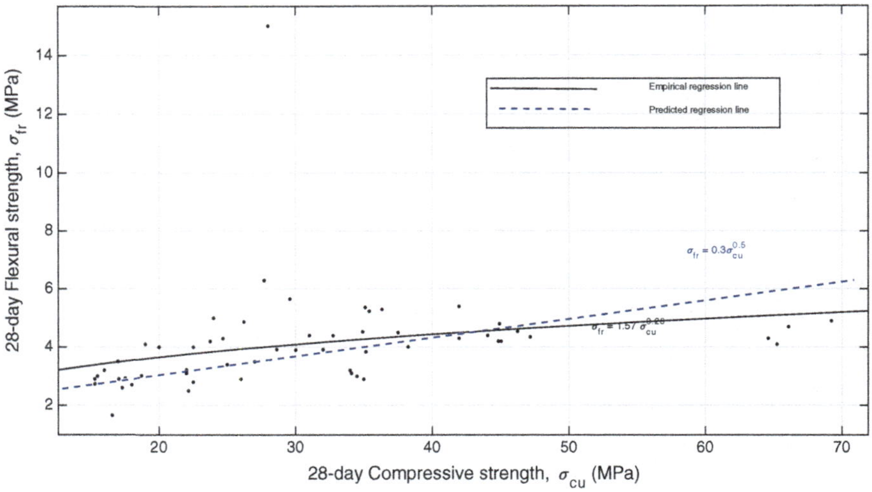

Fig. 12 Relationship between flexural strength and compressive strength of e-plastic concrete >

$$\sigma_{fr} = 0.3\sigma_{cu}^{0.5} \tag{3}$$

$$\sigma_{fr} = 1.57\sigma_{cu}^{0.28} \tag{4}$$

4.1.5 Split Tensile Strength

The split tensile strength is loosely defined as that point where failure is due to the compression load, thereby inducing pure tensile stress along the diameter of the specimen. The replacement of coarse aggregates with e-plastic waste has provided varying results. In some studies, it led to an improvement in the split tensile strength, especially where superplasticizers were used to improve the strength of the concrete while in contrast, in other studies, it demonstrated a negative influence on the split tensile strength as demonstrated in Fig. 13.

The statistical analysis on the split tensile strength of e-plastic concrete demonstrated that a 10% replacement rate of e-plastic waste negatively impacts the split tensile strength of M30 and M40 concrete grade by 8% and 0.1%, respectively. However, in contrast, the e-plastic was found to improve the split tensile strength for M20 and M60 grades by 10% and 29%, respectively.

There is a need for further studies on the impact of e-plastic waste on the split tensile strength. There are contrasting results on the influence of the replacement rate of e-plastic waste on the split tensile strength for M20 concrete as demonstrated in Fig. 14. While studies by Arivalagan (2020), Manjunath (2016) show a decline in the

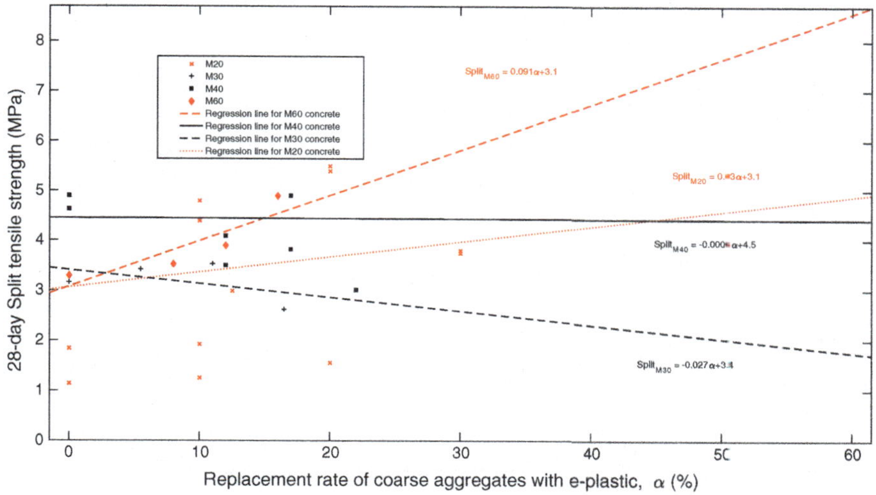

Fig. 13 Influence of e-plastic replacement rate on the split tensile strength for varying concrete grades

split tensile strength by 5% for a 10% replacement rate, other studies have demonstrated improvements in similar M20 concrete between 19 and 37% (Needhidasan et al. 2020a; Santhanam et al. 2020).

The variation of e-plastic waste replacement rate with split tensile strength shows a significant relationship between the water-cement ratio and use of additives on the variation on split tensile strength as demonstrated in Fig. 15. The analysis showed an

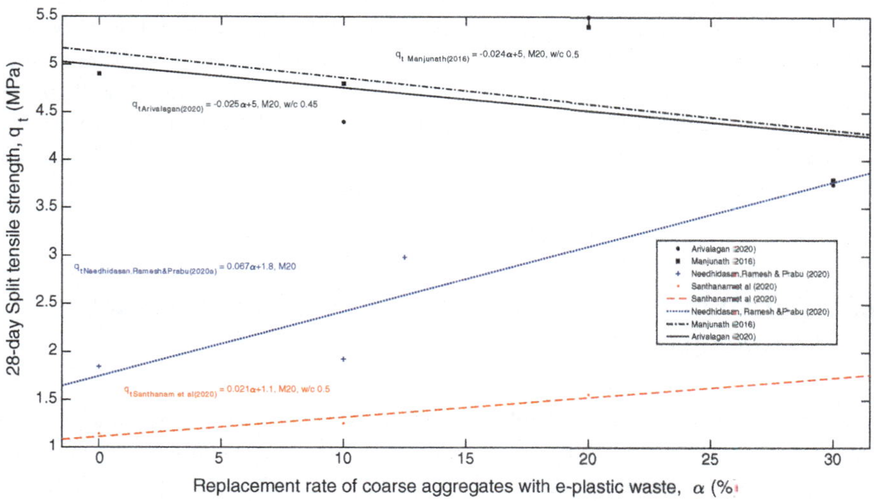

Fig. 14 Influence of e-plastic replacement rate on the split tensile strength from varying studies for ordinary concrete (M20)

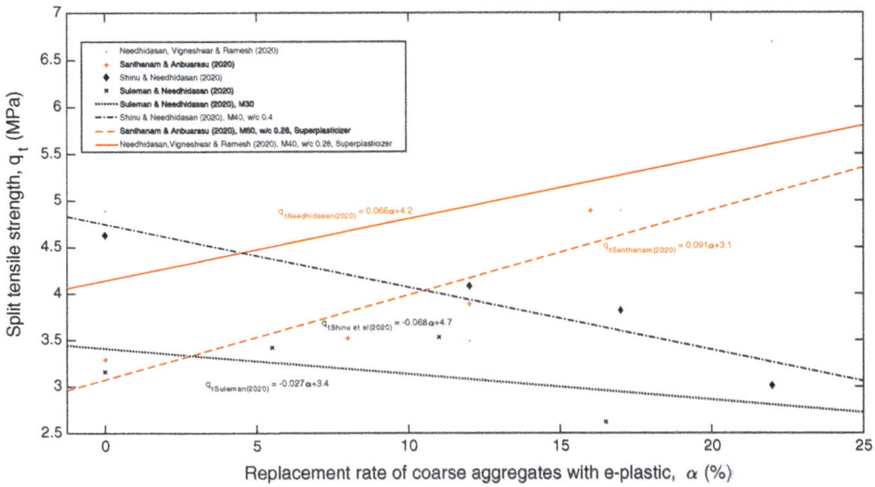

Fig. 15 Influence of e-plastic replacement rate on the split tensile strength from varying studies for standard concrete (M25–M60)

improvement in the split tensile strength by 15–29% where a low water-cement ratio of 0.28 was used alongside a superplasticizer (Needhidasan et al. 2020a; Santhanam and Anbuarasu 2020).

Further to this, there was a correlation between the influence on the split tensile strength and the specific gravity of e-plastic waste as demonstrated in Fig. 16. It was observed that for e-plastics with a higher specific gravity it resulted in a smaller reduction in the split tensile strength for increasing replacement rates as compared to those with lower specific gravities. This can be explained by the fact that aggregates with higher specific gravities are considered stronger than those with lower specific gravities and therefore will have greater contributions to the strength of the concrete. For the e-plastic waste aggregates, e-plastics with specific gravities less than 0.8 were found to result in reductions in the split tensile strength by up to 35% for a 22% replacement rate while those e-plastics with specific gravities greater than 1 reduced the split tensile strength by up to 22% for a 30% replacement rate.

4.2 Durability Properties of Concrete Containing E-Plastic Waste Aggregates

According to IS 456: 2000, durable concrete is one that performs satisfactorily in the working environment during the anticipated exposure conditions while maintaining its desired engineering properties over its service life. Durability is therefore measured by the ability of concrete to resist weathering actions, chemical attack, abrasion, and other processes of deterioration (Luhar and Luhar 2019). Wide research has

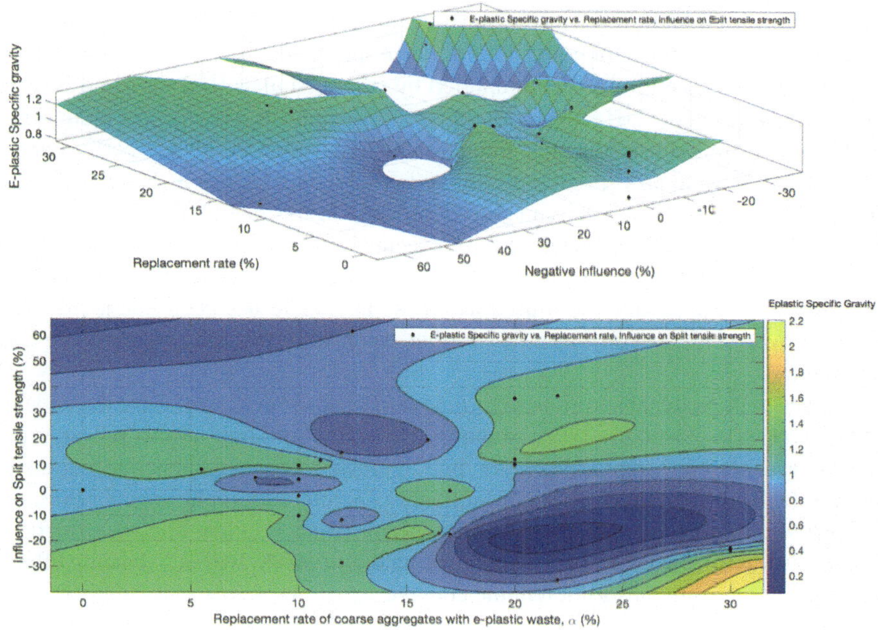

Fig. 16 Influence of e-plastic waste aggregate specific gravity on the split tensile strength

been done testing the durability characteristics of e-plastic concrete based on both the physical and chemical conditions including permeability, shrinkage, sulfate, and chloride resistance.

4.2.1 Permeability

Permeability is a characteristic that governs the flow rate of a liquid into a porous solid (Luhar and Luhar 2019). The effect on permeability can be characterized by the micro-structural properties such as saturated water absorption, porosity, sorptivity, etc. Therefore, the higher the permeability, the lower will be the durability of concrete.

Water Absorption

Water absorption is a measure of the penetrability of water in the pore system of concrete over a specified period of time and is expressed as a percentage increase in the weight of concrete. The factors that affect water absorption include the water mix proportions, type of cement and aggregate used, presence of additives, and length of exposure. The acceptance criteria for durability based on the water absorption levels of concrete is rated as good (>3%), average (3–5%), and worse (<5%) (Bala Rama Krishna and Jagadeesh 2019a, b).

Existing research shows an increasing trend in water absorption with an increase in e-plastic as a replacement of both fine and coarse aggregate. An increase of more than 70% in water absorption has been reported by several studies (Lakshmi and Nagan 2011; Gavhane et al. 2016). According to the acceptance criteria, concrete with e-plastic substituted for coarse aggregate exhibits average durability performance with saturated water absorption rates falling between 3 and 5%, which is lower than that of conventional concrete with water absorption rates below 3% (Lakshmi and Nagan 2011). However, when used as a replacement for fine aggregate, the durability performance of the concrete turned out to be good, with rates below 3%. The increase in absorptivity especially when used as coarse aggregate has been blamed on the inter-particle voids existing in e-plastic that emanates from the dispersion of the aggregates.

However, the water absorption performance of e-plastic concrete is also dependent on the type of e-plastic used. For example, application of High-impact polystyrene (HIPS) granules as a replacement for fine aggregate in self-compacting concrete yielded better performance than conventional concrete with replacement values of up to 30% beyond which a worse durability performance is noted. The water ingress reduction was attributed to the hydrophobic nature of HIPS, with a balancing effect of reducing the water absorption in the concrete.

Porosity and Sorptivity

Whereas sorptivity measures the rate of penetration of water into the pores of concrete by capillarity suction, porosity is a measure of the pore volume occupied by water in a saturated condition, as denoted by the quantity of water removed on drying saturated concrete (Lakshmi and Nagan 2011; Bala Rama Krishna and Jagadeesh 2019a, b). According to Zhang et al. (2020), porosity highly affects the methodology for testing permeability, i.e., it would be inappropriate to test concrete with low porosity and high compactness using the water pressure method. The higher the porosity and sorptivity characteristics of concrete, the lower the durability.

Based on an acceptance criterion for porosity and sorptivity of "excellent" for values less than 9% and 6 mm/$\sqrt{}$hr, respectively, a slight increase in porosity is observed with an increase of e-plastic as a partial replacement for coarse aggregates of up to 10%, beyond which a sharp decrease is noted (Akram et al. 2015; Lakshmi and Nagan 2011). Thus, the durability of e-plastic concrete, though slightly lower than that of conventional concrete, is acceptable. However, when HIPS granules are used especially as replacement of fine aggregate, both porosity and sorptivity are improved with up to 30% level of substitution, exhibiting slightly lower values than conventional concrete (Bala Rama Krishna and Jagadeesh 2019a, b). Higher value addition of HIPS (over 30%), on the other hand, leads to an increase in porosity and hence a decrease in permeability. The increase is due to the inter-particle voids created by the e-plastic aggregates which increase both the porosity and sorptivity aspects of the concrete, thus reducing its durability.

4.2.2 Sulfate Resistance

Sulfate attack denotes an increase in the volume of cement paste in concrete caused by chemical reactions between the products of hydration of cement and solutions containing sulfates (Suchithra et al. 2015). Just like chloride resistance, several studies have shown that e-plastic concrete has better resistance to sulfate attacks than conventional concrete, when e-plastics partially replace either coarse or fine aggregates.

After immersion in a 10% solution of sodium sulfate, e-plastic concrete cubes registered daily reductions of less than 0.5% in weight when substituted for fine aggregates and reductions of 1.5–2.5% when substituted for coarse aggregates. This compares to approximately 1% and 3% weight reductions for fine and coarse aggregate substitutions, respectively, in conventional concrete (Gavhane et al. 2016; Lakshmi and Nagan 2011; Suchithra et al. 2015). Similarly, less reductions in compressive strengths are noted in e-plastic concrete especially with e-plastic replacement values of up to 15%. The addition of 10% fly ash as a partial substitute for cement has also been noted to improve the sulfate resistance properties of e-plastic concrete (Lakshmi and Nagan 2011). Thus, it is clear that e-plastic concrete is not influenced by sulfates, but it is more durable when percentage replacement values of less than 15% of weight are used to replace coarse aggregate.

4.2.3 Chloride Resistance

Chloride-induced corrosion is one of the common mechanisms of deterioration affecting the long-term performance of structures, and thus chloride resistance is one of the most important aspects of concrete durability (Lakshmi and Nagan 2011). Concrete is meant to provide physical and chemical protection to reinforcement steel from both chlorides and sulfate attacks, primarily to avoid corrosion. The chloride resistance, however, is dependent on the porosity of the concrete and the thickness of cover to the reinforcement (AL-Ameeri et al. 2019; Lee et al. 2020). A few studies have looked into the chloride resistance characteristics of concrete incorporating e-plastic as aggregate.

The percentage loss in weight and compressive strength obtained after immersion of the e-plastic concrete cubes in 5% diluted hydrochloric acid is considerably lower than the corresponding loss of weight and compressive strength of conventional concrete cubes (Lakshmi and Nagan 2011; Suchithra et al. 2015). However, an increasing trend in loss of both weight and compressive strength is noted after the substitution levels exceed 15% of coarse aggregate (Lakshmi and Nagan 2011).

The addition of fly ash into the mix as a substitute for cement was also noted to improve the density of concrete, resulting in comparatively lower loss of weight and compressive strength of the e-plastic concrete, thus rendering e-plastic concrete more durable than conventional concrete (Lakshmi and Nagan 2011). This performance has been attributed to the impervious nature of e-plastic aggregates, which present a physical obstacle to chloride ingress.

4.3 Thermal Properties of E-Plastic Concrete

The main properties that form the basis for evaluation of the thermal performance of concrete structures include thermal expansion, thermal conductivity, thermal diffusivity, and specific heat (Sinha 2014). Whereas thermal expansion is a measure of the fire performance of the concrete, thermal conductivity relates to the ability of concrete to conduct heat and is one of the key aspects used to predict temperature variation in elements during hydration. These thermal properties are highly dependent on the moisture content, type of aggregate, the mix proportions, the type of cement, and the temperature of the concrete (Sinha 2014; Kakae et al. 2017; Khalil and Obeidy 2018). Research on the thermal properties of e-plastic concrete is scanty, with a few done on e-waste and plastic concrete generally. Concrete utilizing e-waste or plastics generally has been reported to have lower thermal conductivity values than conventional concrete (Dhanraj et al. 2017; Khalil and Obeidy 2018). The use of e-plastic as a substitute for coarse or fine aggregate in concrete also yields similar results, depicting a reduction in thermal conductivity with an increase in e-plastic content (Lakshmi and Nagan 2011). The decrease in thermal conductivity is mainly attributed to the increase in porosity of the concrete and the lower thermal conductivity value of e-plastic of 0.1175 W/(m K), which is much lower than that of the natural coarse aggregate of 2 W/(m K) (Lakshmi and Nagan 2011).

However, the thickness of the concrete members has been observed to have an impact on the thermal conductivity of the e-concrete as shown in Fig. 17.

It can be noted that there is a general increase in thermal conductivity for both the conventional and e-plastic concrete with an increase in thickness up to a maximum of 100 mm, where the thermal conductivity starts to decrease. However, the e-concrete trendline showed that for every increase in thickness, there was a corresponding

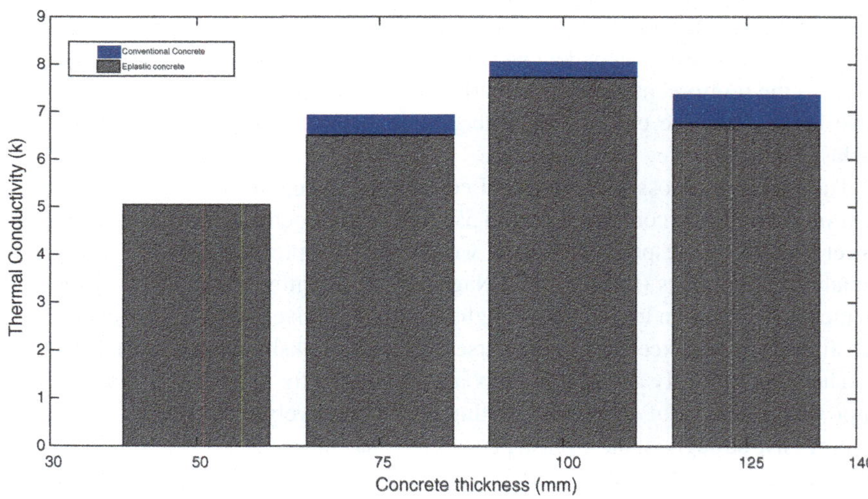

Fig. 17 Comparison of thermal conductivity between e-plastic waste and conventional concrete

increase in thermal conductivity of 0.6242. The resultant R-value (heat resistance) for e-concrete as well as the energy stored was therefore observed to be less than that for conventional concrete.

5 Conclusion

Based on several research findings, the use of e-plastic in concrete as a replacement for fine and coarse aggregate yielded both positive outcomes and criticisms. Some studies have shown that the reuse of plastic e-waste in concrete has many economical, technical, and environmental advantages including the following:

1. Extreme versatility and ability to be tailored to meet specific technical needs.
2. Concrete with e-plastic is comparatively light in weight than the conventional one thus reducing transportation costs.
3. They result in high-impact strength and good abrasion strength.
4. It produces high workable concrete than conventional concrete for the same water-cement ratio, and thus would reduce the high cost of admixture required to produce workable concrete.
5. Better durability and longevity due to its higher resistance to chemicals and water.
6. Better bonding properties as the bond capacity increases as the temperature increases at the melting point.
7. Excellent thermal and electrical insulation properties.
8. Recycle-ability of the elements constructed using e-plastic as a recycled aggregate thus saving natural resources and dumping spaces, which reduces the load caused by waste materials, resulting in a clean environment. In addition, it minimizes the use of natural resources like aggregates, thus preventing the depletion of natural resources.
9. With less production costs when compared to conventional concrete, the utilization of crushed e-plastic waste materials as conventional concrete and other materials in the building construction helps in reducing the cost of concrete manufacturing.

Some of the drawbacks of the utilization of e-plastic waste in the manufacture of concrete included the following:

1. E-plastic waste has low bonding properties making it difficult to obtain a homogeneous mixture thus reducing the compressive, tensile, and flexural strength of concrete especially with replacement values exceeding 10%.
2. When used as a replacement for coarse aggregate, segregation of the concrete is reported to occur, as the water absorption of e-plastic is nearly zero.
3. In addition, it cannot be used in furnaces and other fire places due to its low melting point.

References

Ahirwar S, Malviya P, Patidar V, Singh VK (2016) An experimental study on concrete by using E-waste as partial replacement for coarse aggregate. IJSTE Int J Sci Technol Eng 3(04):7–13. www.ijste.org

Akram A, Sasidhar C, Pasha KM (2015) E-waste management by utilization of E-plastics in concrete mixture as coarse aggregate replacement. Int J Innov Res Sci Eng Technol 04(07):5087–5095. https://doi.org/10.15680/IJIRSET.2015.0407008

AL-Ameeri AS, Imran Rafiq M, Tsioulou O (2019) Influence of carbonation on the resistance of concrete structures to chloride penetration and corrosion. MATEC Web Conf 289:08001. https://doi.org/10.1051/matecconf/201928908001

Alexander M, Mindess S (2010) Aggregates in concrete. In civil engineering. Taylor & Francis

Amuda AG, Uche OA, Amuda A (2014) Physicomechanical characterization of basement rocks for construction aggregate: a case study of Kajuru Area, Kaduna, Nigeria. IOSR J Mech Civ Eng 11(6):46–51. https://doi.org/10.9790/1684-11664651

Arivalagan S (2020) Experimental study on the properties of green concrete by replacement of e-plastic waste as aggregate. Procedia Comput Sci 172(2019):985–990. https://doi.org/10.1016/j.procs.2020.05.145

Bala Rama Krishna CH, Jagadeesh P (2019a) Durability studies on SCC replacing sand partially with HIPS granules. Int J Eng Adv Technol 8(2):72–75

Bala Rama Krishna CH, Jagadeesh P (2019b) Strength and durability assessment of binary blended self-compacting concrete replacing partial sand with electronic plastic waste. Int J Innov Technol Explor Eng 8(5):107–111

Brahmacharimayum M, Shijagurumayum C, Suresh T (2019) E-waste in concrete—a review. 6(6):384–387

Bulut HA, Şahin R (2017) A study on mechanical properties of polymer concrete containing electronic plastic waste. Compos Struct 178(June):50–62. https://doi.org/10.1016/j.compstruct.2017.06.058

Chandra S, Berntsson L (2002) Lightweight aggregate concrete: science, technology, and applications

da Silva TR, de Azevedo ARG, Cecchin D, Marvila MT, Amran M, Fediuk R, Vatin N, Karelina M, Klyuev S, Szelag M (2021) Application of plastic wastes in construction materials: a review using the concept of life-cycle assessment in the context of recent research for future perspectives. Materials 14(13). https://doi.org/10.3390/ma14133549

Dhanabal P, Sushmitha KS, Reddy PN (2021) Study on properties of concrete with electronic waste. Revista Ingeniería De Construcci6n 36(1):48–58

Dhanraj AA, Selvamony C, Joshuva A (2017) An experiment investigation on concrete by using E-waste as fine aggregate and enhanced the thermal insulation and ultrasonic properties. Int J Civ Eng Technol 8(12):392–399

Dhir RK, Brito J de, Mangabhai R, Lye CQ (2017) Production and properties of copper slag. Sustainable construction materials: copper slag. https://doi.org/10.1016/b978-0-08-100986-4.00003-1

Forti V, Balde CP, Kuehr R, Bel G (2020) The global E-waste monitor 2020: quantities, flows and the circular economy potential

Gavhane A, Sutar D, Soni S, Patil P (2016) Utilisation of E-plastic waste in concrete. Int J Eng Res Technol 5(02):594–601. https://doi.org/10.17577/ijertv5is020538

Grigorescu RM, Grigore ME, Iancu L, Ghioca P, Ion RM (2019) Waste electrical and electronic equipment: a review on the identification methods for polymeric materials. Recycling 4(3). https://doi.org/10.3390/recycling4030032

Gull I, Subramanian BM (2014) A new paradigm on experimental investigation of concrete for E-plastic waste management. Int J Eng Trends Technol 10(4):180–186. https://doi.org/10.14445/22315381/IJETT-V10P234

Inamuddin AMI, Lichtfouse E (2021) Water pollution and remediation: organic pollutants. Environmental chemistry for a sustainable world, p 54

Jóźwiak-Niedźwiedzka D, Lessing PA (2019) High-density and radiation shielding concrete. Developments in the formulation and reinforcement of concrete, 2nd edn. Woodhead Publishing Ltd., pp 193–228

Kakae N, Miyamoto K, Momma T, Sawada S, Kumagai H, Ohga Y, Hirai H Abiru T (2017) Physical and thermal properties of concrete subjected to high temperature. J Adv Concr Technol 15(6):190–212. https://doi.org/10.3151/jact.15.190

Kamaruddin MA, Abdullah MMA, Zawawi MH, Zainol MRRA (2017) Potential use of plastic waste as construction materials: recent progress and future prospect. IOP Conf Ser Mater Sci Eng 267:012011. https://doi.org/10.1088/1757-899X/267/1/012011

Khalil W, Obeidy NAl (2018) Some properties of sustainable concrete containing two environmental wastes. MATEC Web Conf 162:1–9. https://doi.org/10.1051/matecconf/201816202029

Kumar R, Malik SV (2016) Experimental study on effective utilization of E-waste in concrete. Int J Innov Res Sci Eng Technol 4(6):5262–5265. http://www.ijirset.com/upload/2015/november/16_Experimental_new.pdf

Kumar S, Baskar K (2014) Response surfaces for fresh and hardened properties of concrete with E-waste (HIPS). J Waste Manag 2014:1–14. https://doi.org/10.1155/2014/517219

Lakshmi R, Nagan S (2011) Investigations on durability characteristics of E-plastic waste incorporated concrete. Asian J Civ Eng 12(6):773–787

Lee Y, Lim S, Lee H (2020) Chloride resistance of Portland cement-based mortar incorporating high aluminate cement and calcium carbonate. Materials 13(2). https://doi.org/10.3390/ma13020359

Luhar S, Luhar I (2019) Potential application of E-wastes in construction industry: a review. Constr Build Mater 203:222–240. https://doi.org/10.1016/j.conbuildmat.2019.01.080

Makri C, Hahladakis JN, Gidarakos E (2019) Use and assessment of "e-plastics" as recycled aggregates in cement mortar. J Hazard Mater 379:120776. https://doi.org/10.1016/j.jhazmat.2019.120776

Manjunath AB (2016) Partial replacement of E-plastic waste as coarse-aggregate in concrete. Procedia Environ Sci 35:731–739. https://doi.org/10.1016/j.proenv.2016.07.079

Masuduzzaman M, Amit SKS, Alauddin M (2018) Utilization of E-waste in concrete and its environmental impact—a review. In: 2018 international conference on smart city and emerging technology, ICSCET 2018, pp 1–4. https://doi.org/10.1109/ICSCET.2018.8537301

Mehta KP, Monteiro PJ (2001) Concrete: microstructure, properties and materials vol 7

Mills W, Tatara R (2016) Potential for reuse of E-plastics through processing by compression molding. Challenges 7(1):13. https://doi.org/10.3390/challe7010013

Needhidasan S, Ramesh B, Joshua Richard Prabu S (2020a) Experimental study on use of E-waste plastics as coarse aggregate in concrete with manufactured sand. Mater Today Proc 22:715–721. https://doi.org/10.1016/j.matpr.2019.10.006

Needhidasan S, Vigneshwar CR, Ramesh B (2020b) Amalgamation of E-waste plastics in concrete with super plasticizer for better strength. Mater Today Proc 22:998–1003. https://doi.org/10.1016/j.matpr.2019.11.253

Neville AM, Brooks JJ (2010) Concrete technology, 2nd edn. Pearson

Parsons LA, Nwaubani SO (2020) E-plastics as partial replacement for aggregates in concrete. Wastes: solutions, treatments and opportunities III, September, pp 275–280. https://doi.org/10.1201/9780429289798-44

Pivnenko K, Astrup TF (2018) E-plastics in Europe: an overview. In: Proceedings CRETE 2018, sixth international conference on industrial & hazardous waste management, pp 1–3. https://doi.org/10.1017/9781316841235.001

Prashant S, Kumar M (2019) Effect of partial replacement of coarse aggregates with E-waste on strength properties of concrete. Sustainable construction and building materials, vol 25. Springer, Singapore, pp 555–560. https://doi.org/10.1007/978-981-13-3317-0_48

Rathore V, Rawat A (2019) Effective utilization of electronic waste in concrete mixture as a partial replacement to coarse aggregates. AIP Conf Proc 2158(September). https://doi.org/10.1063/1.5127161

Sabău M, Vargas JR (2018) Use of e-plastic waste in concrete as a partial replacement of coarse mineral aggregate. Comput Concr 21(4):377–384. https://doi.org/10.12989/cac.2018.21.4.377

Sahajwalla V, Gaikwad V (2018) The present and future of e-waste plastics recycling. Curr Opin Green Sustain Chem 13:102–107. https://doi.org/10.1016/j.cogsc.2018.06.006

Santhanam N, Anbuarasu G (2020) Experimental study on high strength concrete (M60) with reused E-waste plastics. Mater Today Proc 22:919–925. https://doi.org/10.1016/j.matpr.2019.11.107

Santhanam N, Ramesh B, Pohsnem FK (2020) Concrete blend with E-waste plastic for sustainable future. Mater Today Proc 22:959–965. https://doi.org/10.1016/j.matpr.2019.11.204

Shinu M, Needhidasan S (2020) An experimental study of replacing conventional coarse aggregate with E-waste plastic for M40 grade concrete using river sand. Mater Today Proc 22:633–638. https://doi.org/10.1016/j.matpr.2019.09.033

Shinu NMMT, Needhidasan S (2019) An experimental study of replacing conventional coarse aggregate with E-waste plastic for M40 grade concrete using river sand. Mater Today Proc. https://doi.org/10.1016/j.matpr.2019.09.033

Siddique R, Khatib J, Kaur I (2008) Use of recycled plastic in concrete: a review. Waste Manag 28(10):1835–1852

Singh P, Pandey M (2017) E waste management by utilization of E-plastic as coarse aggregate in concrete. Int Res J Eng Technol (IRJET) 4(6):2916–2919. https://irjet.net/archives/V4/i6/IRJET-V4I6716.pdf

Sinha DDA (2014) Thermal properties of concrete. Paripex Indian J Res 3(2):90–91. https://doi.org/10.15373/22501991/feb2014/27

Stenvall E (2013) Electronic waste plastics characterisation and recycling by melt-processing. 1–62

Suchithra S, Kumar M, Indu VS (2015) Study on replacement of coarse aggregate by E-waste in concrete. Int J Tech Res Appl 3(4):266–270

Suleman S, Needhidasan S (2020) Utilization of manufactured sand as fine aggregates in electronic plastic waste concrete of M30 mix. Mater Today Proc 33:1192–1197. https://doi.org/10.1016/j.matpr.2020.08.043

Usuki A (2018) Development of polymer–clay nanocomposites by dispersion of particles into polymer materials. Nanoparticle technology handbook, pp 819–821. https://doi.org/10.1016/B978-0-444-64110-6.00083-4

Zhang Y, Xu S, Fang Z, Zhang J, Mao C (2020) Permeability of concrete and correlation with microstructure parameters determined by 1H NMR. Adv Mater Sci Eng. https://doi.org/10.1155/2020/4969680

Recycling of Mobile Phones: Case Study of the Lithium-Ion Cell Phone Batteries in Brazil

R. G. Silva, M. G. Araújo, J. C. Afonso, C. F. Mahler, and K. R. A. Nunes

Abstract After an intense study of hydrometallurgical processes in acidic environment for the recovery of lithium and cobalt from lithium-ion batteries of cell phones, a life cycle assessment of lithium-ion batteries was carried out, and the impact categories of acids used in chemical treatments and transport were modeled. The modeling confirmed that the hydrochloric acid reached a greater environmental impact in the reuse of the metals in the batteries, for being classified as a strong acid and for having presented the maximum recovery of metals. Hydrofluoric acid achieved the lowest environmental impact in this category as it has positive values in the recovery of these metals. The transport from the reprocessing site to the industrial landfill generated low values in its categories of climate change impacts. Due to the short distance between reprocessing site and industrial landfill, the CO_2 emissions were low. Concerning recovery of cobalt and lithium modeled by the life cycle assessment, the results were not very satisfactory with moderate environmental impact,

R. G. Silva · C. F. Mahler · K. R. A. Nunes (✉)
COPPE/Federal University of Rio de Janeiro - Civil Engineering Program, Centro de Tecnologia - Bloco B - Av. Pedro Calmon, S/N - Cidade Universitária, Rio de Janeiro/RJ CEP 21941-596, Brazil
e-mail: katia.nunes@coc.ufrj.br

R. G. Silva
e-mail: rafael_gundim@yahoo.com.br

C. F. Mahler
e-mail: mahler@coc.ufrj.br

M. G. Araújo
Department of Environmental Health, FIOCRUZ - Oswaldo Cruz Foundation, Rio de Janeiro, Brazil
e-mail: marcelo.araujo@ensp.fiocruz.br

J. C. Afonso
IQ/Department of Analytical Chemistry, Federal University of Rio de Janeiro, Centro de Tecnologia - Bloco A - Av. Pedro Calmon, S/N Cidade Universitária, Rio de Janeiro/RJ, Brazil CEP 21941-596
e-mail: julio@iq.ufrj.br

M. Jawaid and A. Khan (eds.), *Conversion of Electronic Waste in to Sustainable Products*, Sustainable Materials and Technology,
https://doi.org/10.1007/978-981-19-6541-8_7

thus concluding that this technology is not advantageous as a hydrometallurgical treatment.

Keywords Lithium-ion batteries · Metals recovery · Hydrometallurgy · Life cycle assessment

1 Introduction

Lithium-ion batteries are now widely used in portable equipment, photo cameras, electric vehicles, cell phones and other electrical and electronic equipment (EEE) because they have a long service life, low mass, reduced discharge rate and less toxicity compared to other types of batteries (nickel–cadmium, nickel metal-hydride, etc.). The development of new technological versions of these devices and their easier access to the population lead to an inevitable increase in the production and consumption of EEE and, consequently, of their batteries. Due to their high performance compared to other batteries, lithium-ion batteries dominate the portable energy market today (Kang et al. 2010a, b, Chen et al. 2011; Zeng et al. 2012; Wang et al. 2014).

Lithium-ion batteries have in their composition an anode, a cathode, current collectors, a separator, an electrolyte, an outer shell and sealing parts (Shin et al. 2005). The anode is composed of graphite, lithium, carbon and polyvinylidene fluoride (PVDF), which gives these components adhesion to the current collector, which is a copper blade; the electrolyte is typically lithium hexafluorophosphate ($LiPF_6$) in organic solvent, usually ethylene carbonate (CE), diethyl carbonate (DEC), dimethyl carbonate (DMC) or mixtures thereof. The separator is made of polypropylene (PP) or polyethylene (PE). The cathode is made of carbon, mixed $LiCoO_2$ oxide and PVDF, which fixes the components to the cathode current collector and an aluminum blade.

In contrast to the growing consumption, large quantities of waste of electrical and electronic equipment (WEEE) have been produced due to the rapid obsolescence of EEE. According to the GSMA (Group Managed Service Accounts) of UNU (United Nations University) (GSMA 2015) for cell phones, around 189,000 t were discarded worldwide in 2014, of which, approximately 17,000 t in Latin America. In 2014 the total amount of WEEE produced in the world was 3904,000 t. This means that worldwide, WEEE corresponding to cell phones represents less than 0.5% of the total mass of WEEE produced worldwide. However, batteries are a serious environmental problem if disposed of inappropriately. They have high concentrations of dangerous metals in addition to electrolytes. Its presence in landfills can lead to contamination of soil and groundwater (Wang et al. 2014).

The recycling of these batteries is a solution to eliminate the environmental risks arising from their improper disposal. However, until now, the processing of used lithium-ion batteries comes up against problems of high energy consumption, use of toxic and/or corrosive reagents, generation of large amounts of final waste and only moderate recovery of the metals of greatest interest in these batteries that are lithium and cobalt (Pranolo et al. 2010; Lee and Rhee 2002).

In general, leaching of lithium-ion batteries occurs in an acid medium together with a reducing agent (often H_2O_2). Trivalent cobalt is reduced to divalent, facilitating its dissolution. Cobalt is often isolated by fractional precipitation or liquid–liquid extraction, while lithium is often isolated by precipitation of a poorly soluble salt (LiF, Li_3PO_4 and Li_2CO_3) (Zeng et al. 2012).

In this research, based on the concepts of Life Cycle Assessment (LCA), metal recovery processes were investigated using Ecoinvent 3.0 database (Ecoinvent 2015), with the application of the ILCD (International Life Cycle Data System) impact assessment methodology. Hydrometallurgical residues incorporating the environmental impacts related to the recovery of cobalt and the residues generated in the processing of lithium-ion batteries of cellular devices. LCA addresses the environmental impacts of a product, including waste management and, eventually, the management of different waste treatment options. In this study, LCA was used in the decision-making as a tool for the management of electronic waste from cell phone batteries. In the recovery of Co and Li modeled in the LCA, the research activities sought to determine the environmental impacts generated associated with the recovery of cobalt and lithium metals from lithium-ion batteries transformed into waste.

2 Materials and Methods

2.1 Materials

In this research, 50 lithium-ion batteries were used in cell phones from three different manufacturers, produced between 2010 and 2014. The batteries were discharged and placed in a refrigeration chamber at 5 °C. After this procedure, the batteries were opened manually in a hood under exhaustion. The steel shield was removed and placed in a vacuum system for 20 min to remove and condensate the solvent present. After this procedure, it was possible to identify each internal battery component. The separation between them was carried out by mechanical vibration and scraping of the powder. The masses of the combined electroactive components (cathode, anode and electrolyte) were determined using an analytical balance. Figure 1 illustrates the mass of electroactive components after elimination of the solvent and manual separation.

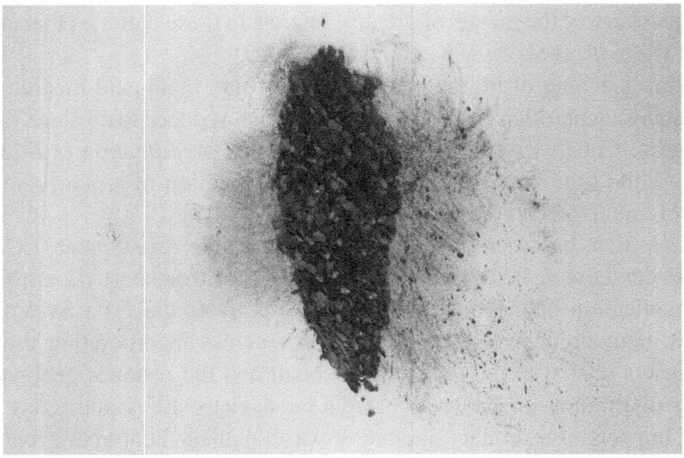

Fig. 1 Electroactive components isolated after elimination of the non-aqueous solvent and manual scraping

2.2 Functional Unit

1 kg homogenized material from the electroactive components of the lithium-ion batteries of cellular devices was used. Of this total weight, 250 g of electroactive components was isolated for each acid used.

2.3 Process Flow

In this chemical reprocessing study, 50 lithium-ion batteries from cell phones of three different brands were analyzed. 1 kg of the materials was recovered from these batteries.

2.4 System Boundaries

Hydrometallurgical treatment in the laboratory, the transport of waste to the industrial landfill and the final disposal of these in the landfill were modeled. The collection of cell phones and the transport of the recovered substances to the place of use were not included as an input in a new production process, as well as estimates of the infrastructure (facilities and equipment) needed to carry out the leaching.

2.5 Leaching LCA

A LCA software (SimaPro 8.3) was used, as well as a Life Cycle Inventory database (Ecoinvent 3.0), with the application of the ILCD impact assessment methodology. The recovery of cobalt and lithium metals in leachate and the disposal of waste in an industrial landfill (class I) was modeled.

The application of the LCA aimed to quantify the potential environmental impacts of the chemical leaching of the lithium-ion batteries of a cellular device on a laboratory scale. By determining the recovery of lithium and cobalt with the LCA software, it was possible to analyze the possible impacts, considering the specific international standards (ABNT 2009a, b). The four technological routes represented by their respective acids are compared (H_2SO_4, HCl, HF, HCOOH) in the leaching process, describing the recovery of lithium and cobalt from lithium-ion batteries, and forwarding their waste to the industrial landfill located in the municipality of Belford Roxo (Rio de Janeiro—Brazil).

3 Results and Discussion

3.1 General Composition of Batteries

Table 1 describes the average percentage composition (m/m) of the 50 lithium-ion batteries used in this research.

The electroactive mass (anode, cathode and electrolyte) corresponds to 40% of the total battery mass. In this mass, the presence of cobalt, lithium, aluminum and copper was detected in FRX as the most abundant metals. Other elements were identified in small amounts: calcium, nickel and phosphorus; the latter data suggest that the electrolyte is lithium hexafluorophosphate ($LiPF_6$). The non-aqueous solvent (propylene

Table 1 Average composition of lithium-ion batteries

Components	Percentages (m/m)
Cathode + anode + electrolyte*	40
External plastic housing	23
Steel shield	10.5
Non-aqueous solvent	4.5
Copper blade	8.2
Aluminum foil	6.1
Polymeric separator	5.1
Electrical contacts	2.6

* Carbon corresponds to 30–32% by weight of these components (12–13% by weight of the total sample)

carbonate, 1,3-dioxolane and dimethoxymethane) condensed after vacuum removal represents approximately 4.5% w/w of the stack composition. The composition data presented are in accordance with the values reported in the literature (Nazri and Pistoia 2003; Varela et al. 2002).

3.2 Application of LCA to Chemical Reprocessing of Cell Phone Batteries

3.2.1 Life Cycle Inventory (LCI)

The compilation of the LCI is directly linked to the formulation of the study that defines the objectives and scope of the acid routes constituted by primary and secondary LCA data studied on the acid leaching of lithium-ion batteries. This inventory analysis showed primary data collection in the chemistry laboratory to measure the inputs and outputs belonging to the chemical reprocessing system.

Table 2 presents, respectively, the values of inputs and outputs of processes based on the routes of acid leaching of sulfuric acids, hydrochloric acid and hydrofluoric acid and formic acid.

In Table 2, the highest concentration of cobalt and lithium in the leachate is present in route 2 with values of 0.003969 and 0.00048 kg. The LCI data related to H_2SO_4, HCl, HF, HCOOH acids come from the Ecoinvent database (Ecoinvent 2015).

3.2.2 Electric Power from Reprocessing

The electrical energy consumption estimated in the chemical reprocessing analysis of lithium-ion batteries was based on the Brazilian energy grid, thus, the existing energy matrix in SimaPro 8.3 was adjusted to the Brazilian energy reality. Secondary data for energy sources were obtained through the Ecoinvent bank (2013–2014). Table 3 shows the main sources of energy according to the national energy balance of 2014.

The electrical energy used in the chemical reprocessing of lithium-ion batteries is 0.21 kWh for the equipment (electronic scale, muffle, stove, and exhaust hood) for 3 h in active mode for 60 days.

3.2.3 Transport System

For the calculation of emission, the average distance of 52 km from UFRJ (Federal University of Rio de Janeiro) to the nearest industrial landfill was considered, considering round trip. Table 4 shows the modeling and transportation of lithium-ion battery waste considering the distance from the collection point to the destination of the waste.

Table 2 Inputs and outputs of routes 1, 2, 3 and 4 (H_2SO_4, HCl, HF and HCOOH)

Inputs Routes 1, 2, 3 and 4	Unit	Quantity
Batteries (electroactive components)	kg	1.00E+00
Acids on each route (H_2SO_4, HCl, HCOOH and HF)	kg	5.00E-03
Hydrogen peroxide on each route (H_2O_2) there is no for HCOOH (it is just water)	kg	5.00E-03
Inorganic material	kg	8.00E-04
Distilled water for dilution on each route (H_2O)	kg	4.00E-01
Outputs Route 1 (H_2SO_4)		
Cobalt	kg	3.20E-03
Lithium	kg	3.90E-04
Waste (electroactive components)	kg	1.50E-01
Effluents	kg	9.00E-02
Other materials (metals)	kg	1.70E-04
Output Route 2 (HCl)		
Cobalt	kg	3.96E-03
Lithium	kg	4.30E-04
Waste (electroactive components)	kg	1.60E-01
Effluents	kg	8.50E-02
Other materials (metals)	kg	1.90E-04
Output Route 3 (HF)		
Cobalt	kg	3.83E-03
Lithium	kg	4.30E-04
Waste (electroactive components)	kg	1.40E-01
Effluents	kg	1.00E-01
Other materials (metals)	kg	1.90E-04
Output Route 4 (HCOOH)		
Cobalt	kg	2.46E-03
Lithium	kg	2.80E-04
Waste (electroactive components)	kg	1.30E-01
Effluents	kg	1.10E-01
Other materials (metals)	kg	1.40E-04
Outputs—Waste	*Final destination*	
Type I—hazardous	Industrial landfill	
Type II—non-hazardous	Industrial landfill	
Type II—non-hazardous: scrap metal	Industrial landfill	

Table 3 Brazilian energy supply (EPE 2014)

Energy matrix	%
Hydroelectric	70.6
Natural gas	11.3
Biomass	7.6
Oil derivatives	4.4
Coal derivatives	2.6
Nuclear energy	2.4
Wind energy	1.1

Table 4 Data for modeling and transporting lithium-ion battery waste

Exit location	Arrival place	Distance km (round trip)	Means of transport
Solid waste laboratory IQ/UFRJ	Belford Roxo industrial landfill	52 km	Light commercial vehicle

After data collection at the solid waste laboratory at IQ/UFRJ, the SimaPro 8.3 software was used to model the data, being considered a functional unit of 1 kg.

3.2.4 Landfill Disposal Process

In this scenario, the residues from chemical reprocessing were disposed of directly at the industrial landfill. The materials disposed of in the industrial landfill are: external plastic housing, steel shielding, non-aqueous solvent, copper foil, aluminum foil, polymeric separator and electrical contacts.

3.3 Assessment of Environmental Impacts in the Life Cycle

Based on the results obtained in acid leaching, the most environmentally efficient procedure for recovering cobalt and lithium originating from obsolete cell phone batteries was chosen. The lower the environmental impact in analogy to the scenarios, the better the procedure developed for recovering cobalt and lithium.

According to the data obtained, the acids named for the method of assessing the potential impacts on the life cycle in leaching were sulfuric, hydrochloric, hydrofluoric and formic acids because they provide a high percentage of extraction of cobalt and lithium in acid leaching using lower concentrations. To choose the best acid among the four from the environmental point of view, an environmental assessment was carried out based on the LCA methodology of the four acids, adopting the environmental impact assessment methodology ILCD 2011 "Midpoint + V1.09/EC-JRC

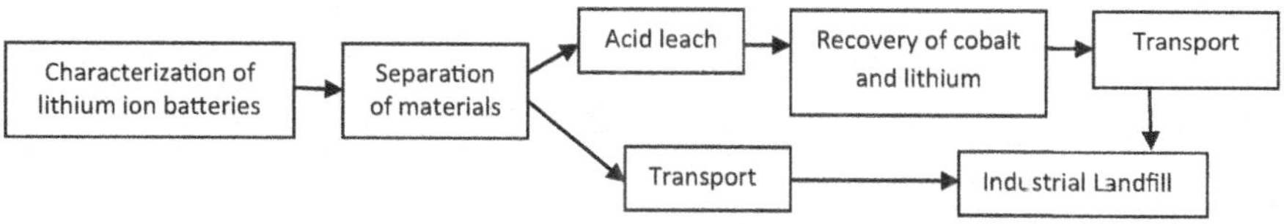

Fig. 5 Acid leach route

Global", equal weighting in the LCA software, where the environmental impacts of nine different categories are assessed.

In Fig. 5, the structure of the acid leaching system in the life cycle was demonstrated through the SimaPro 8.3 program, through the generation of a flowchart. Thus, for the formation of all cycles, data obtained in the solid waste laboratory at UFRJ were used, through chemical analysis. The environmental impacts were separated by acids (H_2SO_4, HCl, HF and HCOOH) and each acid generated a different scenario.

3.4 Potential Environmental Impacts of Acids on Acid Leaching

Initially, the data for each acid were used to compare the treatment routes in the acid leaching process, based on the evaluation of the impacts on the life cycle of lithium-ion batteries. The data used for the impact assessment were obtained through item 2.2, which represents the acid leaching process in the recovery of cobalt and lithium in the chemical reprocessing of batteries, thus calculating the potential impacts of the acids used in the respective chemical treatment. Table 5 presents the acid routes that determine the potential impacts of each acid across its categories.

It should be noted that the values in the impact categories for hydrogen peroxide are the same in all routes observed in Table 5, except for formic acid that already has a reducing characteristic, and does not require peroxide. The main reason why hydrogen peroxide has similar potential impact values is explained by the fact that it is used in the same amount for the three acid routes, being only an aid in reducing and recovering the metals used in acid leaching.

The chemical reprocessing routes can be compared with each other. The results indicate that the most relevant categories are climate change, ozone depletion and depletion of mineral and fossil resources.

In route 1, the values expressed for the climate change impact category in the analysis with sulfuric acid are responsible for generating 7.86E-04 kg CO_2 eq. In Routes 2 and 3, in the processes with hydrochloric and hydrofluoric acids present values of 8.12E-03 and 1.72E-02 kg CO_2 eq, being observed that Route 3 has a higher value than Route 2. Route 4 which represents leaching with formic acid, the value of 3.30E-02 kg CO_2 eq has the highest value in relation to other acids studied in chemical reprocessing in the present study.

Table 5 Comparison between the potential environmental impacts of acids used in acid leaching by the method ILCD 2011 Midpoint

Impact category	Route 1		Route 2		Route 3		Route 4	Unit
	H_2SO_4	H_2O_2	HCl	H_2O_2	HF	H_2O_2	HCOOH	
Climate changes	7.86E-04	5.86E-03	8.12E-03	5.86E-03	1.72E-02	5.86E-03	3.30E-02	kg CO_2 eq
Ozone depletion	3.30E-10	5.55E-10	3.57E-09	5.55E-10	2.64E-09	5.55E-10	4.20E-09	kg CFC-11 eq
Human toxicity, non-carcinogenic	2.02E-09	1.75E-09	3.24E-09	1.75E-09	1.55E-08	1.75E-09	6.16E-09	CTUh
Human toxicity, carcinogens	1.15E-10	1.73E-09	5.07E-10	1.73E-09	1.41E-09	1.73E-09	1.18E-09	CTUh
Particulate material	2.53E-06	4.51E-06	9.51E-06	4.51E-06	2.88E-05	4.51E-06	3.28E-05	kg PM2.5 eq
HH ionizing radiation	1.48E-04	5.71E-04	8.43E-04	5.71E-04	2.50E-03	5.71E-04	2.57E-03	kBq U235 eq
E ionizing radiation (interim)	8.56E-10	1.90E-09	3.16E-09	1.90E-09	1.04E-08	1.90E-09	1.30E-08	CTUe
Photochemical formation of ozone	9.00E-06	1.75E-05	2.27E-05	1.75E-05	8.97E-05	1.75E-05	8.92E-05	kg NMVOC eq
Acidification	4.25E-05	3.28E-05	5.36E-05	3.28E-05	3.30E-04	3.28E-05	2.17E-04	molc H + eq
Terrestrial eutrophication	2.55E-05	4.86E-05	8.17E-05	4.86E-05	2.85E-04	4.86E-05	2.80E-04	molc N eq
Freshwater eutrophication	1.01E-06	2.27E-06	4.58E-06	2.27E-06	1.29E-05	2.27E-06	1.07E-05	kg P eq
Marine eutrophication	2.33E-06	5.05E-06	8.44E-06	5.05E-06	2.72E-05	5.05E-06	2.71E-05	kg N eq
Freshwater ecotoxicity	3.73E-02	6.90E-02	7.73E-02	6.90E-02	3.24E-01	6.90E-02	1.58E-01	CTUe
Land use	5.47E-03	6.12E-03	8.58E-03	6.12E-03	4.39E-02	6.12E-03	6.46E-02	kg C deficit
Depletion of water resources	1.64E-05	3.28E-05	-1.46E-06	3.28E-05	5.50E-05	3.28E-05	1.88E-05	m^3 water eq
Depletion of mineral and fossil resources	4.34E-06	3.00E-07	5.06E-07	3.00E-07	5.28E-05	3.00E-07	7.87E-07	kg Sb eq

Ozone depletion in formic acid has the highest value in this impact category of 4.20E-09 kg CFC-11 eq, while sulfuric acid has a value of 3.30E-10 kg CFC-11, which is lower if purchased with the other routes, being considered the lowest value among the acids used. This category is related to the impacts associated with the decrease in ozone levels in the stratospheric layer of the atmosphere because of emissions of substances such as CFCs.

Route 4 that represents formic acid has high impact categories when compared to the other routes used in this acid leaching research. The high value represented in the mineral resource depletion category is related to the depletion of natural mineral resources. Route 3 has the lowest values in most of its impact categories. The increasing order of the potential impacts of acids in relation to the routes above: HF > HCOOH > HCl > H_2SO_4 (Hischier et al. 2007).

3.5 Potential Impacts on the Recovery of Cobalt and Lithium

The recovery of cobalt and lithium was calculated using SimaPro 8.3 (Table 6), and the categories of environmental impacts generated in this recovery were also evaluated. The evaluation of the impacts on the life cycle of chemical reprocessing serves to measure whether the treatment used is advantageous or not in the recovery of cobalt and lithium contained in lithium-ion batteries.

In Table 6, it was observed that most categories of potential environmental impacts presented negative values that correspond to the environmental credit in the recovery of these two metals in each technological route. The impact category of depletion of water resources had a positive impact for lithium recovery, which means that there was a greater expenditure of water in chemical reprocessing for lithium recovery.

The interpretation of the results of the characterization of the environmental impacts in 2011 does not allow us to visualize which are the most important processes in terms of the different environmental impacts. This can be done with the weighting of impacts in the ILCD Midpoint + V1.09/EC-JRC Global method, which converts impacts in the same unit—μPt through weighting factors.

3.6 Comparison of Potential Environmental Impacts

Table 7 presents the process-weighted results for the hydrochloric acid route. This result shows that transport to the landfill generates environmental impacts in the order of 10 of the total sulfuric acid routes for the distance of 52 km. However, it would practically cancel the environmental credit for recycling if the distance was increased to 520 km, emphasizing the relevance of the environmental impacts of logistical processes in recycling.

Table 6 Potential impacts on the recovery of cobalt and lithium through acid leaching by the ILCD 2011 Midpoint method

Impact Category	H_2SO_4		HCl		HF		HCOOH		Unit
	Cobalt	Lithium	Cobalt	Lithium	Cobalt	Lithium	Cobalt	Lithium	
Climate changes	−1.34E-01	−9.16E-02	−1.66E-01	−1.13E-01	−1.60E-01	−1.01E-01	−1.03E-01	−6.58E-02	kg CO_2 eq
Ozone depletion	−1.24E-08	−1.15E-08	−1.53E-08	−1.41E-08	−1.48E-08	−1.26E-08	−9.50E-09	−8.22E-09	kg CFC-11 eq
Human toxicity, non-carcinogenic	−5.89E-08	−3.20E-08	−7.27E-08	−3.94E-08	−7.04E-08	−3.53E-08	−4.53E-08	−2.30E-08	CTUh
Human toxicity, carcinogens	−7.71E-09	−7.87E-09	−9.51E-09	−9.68E-09	−9.21E-09	−8.67E-09	−5.92E-09	−5.65E-09	CTUh
Particulate material	−1.54E-04	−9.71E-05	−1.90E-04	−1.20E-04	−1.84E-04	−1.07E-04	−1.18E-04	−6.97E-05	kg PM2.5 eq
HH ionizing radiation	−1.22E-02	−1.05E-02	−1.51E-02	−1.30E-02	−1.46E-02	−1.16E-02	−9.38E-03	−7.56E-03	kBq U235 eq
E ionizing radiation (interim)	−4.87E-08	−3.71E-08	−6.01E-08	−4.57E-08	−5.82E-08	−4.09E-08	−3.74E-08	−2.67E-08	CTUe
Photochemical formation of ozone	−1.36E-03	−2.65E-04	−1.68E-03	−3.26E-04	−1.63E-03	−2.92E-04	−1.05E-03	−1.90E-04	kg NMVOC eq
Acidification	−1.77E-03	−6.25E-04	−2.19E-03	−7.70E-04	−2.12E-03	−6.89E-04	−1.36E-03	−4.49E-04	molc H+eq
Terrestrial eutrophication	−6.68E-03	−1.16E-03	−8.24E-03	−1.43E-03	−7.98E-03	−1.28E-03	−5.13E-03	−8.32E-04	molc N eq
Freshwater eutrophication	−5.58E-05	−5.68E-05	−6.89E-05	−6.99E-05	−6.67E-05	−6.26E-05	−4.29E-05	−4.08E-05	kg P eq
Marine eutrophication	−5.34E-04	−1.45E-04	−6.59E-04	−1.79E-04	−6.38E-04	−1.60E-04	−4.10E-04	−1.04E-04	kg N eq
Freshwater ecotoxicity	−1.30E+00	−7.30E-01	−1.61E+00	−8.99E-01	−1.55E+00	−8.05E-01	−1.00E+00	−5.24E-01	CTUe
Land use	−6.30E-01	−2.08E-01	−7.78E-01	−2.56E-01	−7.53E-01	−2.30E-01	−4.84E-01	−1.50E-01	kg C deficit
Depletion of water resources	−1.83E-05	7.61E-05	−2.25E-05	9.36E-05	−2.18E-05	8.39E-05	−1.40E-05	5.46E-05	m^3 water eq
Depletion of mineral and fossil resources	−4.38E-04	−3.91E-05	−5.40E-04	−4.82E-05	−5.23E-04	−4.32E-05	−3.37E-04	−2.81E-05	kg Sb eq

Table 7 Results of assessments of potential environmental impacts for the hydrometallurgical route with hydrochloric acid by the method ILCD 2011 Midpoint + V1.09/EC-JRC Global

Inputs/processes	Impact	(μPt)
Water	6.27E-02	−0.1
Other chemicals	2.04E+00	−2.5
Hydrochloric acid	6.32E+00	−7.9
Hydrogen peroxide	1.18E+01	−14.7
Transport	7.92E+00	−9.9
Electricity	2.22E+01	−27.7
Cobalt	−3.17E+02	395.3
Lithium	−1.11E+02	138.5
Industrial landfill	2.97E+02	−370.9
Total	−8.01E+01	100.0

The table shows that the most relevant environmental impacts are related to the recovery of Co and Li (credits of 395.3 and 138.5) and the disposal of waste and effluents from recycling in an industrial landfill (impact of 370.9).

Table 8 presents the results for the four acid routes by the ILCD Midpoint methodology in a single Pt score. The results of the evaluation of the potential environmental impacts of the recovery processes of Cobalt and Lithium by hydrometallurgical routes by different weighted acids allow to be analyzed in different ways.

Initially, if the total values are compared, it appears that the route with hydrochloric acid has greater environmental credit (negative values), therefore it is the one that allows greater reduction of environmental impacts with the recovery of cobalt and lithium. It is noticed that the carcinogenic human toxicity category has the highest impact values for all four acid routes, almost equaling the values of environmental credits (negative in the mineral resource depletion category) related to the recovery of the two metals. The formic acid route has no negative environmental impact (environmental credit), which means that this recovery process is not environmentally indicated. It should be noted that the route through sulfuric acid represents 18 of the routes through hydrochloric acid and hydrofluoric acid 38, which are not very relevant differences (Hischier et al. 2005).

3.7 *Sensitivity Analysis*

The objective was to perform a sensitivity analysis to assess the veracity of the results obtained in the study. In review, the evaluation already carried out was decided to address only the energy parameters in the stages of chemical reprocessing simulated in the sensitivity analysis. The Brazilian Grid and the global Grid and their respective impact categories were compared through their weighting.

As can be seen, in Table 9, the differences in relation to the impact categories depending on the type of electrical energy used. Comparing the Brazilian grid with

Table 8 Results of assessments of potential environmental impacts in a single score by the ILCD 2011 Midpoint + V1.09/EC-JRC Global method

Impact categories (μPt)	Sulfuric	Formic	Hydrochloric	Hydrofluoric
Climate changes	−1.65E-01	8.54E-01	−3.20E-01	−1.11E-01
Ozone depletion	−3.47E-02	4.59E-02	−1.61E-02	−1.25E-02
Human toxicity, non-carcinogenic	−1.30E+01	3.94E-01	−1.85E+01	−1.09E+01
Human toxicity, carcinogens	1.68E+02	1.85E+02	1.55E+02	1.62E+02
Particulate material	2.64E-02	1.41E+00	−3.75E-01	5.84E-02
HH ionizing radiation	−2.55E+00	2.62E-01	−3.07E+00	−2.16E+00
Photochemical formation of ozone	−9.81E-01	−2.75E-03	−1.18E+00	−9.81E-01
Terrestrial eutrophication	−2.09E+00	−9.16E-01	−2.46E+00	−2.23E+00
Freshwater eutrophication	−1.25E-01	2.57E-01	−3.26E-01	−1.63E-01
Marine eutrophication	−9.31E-01	−3.66E-01	−1.09E+00	−9.77E-01
Freshwater ecotoxicity	−1.71E+00	1.05E+01	−6.62E+00	−2.30E-01
Land use	−2.85E-05	4.56E-03	−1.00E-03	−1.05E-04
Depletion of water resources	3.40E-01	2.87E-01	3.31E-01	3.75E-01
Depletion of mineral and fossil resources	−1.60E+02	1.22E+02	−2.00E+02	−1.74E+02
Total	−1.45E+01	7.55E+01	−8.01E+01	−3.07E+01

the global grid, it can be noted that the impact categories vary by 25 in relation to energy expenditure.

The European electricity grid has higher values in all its categories compared to the Brazilian grid. This variation occurs due to the difference in the energy sources of these regions. In Brazil, there is a large share of renewable energy sources such as hydroelectric power plants.

Therefore, there is relevance of electricity for the decision to recycle Co and Li recovery from cell phone batteries in hydrometallurgical routes with the acids used, depending on the specificity of the local grid.

To analyze the result with a different methodology, the four acid leaching routes were modeled with the application of the ReCiPe End Point Methodology methodology. It is an impact assessment methodology aimed directly at damage, which may have different results from the ILCD MidPoint methodology, which is focused on problems. The results using the ReCiPe methodology confirm the interpretation of the previous item by the ILCD methodology, as can be seen in Table 10.

It can be noted that the hydrochloric acid route is the one that generates the least environmental impact. It is also observed that the environmental credits from the recovery of cobalt and lithium are supplanted by other impacts of resource depletion of the processes necessary for this treatment in all four routes. Finally, as all routes have positive total environmental impacts, there are not enough environmental credits

Table 9 Comparison of sensitivity analysis of potential impacts between the Brazilian and global grid for modeling sulfuric acid

Impact Categories		Hydrochloric Acid	
		Brazil	Europa
Climate changes	kg CO_2 eq	−3.40E-02	2.53E+00
Ozone depletion	kg CFC-11 eq	−2.94E-09	1.41E-07
Human toxicity, non-carcinogenic	CTUh	−4.30E-08	7.88E-07
Human toxicity, carcinogens	CTUh	2.87E-08	1.32E-07
Particulate material	kg PM2.5 eq	−2.85E-05	2.74E-03
HH ionizing radiation	kBq U235 eq	−1.11E-02	4.04E-01
E ionizing radiation (interim)	CTUe	−3.67E-08	1.31E-06
Photochemical formation of ozone	kg NMVOC eq	−7.99E-04	5.17E-03
Acidification	molc H+ eq	−1.27E-03	1.47E-02
Terrestrial eutrophication	molc N eq	−6.04E-03	1.60E-02
Freshwater eutrophication	kg P eq	−3.20E-05	1.51E-03
Marine eutrophication	kg N eq	−4.98E-04	1.80E-03
Freshwater ecotoxicity	CTUe	−3.71E-01	3.66E+02
Land use	kg C deficit	−7.81E-02	1.42E+00
Depletion of water resources	m^3 water eq	3.42E-04	−2.29E-3
Depletion of mineral and fossil resources	kg Sb eq	−5.78E-04	1.47E-04

Table 10 Results of the LCA for different acids in the leaching of Co and Li batteries by the ReCiPe Endpoint methodology

Impact Category (mPt)	Sulfuric	Formic	Hydrochloric	Hydrofluoric
Human health	−4.84E-01	5.70E+00	−2.09E+00	−5.26E-01
Ecosystems	1.02E+00	1.47E+00	9.18E-01	9.98E-01
Resources	8.63E+00	1.60E+01	7.88E+00	9.61E+00
Total	9.16E+00	2.32E+01	6.72E+00	1.01E+01

in recovery to recommend hydrometallurgical treatment for these routes (Ekvall and Weidema 2004).

4 Conclusions

The leaching of lithium and cobalt from lithium-ion batteries in an acid medium showed that the weak acids tested showed recovery comparable (HF) or lower (HCOOH) to those obtained with traditional strong acids (HCl and H_2SO_4). The presence of H_2O_2 is essential for the good conduct of the process, being in this respect a better Co(III) reducer than formic acid.

The increase in time and temperature favored leaching, reaching a constant level with little energy expenditure (~40 °C) after 3 h of reaction for all acids. The insoluble fraction after leaching is mainly composed of carbon (graphite) from the cathode of the batteries. The four processes tested generated similar residues.

In the results of environmental impacts by the ILCD MidPoint LCA methodology, the hydrochloric acid leach route had the least environmental impact among the other acid leach routes for the recovery of cobalt and lithium. However, it should be noted that the difference between the two extremes is of the same order of magnitude as the electricity grid, transport over long distances.

The importance of the disposal process in an industrial landfill, which presents relevant environmental impacts, for the category of human carcinogenic toxicity is emphasized. It is because the impacts are calculated from an average of the waste disposed of in an industrial landfill according to the Ecoinvent dataset. However, not necessarily, the residues of these recovery processes by hydrometallurgy are effectively toxic as the average of the disposed residues.

The study points to the need to reassess the model of mineral exhaustion, which considers that it does not consider that the demand for cobalt and lithium will increase appreciably in the coming decades due to the increasing use of batteries for the accumulation of alternative renewable energies that are not constant, such as wind power and solar panels, requiring accumulation batteries that are currently produced with cobalt and lithium.

It is worth noting that the data obtained are for procedures developed in the laboratory, and it is expected through this research a likely decrease in the environmental impacts related to the recovery of metals in relation to the reference routes.

It can be concluded that the LCA tool is an appropriate and relevant tool in the decision-making process of recycling, collaborating to obtain the least possible environmental impact.

Acknowledgements To CNPq (Conselho Nacional de Desenvolvimento Científico e Tecnológico), CAPES (Coordenação de Aperfeiçoamento de Pessoal de Nível Superior) and FAPERJ (Fundação de Amparo à Pesquisa do Estado do Rio de Janeiro) for the financial support. To the Brazilian Mineral Technology Center (CETEM) for the X-ray diffraction analysis. To the Institute of Nuclear Engineering (IEN) for the X-ray fluorescence analysis.

References

ABNT (2009a) NBR ISO 14040: Gestão Ambiental – Avaliação do ciclo de vida – Princípios e estrutura. Rio de Janeiro

ABNT (2009b) NBR ISO 14044: Gestão Ambiental – Avaliação do ciclo de vida – Requisitos e orientações. Rio de Janeiro

Chen L, Tang X, Zhang Y, Li L, Zeng Z, Zhang Y (2011) Process for the recovery of cobalt oxalate from spent lithium-ion batteries. Hydrometallurgy 180(1–2):80–86

Ecoinvent (2015) About us. http://www.ecoinvent.org. Accessed Sept 2015
Ekvall T, Weidema BP (2004) System boundaries and input data in consequential life cycle inventory analysis. Int J Life Cycle Assess 9(3):161–171
EPE (Energy research company) (2014) Brazilian Energy Balance 2014. Brasília. Available at: https://www.epe.gov.br/en/publications/publications/brazilian-energy-balance/brazilian-energy-balance-2014 (accessed 10.09.2022)
GSMA (Group Managed Service Accounts) (2015) eWaste in Latin America: statistical analysis and policy recommendations. United Nations University, UNU-IAS Institute for the Advanced Study for the Sustainability
Hischier R, Waeger P, Gauglhofer J (2005) Does WEEE recycling make sense form an environmental perspective? The environmental impacts of the Swiss take-back and recycling systems for waste electrical and electronic equipment. Environ Impact Assess Rev 25:525–539
Hischier R, Classen M, Lehmann M, Scharnhorst W (2007) Life cycle inventories of electric and electronic equipment: production, use and disposal. Data v 2.0, Ecoinvent report no.18. Swiss Centre for Life Cycle Inventories. EMPA TSL, Dübendorf, Swiss
Kang JG, Sohn G, Chang H, Senanayake G, Shin SM (2010a) Recovery of cobalt sulphate from spent lithium-ion batteries by reductive leaching and solvent extraction with Cyanex 272. Hydrometallurgy 100:168–171
Kang JG, Sohn G, Chang H, Senanayake G, Shin SM (2010b) Preparation of cobalt oxide from concentrated cathode material of spent lithium-ion batteries by hydrometallurgical method. Adv Power Technol 21(2010b):175–179
Lee CK, Rhee KI (2002) Preparation of LiCoO from spent lithium-ion batteries. J Power Sources 109:17–21
Nazri GA, Pistoia O (2003) Lithium batteries: science and technology. Kluwer Academic Publishers
Pranolo Y, Zhang W, Cheng CY (2010) Recovery of metals from spent lithium-ion battery leach solutions with a mixed solvent extractant system. Hydrometallurgy 102(1–4):3742
Shin MS, Kim NH, Sohn JS, Yang DH, Kim YH (2005) Development of a metal recovery process from ion battery wastes. Hydrometallurgy 79:172–181
Varela FH, Huguenin M, Malta R, Torresi M (2002) Materiais para cátodo de baterias secundárias de lítio. Quimica Nova 25(2):287–299
Wang X, Gaustad G, Babbitt CW, Bailey C, Ganter MJ, Landi BJ (2014) Economic and environmental characterization of an evolving Li-ion battery waste stream. J Environ Manag 135:126–134
Zeng GS, Deng XR, Luo SL, Luo XB, Zou JP (2012) A copper-catalysed bioleaching process for enhancement of cobalt dissolution from spent lithium-ion batteries. J Hazard Mater

A Bibliometric Approach to the Current State of the Art of Risks in E-waste Supply Chains

Pedro Senna, Lino Guimarães Marujo, Augusto da Cunha Reis, and Ana Carla de Souza Gomes dos Santos

Abstract E-waste management is becoming a very challenging subject since it is very multidisciplinary, encompassing concepts such as circular economy, closed-loop supply chains, supply chain risk management, and supply chain resilience. These pillars must be strongly supported by a huge amount of quality data, therefore, opening an important interface with the concept of smart cities. Among the main challenges, is the need to motivate the customers to collaborate, creating a culture of reusing and recycling end-of-life products. In addition, it is crucial to develop reverse recycling channels suitable to each client and enabled by an effective logistics network design. Despite the relevance of this topic, many significant gaps can be pointed out, since there are still few relevant papers. Thus, there is a substantial necessity of understanding what have been done by scholars to establish the state of the art. In this sense, this paper has the objective of mapping the state of the art of e-waste management through a bibliometric study. Among the main results, we highlight the mapping of the state of the art composed of the literature statistics.

Keywords Bibliometrics · E-waste management · E-waste supply chains · Supply chain management

---

P. Senna (✉) · A. da Cunha Reis
Federal Centre for Technological Education Celso Suckow da Fonseca (CEFET/RJ), Postgraduate Program in Production Engineering and Systems (PPPRO), Rio de Janeiro/RJ, Brazil
e-mail: pedro.senna@cefet-rj.br

A. da Cunha Reis
e-mail: augusto.reis@cefet-rj.br

P. Senna · L. G. Marujo
Production Engineering Program, PEP/COPPE-UFRJ, Rio de Janeiro/RJ, Brazil
e-mail: lgmarujo@ufrj.br

A. C. de Souza Gomes dos Santos
Industrial Production Management Department and Graduate Program in Production Engineering and Systems (PPPRO), Federal Institute of Education, Science, and Technology of Rio de Janeiro (IFRJ) and Federal Centre for Technological Education Celso Suckow da Fonseca (CEFET/RJ), Rio de Janeiro/RJ, Brazil
e-mail: ana.carla@ifrj.edu.br

M. Jawaid and A. Khan (eds.), *Conversion of Electronic Waste in to Sustainable Products*, Sustainable Materials and Technology,
https://doi.org/10.1007/978-981-19-6541-8_8

1 Introduction

E-waste is increasingly becoming a crucial issue for managers and all societies, motivated by a considerable increase in demand for manufactured products and incorrect discarding (Cai and Choi 2021). The management of hazardous electronic wastes contributes to reducing environmental risks and provides an alternative source of scarce raw materials. In this sense, research on the topic has been growing and the idea is being associated with circular economy and supply chain management.

Among the main challenges, is the urge to convince the customers to collaborate, creating a culture of reusing and recycling end-of-life products. In addition, each client is used to different types of channels, in this sense, it is crucial to create reverse recycling channels suitable to each client and enabled by effective logistics network design.

In order to address an effective e-waste management, managers must consider some important building blocks such as (i) circular economy, (ii) supply chain management, (iii) supply chain risk management, (iv) supply chain resilience, and (v) closed-loop supply chains.

The most relevant challenges considering e-waste management include (i) lack of infrastructure to collect and recycle discarded consumer goods (Tansel 2020) and (ii) pressures related to environmental agencies, government, and original equipment manufacturers to adopt, develop, and innovate environment-friendly e-waste mitigation strategies (Garg 2021).

Although there is a considerable variety of studies, to the best of our knowledge, no study establishes the state of the art considering e-waste management. In this regard, this paper has the objective of mapping state of the art of e-waste management through a bibliometric study. Moreover, this study aims to answer four research questions: (RQ1) Which are the leading research countries considering e-waste management? (RQ2) Which are the most relevant sources and main topics covered by the most cited papers? (RQ3) Which are the main concepts associated with e-waste management? (RQ4) Which are the main research gaps considering e-waste management?

In addition to this introductory section, Sect. 2 presents a brief literature review, Sect. 3 presents the methodology, Sect. 4 presents the bibliometric results that answer each of the RQs, Sect. 5 presents the discussion, and Sect. 6 closes the paper with the conclusions, limitations, and further research opportunities.

2 Literature Review

2.1 E-Waste Management

E-waste is a major environmental issue and can be minimized by motivating values recovery from the waste stream through reverse supply chains (Rahman and Subramanian 2012). In a very challenging reality including scarce resources, rising costs, and severe environmental hazards, companies must begin to understand returns as value streams and maximize the revenue from proper refurbishment, and prompt resale through the appropriate channels (Guide et al. 2005). In the last decades, e-waste has been treated mainly as a cost factor in production (Zoeteman et al. 2010). Regarding e-waste, there is an academic consensus that landfill of waste electronic and electric equipment is not an acceptable end-of-use management option (Geyer and Blass 2010).

Moreover, businesses must understand that, in addition to selling products, understanding and providing solutions to the customers' needs (Mont et al. 2006). Guide et al. (2005) identified five critical processes which form the backbone of reverse the supply chain: (i) product acquisition, (ii) reverse logistics, (iii) inspection and disposition, (iv) refurbishment or remanufacturing, and (v) sales and marketing. To deal with this new reality, managers must discuss changes in product design and the supply chain (Mont et al. 2006). A proactive approach concerning sustainable supply chain strategies maximizes resource productivity and improves industrial and environmental performance (Kumar et al. 2012).

Extended producer responsibility (EPR) holds the producer responsible for the environmental impacts of end-of-life products and is increasingly becoming an important strategy to deal with e-waste (Jacobs and Subramanian 2012). Nevertheless, Jacobs and Subramanian (2012) highlight that an effective solution regarding e-waste must involve all supply chain stakeholders and well-designed reverse channels. Additionally, Zoeteman et al. (2010) affirm that the original equipment manufacturers, instead of legislators, must be in the "driver seat".

Circular economy (CE) requires managers to redesign their supply chains and business models (Lüdeke-Freund et al. 2019). Moreover, CE can be considered an ongoing process to achieve greater resource efficiency and effectiveness (Lüdeke-Freund et al. 2019). Closed-loop supply chain (CLSC) and reverse logistics (RL) are essential parts of the waste management process (Islam and Huda 2018). Using the reverse channels effectively is an important strategy to increase the amount of materials recovered from the waste stream and minimize environmental damage (Rahman and Subramanian 2012). A reverse logistics network may be either: (i) a closed loop—sources (origin) and sinks (destination) coincide and flow form cycles in a given supply chain; or (ii) an open loop system—flows enter a company and leave at another (Rahman and Subramanian 2012). Implementing e-waste CE requires closing the loops, which requires the design of reverse channels that is supported by good quality data. In this sense, industry 4.0 technologies provide a strong support backbone to implement such concepts (Nascimento et al. 2019). Nevertheless, closing the loops in the e-waste supply chain is not an easy challenge and must be supported by a cultural change in order to be long-lasting and profitable (Kumar et al. 2012).

2.2 Supply Chain Challenges

The last 20 years consolidated supply chain constructs such as internal and external process integration (Croxton et al. 2001), supply chain modeling (Saen et al. 2016; Lehoux et al. 2016; De Meyer et al. 2016), and key performance indicators (Cai et al. 2009). In the early 2000s, supply chain management was defined as integration and coordination of key supply chain business processes going through all suppliers and manufacturers, to end-user providing products, services, and information for customers and all stakeholders (Croxton et al. 2001; Lambert and Cooper 2000). By understanding a supply chain as a set of processes built the path to understand that some risks have the potential to generate hazards to a whole supply chain. Bowersox et al. (1999) recommended Lean Launch of products to mitigate the risks of higher inventory levels that a push strategy would generate. Lonsdale (1999) presented outsourcing practices as a means to mitigate risks Zsidisin et al. (2000) studied inbound supply risks such as quality, design, cost, availability, manufacturability, supplier, legal, environmental, health, and safety. Hallikas et al. (2002) presented the assessment of networking risks in a network composed of a supplier and a buying company. In addition, Hallikas et al. (2004) expanded the study of Hallikas et al. (2002) presenting methods for risk management in more complex network environments. At this point, the backbone of the supply chain risk literature was complete, which led the scholars like Juttner et al., Norrman and Jansson (2004) to tailor the concept of Supply Chain Risk Management (SCRM).

A concept which is intrinsically connected to SCRM is supply chain resilience (SCRes).

Shuai et al. (2011) affirm that managers must emphasize resilience generation to prevent the vulnerability of supply chain systems. Supply chain resilience can be understood as an adaptive capability of the supply chain to prepare for undesirable events, respond to disruptions, and recover from them while maintaining continuity of operations at the desired level of connectedness and control over structure and function (Ponomarov and Holcomb 2009). Additionally, SCRes is concerned with the system's ability to return to its original state or a new, more desirable one after experiencing a disturbance, and avoiding the failure modes (Carvalho and Cruz Machado 2007) and it is considered a matter of survival for supply chains Barroso et al. (2011) and Carvalho et al. (2012). Additionally, SCRes can react to the adverse effects brought by disturbances that occur in a given moment to maintain SC's objectives (Barroso et al. 2010).

In this sense, the last decade consolidated academic understanding of supply chain risk management and supply chain resilience. Nevertheless, studies that discuss the importance of creating a resilience culture are still lacking. Christopher and Peck (2004) propose the following principles to design resilient Supply Chain Collaboration, Supply Chain Agility, Supply Chain Visibility, and SCRM Culture.

Concepts such as circular economy and closed-loop supply chains are connected to SCRes, since they present ways of generating resilience for all society. There are human factors that help building an SCRes culture: collaboration, mutual trust, openness, communication, training, learning, and good customer service (Senna et al. 2020a).

In this sense, supply chains must be resilient in order to close the loops, expanding the resilience culture to motivate end-users to collaborate with e-waste returns. The creation of this culture benefits the companies, that recover valuable materials, tax incentives from the government, and avoid the discard of hazardrous materials in the environment.

The creation of a resilience and a recycling culture is important to involve end-users that already spend considerable amounts of time in traffic and working. With the several lockdowns to slow down COVID-19 infections, people and material flows decreased drastically, nevertheless, with the resumption of most activities, the traffic increases and once again become a considerable amount of time in the day of an average worker (Senna et al. 2020b). In this sense, reverse flows must be designed to offer advantages to those customers that are willing to return end-of-life (EOL) products. The motivation must be created by a joint action of all stakeholders, including government, that may create incentives to companies that effectively draw in the clients to a recycling culture.

2.3 Bibliometry

Bibliometry is an important technique that helps to identify the current state of the art of a given research stream. Bibliometry can be defined as the collection, handling, and analysis of quantitative bibliographic data from scientific publications (Verbeek et al. 2002). Ardito et al. (2019) argue that bibliometric analysis is fundamental to analyze the connections from the citations of articles in research areas. Bibliometry encompasses the identification of authors, publications, journals (Wu and Wu 2017), and the co-citation of documents (Fahimnia et al. 2015; Appio et al. 2016). Bibliometrics can be used to assess network connections and relevance of a research stream applying the network theory (Liu et al. 2015; Silva et al. 2020) and identifying citations about similar research areas (Hjørland 2013). Therefore, bibliometric analysis connects publications that attribute development to the research field under analysis (Di Stefano et al. 2010).

3 Methodology

Our research methodology is presented in the workflow seen in Fig. 1.

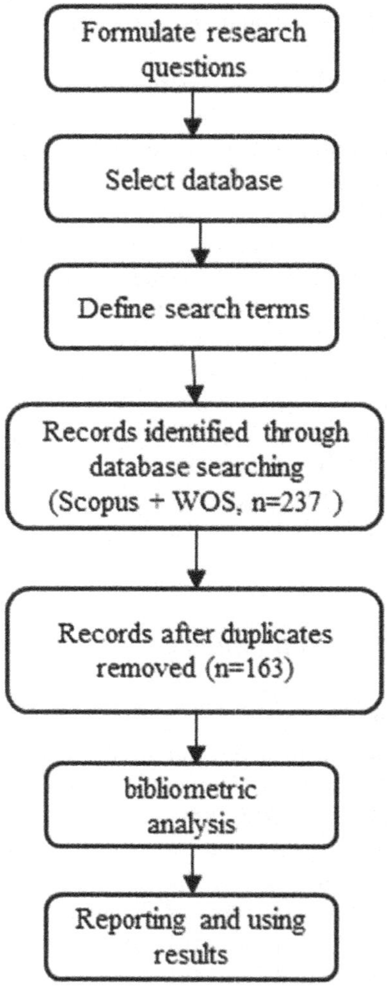

Fig. 1 Research methodology

3.1 Formulate Research Question

The research questions must be clear and highlight an objective research stream, in this sense, this paper proposes four research questions: RQ1) Which are the leading research countries considering e-waste management? RQ2) Which are the most relevant sources and main topics covered by the most cited papers? RQ3) Which are the main concepts associated with e-waste management? RQ4) Which are the main research gaps considering e-waste management?

Table 1 Search terms

STR	Strings	Scopus	WOS
STR1	*(E-waste AND "supply chain risk management") OR ("eletronic waste" AND "supply chain risk management")*	0	0
STR2	*Refurbishment AND "supply chain risk management"*	0	0
STR3	*Refurbishment AND "supply chain risk"*	1	0
STR4	*(E-waste AND "supply chain") OR ("eletronic waste" AND "supply chain")*	97	102
STR5	*Refurbishment AND "supply chain"*	29	25

3.2 Database Selection

We chose Scopus and Web of Science databases in this study, Scopus database is the largest searchable citation and abstract source, while WOS was the only citation database and publication covering all science domains for many years (Chadegani et al. 2013). Both bases are among the most relevant scientific databases in the world (Wang et al. 2019). Vieira and Gomes (2009) affirm that about 2/3 of the studies can be found in both databases and 1/3 only in one database. Furthermore, both databases provide ". bib" files providing full search data (references, article authors, etc.) that can be further analyzed in bibliometric packages. In addition, Scopus and WOS as search engines allow locating papers of Science Direct, Taylor and Francis, Emerald, Springer, and other important bases. The search covered January 2005–July 2021.

3.3 Search Terms Definition

3.4 Bibliometric Analysis

This paper conducted a bibliometric analysis to map the state of the art of this segment. In this sense, we formulated four objective research questions that are answered with bibliometric statistics and discussion regarding well-cited papers.

4 Results of Bibliometric Study

4.1 (RQ1) Which are the Leading Research Countries Considering E-waste Management?

This RQ aims to investigate the leading research countries on the topic. Our criteria to rank the countries is the number of papers, which constitutes scientific evidence of the importance that the topic is being given in each country. Figure 2 shows the paper distribution along the years.

Between 2005 and 2015, the number of papers oscillated. However, from 2016 to 2020, we see a clear tendency of publication growth. It is important to highlight that the bibliographic Search did not include the full year 2021. Figure 3 shows the most productive countries considering paper production.

China and Índia outstand in terms of paper production. Both countries comprise about 35% of the world's population and, therefore, a considerable amount of e-waste production.

In 2019, China produced 10,129 kt of e-waste, having a 16% of collection rate. India produced 3,230 kt recovering only 1% of e-waste (Baldé et al. 2019). The data shows that both countries are responsible for a relevant share of the world's e-waste,

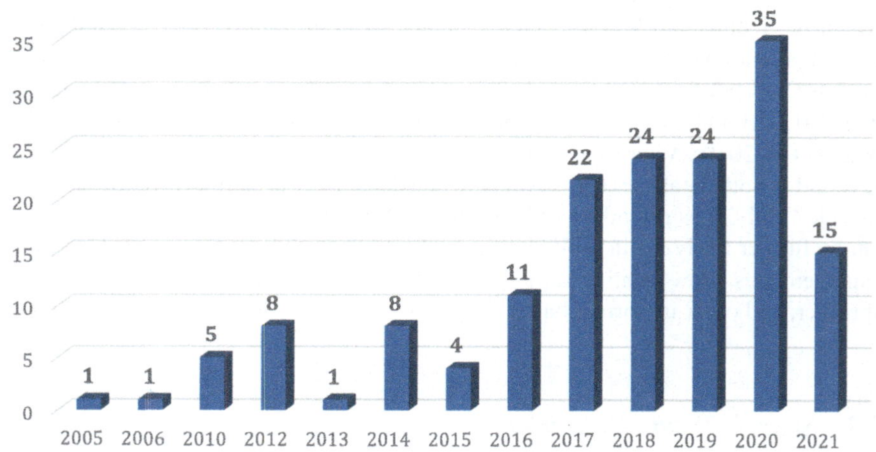

Fig. 2 Paper distribution

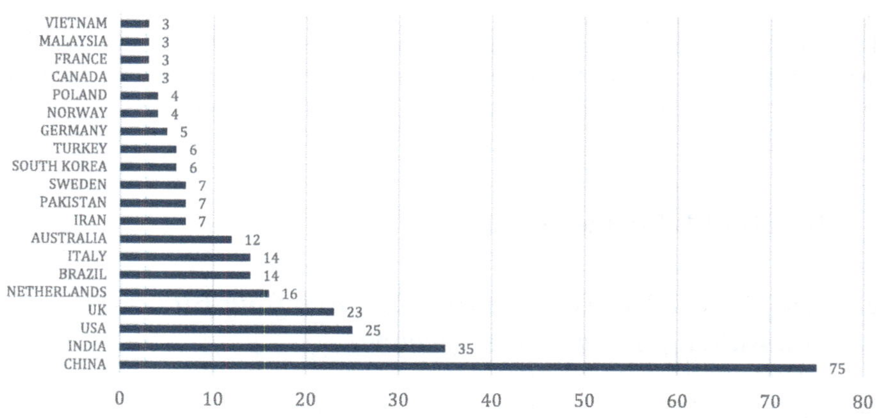

Fig. 3 Country's production

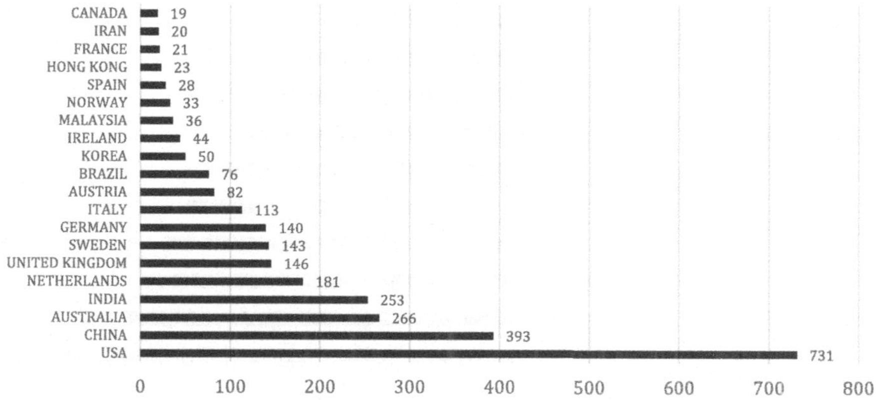

Fig. 4 Countries' citations

in this sense, researchers from both countries are leading paper production as shown in Fig. 3.

Figure 4 shows the most cited countries. We note a significant change in the rank regarding the number of citations. The USA appears second in Fig. 3 and first in Fig. 4. We also highlight Australia, in which 12 publications led to 266 citations.

4.2 *(RQ2) Which are the Most Relevant Sources and Main Topics Covered by the Most Cited Papers?*

This research question investigates the topics covered by the most cited papers. Considering the combination of e-waste management and supply chain management, the Journal of Cleaner Production is the most relevant source (22 papers) as shown in Fig. 5.

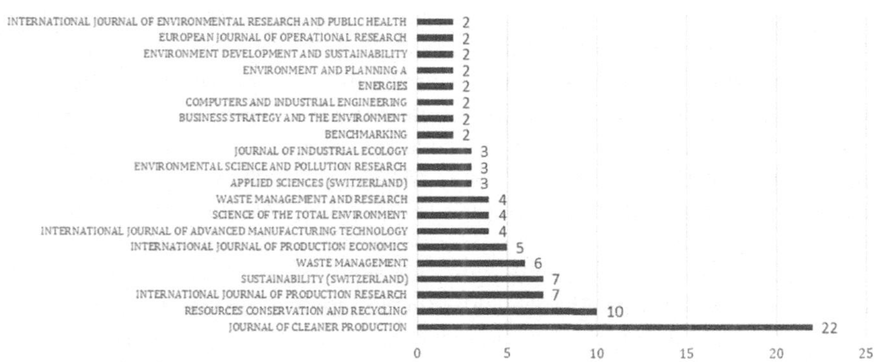

Fig. 5 Most relevant sources

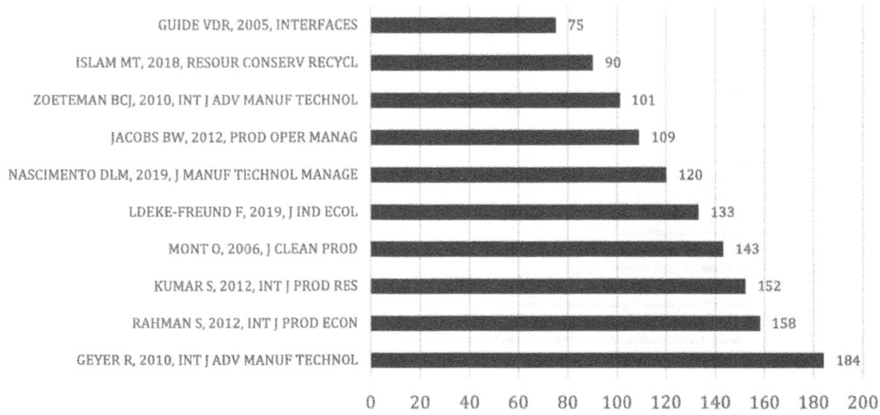

Fig. 6 Most cited papers

Figure 6 shows the ten most cited papers.

The papers are 1–Hewlett-Packard Company Unlocks the Value Potential from Time-Sensitive Returns, 2–Reverse logistics and closed-loop supply chain of Waste Electrical and Electronic Equipment (WEEE)/E-waste: A comprehensive literature review, 3–Handling WEEE waste flows: on the effectiveness of producer responsibility in a globalizing world, 4–Sharing Responsibility for Product Recovery Across the Supply Chain, 5–Exploring Industry 4.0 technologies to enable circular economy practices in a manufacturing context A business model proposal, 6–A Review and Typology of Circular Economy Business Model Patterns, 7–A new business model for baby prams based on leasing and product remanufacturing, 8–A green supply chain is a requirement for profitability, 9–Factors for implementing end-of-life computer recycling operations in reverse supply chains, and 10–The economics of cell phone reuse and recycling. Table 2 shows a brief analysis of these papers.

We organized the analysis by highlighting the main approach/technique, the main literature constructs mentioned by the papers, the variables that build these constructs, and the main findings. As the main findings, we note that Kumar et al. (2012) and Guide et al. (2005) provide roadmaps and processes that can be further investigated, detailed, and tested in case studies.

All of these papers present unique contributions, providing different milestones to this research field. This set of papers forms a timeline where Guide et al. (2005) conduct a real case study providing a guide for other companies and scholars to implement circular supply chain measures. Mont et al. (2006) carry a study on baby prams, a product with concise usage, but very durable, with untamed potential for EOL extension. The authors discuss barriers to the implementation of a circular supply chain concerning product design. This is a very significant result, meaning that product design affects all stakeholders in the supply chain and the capability of reusing and remanufacturing a given product, affecting the whole supply chain and the environment. In this sense, product design should have a more holistic approach. Zoeteman et al. (2010) go further and broadens the stakeholders whole in waste

Table 2 Brief analysis of the 10 most cited papers

Reference	Approach	Literature construct	Variables	Main findings
Guide et al. (2005)	Mathematical Programming	Circular supply chain	Refurbishment, resale channels, remarketing, reverse flows, reverse logistics, inspection, remanufacturing	The authors used a mathematical programming model to recover value from the returned products The paper constructs a roadmap to reverse SC: (i) Treat returns as a value stream, not a waste stream (ii) Consider the entire reverse supply chain (iii) Identify and develop the right performance metrics using simple visual scorecards (iv) Construct simple models based on the right information (v) Use the models to analyze the economic impact of alternative designs and operational policies (vi) Align the organizational structure and the reward systems to motivate economic profit from the reverse supply chain
Mont et al. (2006)	Conceptual design for supply chain	Circular supply chain	Reverse logistics, refurbishment, remanufacturing, Product redesign for reconditioning	The study presents a case study on baby prams. It discusses potential barriers and necessary changes in product design and the supply chain to make it work The new model may provide customers with a high-quality pram in a "like new" condition for the period they need it

(continued)

Table 2 (continued)

Reference	Approach	Literature construct	Variables	Main findings
Zoeteman et al. (2010)	Semi-structured interviews, surveys, and workshops. And literature analysis	E-waste management	Legislating, promoting extended producer responsibility, recovery system for e-waste, recycling	The authors argue that putting the original equipment manufacturers, instead of legislators in the "driver seat", will strengthen the opportunities for high-level recovery
Geyer and Blass (2010)	Mathematical analysis	Circular supply chain	Voluntary or mandatory take-back, collection programs, recycling, reverse logistics, return incentive, shipping, inspection, sorting	The objective of the paper is to investigate and quantify the performance of the reverse logistics, reuse, and recycling operations for end-of-use cell phones
Rahman and Subramanian (2012)	DEMATEL	Circular supply chain	Legislation, Customer demand Customer, Strategic cost/benefit, Environmental concern, Volume and Quality, Incentive, Resource Integration and coordination	The paper identifies critical factors for implementing EOL computer recycling operations and investigates the relationship among the factors influencing computer recycling operations in reverse supply chains using the cognition mapping process DEMATEL
Jacobs and Subramanian (2012)	Mathematical analysis	Recycling operations	Regulation, supply chain profits, EPR, sharing of responsibility for product recovery	The paper highlights the need to involve all actors in the supply chain in order to best achieve the aims of EPR

(continued)

Table 2 (continued)

Reference	Approach	Literature construct	Variables	Main findings
Kumar et al. (2012)	Conceptual analysis	Green supply chain	Global warming, International regulation and legislation, Brand reputation Stakeholders' increasing awareness, commodity prices, Potential value creation in green supply chain, More integrated and better-managed supply chains, trained workforce, controlled processes, continuous improvement	The paper identifies the most important processes regarding green supply chains: (i) Product/process/package design (ii) Manufacturing (iii)Transportation and shipping (iv) Consumption and disposal
Islam and Huda (2018)	Literature review	E-waste management	Reverse logistics	The article summarizes concepts such as reverse logistics, CLSC, and OLSC and the main techniques that are being used in the literature. In addition, the authors affirm that there is a need for investigation of other alternatives—reuse and repair in the network. No single research was found that considered recycling, remanufacturing, reuse, and repair in an integrated manner
Nascimento et al. (2019)	Conceptual analysis	Circular economy, Industry 4.0	Reuses, recycle, reverse logistics, additive manufacturing	The purpose of this paper is to explore how Industry 4.0 can be integrated with circular economy

(continued)

Table 2 (continued)

Reference	Approach	Literature construct	Variables	Main findings
Ludeke-Freund et al. (2019)	Literature review	Circular economy	Repair, maintenance, Reuse, distribution, Refurbishment, manufacturing, Recycling, Cascading, and repurposing, Biochemical feedstock extraction	The paper identifies 26 circular economy business models from the literature

management, affirming that OEMs should have a say in the legislation since they will probably carry on the remanufacturing process. This result highlights the urge for integration in a supply chain, moreover, not only supply chain integration but also stakeholders integration including governments. Geyer and Blass (2010) conduct a study that is able to quantify the performance of the reverse logistics, reuse, and recycling operations which is an invaluable tool to convince shareholders of the potential earnings within the reverse channels. Rahman and Subramanian (2012) map the critical factors for implementing EOL recycling operations. The use of DEMATEL methodology can be tested in other case studies, therefore, it helps the understanding of which are the relevant factors for which business. Jacobs and Subramanian's (2012) study presents similar conclusions compared with Zoeteman et al. (2010), highlighting the need of stakeholders' integration to achieve better results regarding EPR. Kumar et al. (2012) detail the processes needed to achieve a green supply chain. To the best of our knowledge, it is the first paper to conduct such study. Islam and Huda's (2018) study deepens the conceptual understanding by discussing the differences in reverse logistics, open-loop supply chains, and closed-loop supply chains. The authors also highlight the urge for an integrated methodology/framework that integrates recycling, remanufacturing, reuse, and repair. Nascimento et al. (2019) is a conceptual paper that aims to integrate additive manufacturing and industry 4.0 technologies within circular economy practices. Lüdeke-Freund et al. (2019) carry a study that summarizes 26 circular economy business models from the literature consolidating circular economy knowledge up to the date of the study.

4.3 (RQ3) Which are the Main Concepts Associated with E-waste Management?

In this section, we aim to analyze the most popular concepts related to e-waste management. Figure 7 shows the word cloud.

There is no surprise regarding the most cited concepts, including supply chains, recycling, and waste management. Nevertheless, the least cited words reveal with which concepts these main keywords are related to. Based on the analysis of the ten most cited papers and further reading, we selected some concepts to analyze: (i) Extended producer responsibility (EPR). This concept is becoming a significant trend considering e-waste management. The producers are increasingly being held responsible for the products even after they are sold to final customers. The main point of the discussion is the responsibility lies only with the producer or should it be shared among all the stakeholders including the customers? (ii) Product design——Scholars have analyzed that product design is strictly connected with the supply chain returns strategy. Products should be conceived in a way that facilitates remanufacturing, reuse, refurbishment, and recycle. (iii) Optimization—Optimization models should consider that EOL products may reenter the supply chain loops as remanufactured products and/or as a new source of raw materials. Data should be collected and optimization heuristics should be integrated with machine learning techniques. Figure 8 shows the co-word analysis. We note four main clusters that we further discuss.

Considering the co-word analysis, the keyword network is composed of four main clusters, namely: (i) C1—Waste management, (ii) C2—Supply Chain management, (iii) C3—e-waste, and (iv) C4—Circular economy and CLSC. Cluster C1 is mainly a cluster with concepts related to waste management in general and has related concepts

Fig. 7 Wordcloud

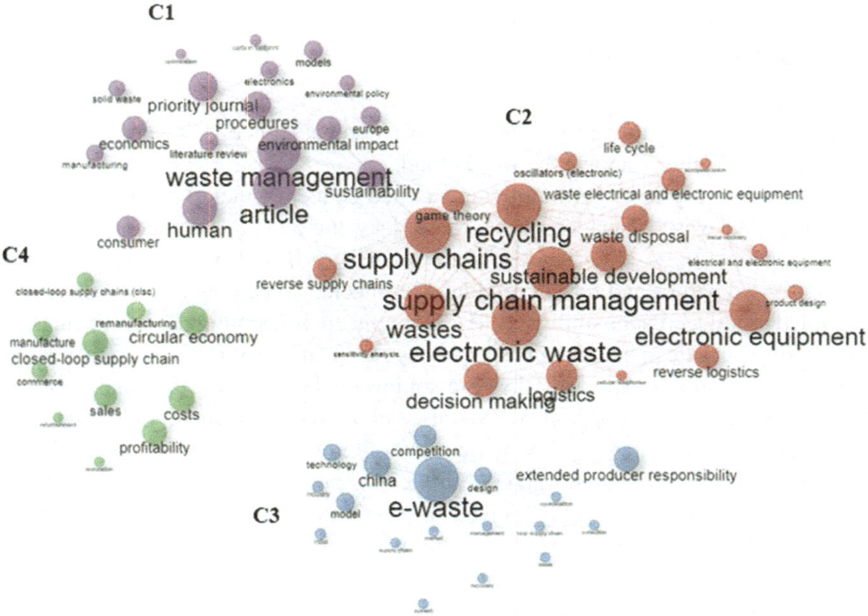

Fig. 8 Keyword network

such as consumer, manufacturing, sustainability, and procedures to deal with waste. C2 is the supply chain management cluster. Supply chain management is crucial to manage waste, because it involves many different companies, customers, and the reverse flows have the task of enabling EOL collection. C2 mentions product design, reverse logistics, and decision-making as key nodes. C3 mentions e-waste, which is connected with design, competition, and EPR. China is also mentioned by many scholars since it is a huge producer of e-waste. Lastly, C4 mentions concepts that enable a circular economy such as closed-loop supply chains and remanufacturing. Figure 9 shows the three fields' plot analysis.

Figure 9 reveals that Chinese and Indian authors outstand in this field, publishing papers that comprise the most relevant keywords in the field, therefore, showing that scholars are interested in this problem.

4.4 (RQ4) Which are the Main Research Gaps Considering E-waste Management?

Although we see a growingly importance of this research stream, there are still some considerable research gaps that should be addressed by researchers to complement the field. The keyword analysis reveals that the risks of these processes are not being considered by the researchers, which is confirmed by keyword graphs and by the

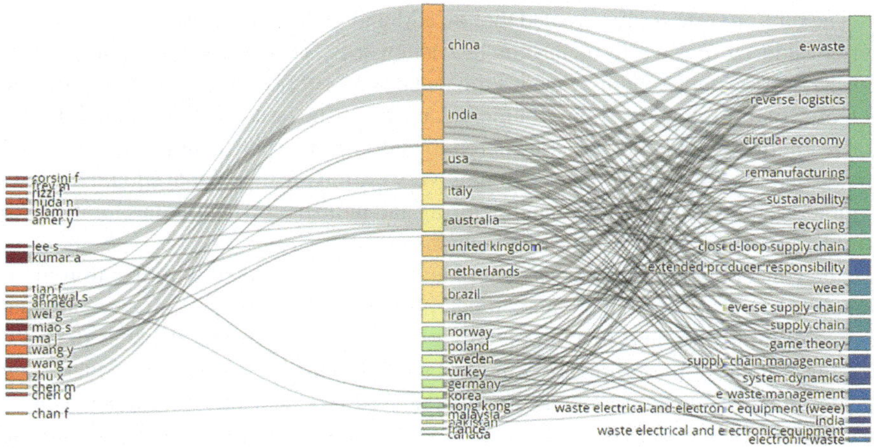

Fig. 9 Three fields plot

reading of the papers. The matter is not even related to future trends by researchers. Risks that comprise not only a company but the supply chain as a whole form a research field known as supply chain risk management (SCRM). The uncertainties of the reverse flows, as well as the demand for remanufactured products, are examples that are only the tip of the iceberg of supply chain risks. Supply chain resilience, which is the capability that a supply chain has of going through undesirable effects and returning to its original state, is a concept strictly connected with SCRM, and it is also not mentioned by the papers. The least cited words of the keyword cloud often provide important analysis, by showing concepts that are insufficiently studied, in this sense, the smaller nodes of the keyword cloud should be analyzed carefully.

5 Discussion

This research presents a scanning of the literature to serve as a guide to researchers and scholars to summarize the state of the art of the subject. Therefore, the study identified the main authors, the top-cited papers, co-citation between papers, as well as other statistics that help to create a panorama of the literature. This research raised some important points that are further discussed in this section. Kumar et al. (2012) mapped some important processes. In this regard, some essential questions remain: (i) Since the paper is from 2012, are the processes still those? (ii) Are these processes the same for all supply chains? What are the main differences? Which processes are common? (iii) How to empirically validate these processes? (iv) Would it be possible to create a generic framework that encompasses a process mapping and validation methodology that is useful to all supply chains? We understand that these are important questions that need further investigation.

The bibliometric study revealed that SCRM and SCRes are constructs that are not yet considered in e-waste management frameworks. This is a very important gap, since e-waste supply chains, besides the usual risks supply chains present, have the potential of unfolding many environmental and health hazards. SCRes methodologies and techniques should also be considered to build strong, robust, and resilient supply chains that are less prone to risks. Another very promising theme that should be more explored is the use of additive manufacturing and industry 4.0 techniques jointly with circular economy and e-waste supply chains.

The concept of smart cities has the potential of being related with the aforementioned concepts of this paper, since it can provide vast amounts of quality data that enable optimization and machine learning techniques that support better decision making. In addition, optimization models should consider that EOL products may reenter the supply chain loops as remanufactured products and/or as a new source of raw materials. Data should be collected and optimization heuristics should be integrated with machine learning techniques.

Each customer finds it easy to use a different channel. In this sense, an integrated sales/returns channel approach, namely omnichannel, can be a vital asset to motivate and integrate all stakeholders in the circular supply chain processes. Therefore, creating a culture of returns and integrating stakeholders and channels is a key feature and finds support in Kumar et al. (2012). In terms of managerial implications, managers and companies that are related to a supply chain must redesign processes to include circular e-waste supply chain processes and activities along the SC and systematically map supply chain risks.

Our keyword cloud analysis reveals that the smaller nodes have unfolded potential for research. The co-word analysis presented four main clusters: C1, C2, C3, and C4. Nevertheless, our analysis compiled in Table 2 discloses other constructs besides C1–C4. In Table 2, we identified some of the variables that compose these constructs, however, we were not exhaustive. Additionally, the analysis summarized in Table 2 must be broadened by a full content analysis of all papers found in this research.

6 Conclusions

This paper had the objective of mapping the state of the art of e-waste management through a bibliometric study. Therefore, bibliometric analysis was the chosen technique to accomplish this objective. We addressed this objective by investigating four research questions. We answered each RQ with the appropriate methods and presented a final discussion summarizing the findings and implications. We did not find any similar bibliometric study in our search, in this sense, this paper offers a unique contribution. As the main limitations, we highlight that this paper is a product of our method of choice and other methods in other supply chains could obtain different results. Our bibliometric analysis included Scopus and Web of Science databases and considered only published papers in English, excluding doctoral theses, congress papers, and documental analyzes. This paper accomplished the first literature scan concerning e-waste management and supply chain management. Since we mapped the main papers, we recommend a thorough Systematic Literature Review with rigorous content analysis as a future research. Finally, some important research questions arise: Which are the main constructs of circular e-waste management? Which are the risks that affect these supply chains? Which techniques are already being applied to identify, assess and mitigate supply chain risks?

References

Appio FP, Martini A, Testa S, Massa S (2016) Unveiling the intellectual origins of social media-based innovation: insights from a bibliometric approach. Scientometrics 108:355–388

Ardito L, Scuotto V, Del GM, Petruzzelli AM (2019) A bibliometric analysis of research on big data analytics for business and management. Manag Decis 57:1993–2009

Baldé CP, Forti V, Gray V, Kuehr R, Stegmann P (2019) The global e-waste monitor. 2019, United Nations University (UNU), International Telecommunication Union (ITU) & International Solid Waste Association (ISWA), Bonn/Geneva/Vienna

Barroso AP, Machado VH, Barros AR, Machado VC (2010) Toward a resilient supply chain with supply disturbances. 245–249

Barroso AP, Machado VH, Machado VC (2011) The resilience paradigm in the supply chain management: a case study. 928–932

Bowersox DJ, Stank TP, Daugherty P (1999) Lean launch: managing product introduction risk through response based logistics. J Prod Innov Manag 16:557–568

Cai YJ, Choi TM (2021) Extended producer responsibility: a systematic review and innovative proposals for improving sustainability. IEEE Trans Eng Manag 68:272–288. https://doi.org/10.1109/TEM.2019.2914341

Cai J, Liu X, Xiao Z, Liu J (2009) Improving supply chain performance management: a systematic approach to analyzing iterative KPI accomplishment. Decis Support Syst 46:512–521

Carvalho H, Cruz-Machado V, Tavares JG (2012) A mapping framework for assessing supply chain resilience. Int J Logist Syst Manag 12:354–373

Carvalho H, Cruz Machado V (2007) Designing principles to create resilient supply chains. Proc 2007 Ind Eng Res Conf Nashv

Chadegani AA, Salehi H, Yunus M et al (2013) A comparison between two main academic literature collections: web of science and scopus databases. Asian Soc Sci 9:18–26

Christopher M, Peck H (2004) Building the resilient supply chain. Int J Logist Manag 15:1–13. https://doi.org/10.1108/09574090410700275

Croxton K, Dastugue S, Lambert D, Rogers D (2001) The supply chain management processes. Int J Logist Manag 12:13–36

da Silva JBN, Senna P, Chousa A, Coelho O (2020) Data mining and operations research techniques in supply chain risk management: a bibliometric study. Brazilian J Oper Prod Manag 17:1–14. https://doi.org/10.14488/bjopm.2020.029

De Meyer N, Snoeck M, Cattrysse D, Van Orshovena J (2016) A reference data model to support biomass supply chain modelling and optimization. Environ Model Softw 83:1–11

Di Stefano G, Peteraf M, Verona G (2010) Dynamic capabilities deconstructed: a bibliographic investigation into the origins, development, and future directions of the research domain. Ind Corp Chang 19:1187–1204

Fahimnia B, Sarkis J, Davarzani H (2015) Green supply chain management: a review and bibliometric analysis. Int J Prod Econ 162:101–114

Garg CP (2021) Modeling the e-waste mitigation strategies using grey-theory and DEMATEL framework. J Clean Prod 281:124035. https://doi.org/10.1016/j.jclepro.2020.124035

Geyer R, Blass VD (2010) The economics of cell phone reuse and recycling. Int J Adv Manuf Technol 47:515–525. https://doi.org/10.1007/s00170-009-2228-z

Guide VDR, Muyldermans L, Van Wassenhove LN (2005) Hewlett-packard company unlocks the value potential from time-sensitive returns. Interfaces (Providence) 35:281–293. https://doi.org/10.1287/inte.1050.0145

Hallikas J, Virolainen VM, Tuominen M (2002) Risk analysis and assessment in network environments: a dyadic case study. Int J Prod Econ 78:45–55. https://doi.org/10.1016/S0925-5273(01)00098-6

Hallikas J, Karvonen I, Pulkkinen U et al (2004) Risk management processes in supplier networks. Int J Prod Econ 90:47–58. https://doi.org/10.1016/j.ijpe.2004.02.007

Hjørland B (2013) Citation analysis: a social and dynamic approach to knowledge organization. Inf Process Manag 49:1313–1325

Islam MT, Huda N (2018) Reverse logistics and closed-loop supply chain of waste electrical and electronic equipment (WEEE)/E-waste: a comprehensive literature review. Resour Conserv Recycl 137:48–75. https://doi.org/10.1016/j.resconrec.2018.05.026

Jacobs BW, Subramanian R (2012) Sharing responsibility for product recovery across the supply chain. Prod Oper Manag 21:85–100. https://doi.org/10.1111/j.1937-5956.2011.01246.x

Kumar S, Teichman S, Timpernagel T (2012) A green supply chain is a requ3. supply chacin and supply chain management theoriesirement for profitability. Int J Prod Res 50:1278–1296

Lambert DM, Cooper MC (2000) Issues in supply chain. Management 83:65–83

Lehoux N, LebelL L, Elleuch M (2016) Benefits of inter-firm relationships: application to the case of a five sawmills and one paper mill supply chain. Inf Syst Oper Res 54:192–209

Liu Z, Yin Y, Liu W, Dunford M (2015) Visualizing the intellectual structure and evolution of innovation systems research: a bibliometric analysis. Scientometrics 103:135–158

Lonsdale C (1999) Effectively managing vertical supply relationships: a risk management model for outsourcing. Supply Chain Manag An Int J 4:176–183. https://doi.org/10.1108/13598549910284499

Lüdeke-Freund F, Gold S, Bocken NMP (2019) A review and typology of circular economy business model patterns. J Ind Ecol 23:36–61. https://doi.org/10.1111/jiec.12763

Mont O, Dalhammar C, Jacobsson N (2006) A new business model for baby prams based on leasing and product remanufacturing. J Clean Prod 14:1509–1518. https://doi.org/10.1016/j.jclepro.2006.01.024

Nascimento DLM, Alencastro V, Quelhas OLG et al (2019) Exploring Industry 4.0 technologies to enable circular economy practices in a manufacturing context: a business model proposal. J Manuf Technol Manag 30:607–627. https://doi.org/10.1108/JMTM-03-2018-0071

Norrman A, Jansson U (2004) Ericsson's proactive supply chain risk management approach after a serious sub-supplier accident. Int J Phys Distrib Logist Manag 34:434–456. https://doi.org/10.1108/09600030410545463

Ponomarov SY, Holcomb MC (2009) Understanding the concept of supply chain resilience

Rahman S, Subramanian N (2012) Factors for implementing end-of-life computer recycling operations in reverse supply chains. Int J Prod Econ 140:239–248. https://doi.org/10.1016/j.ijpe.2011.07.019

Saen FR, Fisher R, Mahdiloo M (2016) Sustainable supply chain modeling and optimization. Transp Res Part D Transp Environ 48:409–410

Senna P, Da Cunha RA, Castro A, Dias AC (2020a) Promising research fields in supply chain risk management and supply chain resilience and the gaps concerning human factors: a literature review. Work 67:487–498. https://doi.org/10.3233/WOR-203298

Senna P, Reis A, Santos IL et al (2020b) A systematic literature review on supply chain risk management: is healthcare management a forsaken research field? Benchmarking 28:926–956. https://doi.org/10.1108/BIJ-05-2020-0266

Shuai Y, Wang X, Zhao L (2011) Research on measuring method of supply chain resilience based on biological cell elasticity theory. 264–268

Tansel B (2020) Increasing gaps between materials demand and materials recycling rates: a historical perspective for evolution of consumer products and waste quantities. J Environ Manage 276:111196. https://doi.org/10.1016/j.jenvman.2020.111196

Verbeek A, Debackere K, Luwel M, Zimmermann E (2002) Measuring progress and evolution in science and technology—I: the multiple uses of bibliometric indicators. Int J Manag Rev 4:179–211

Vieira ES, Gomes JANF (2009) A comparison of scopus and web of science for a typical university. Scientometrics 81:587–600

Wang M, Liu P, Gu Z, et al (2019) A scientometric review of resource recycling industry. Int J Environ Res Public Health 16. https://doi.org/10.3390/ijerph16234654

Wu Y-CJ, Wu T (2017) A decade of entrepreneurship education in the Asia Pacific for future directions in theory and practice. Manag Decis 55:1333–1350

Zoeteman BCJ, Krikke HR, Venselaar J (2010) Handling WEEE waste hows: on the effectiveness of producer responsibility in a globalizing world. Int J Adv Manuf Technol 47:415–436. https://doi.org/10.1007/s00170-009-2358-3

Zsidisin GA, Panelli A, Upton R (2000) Purchasing organization involvement in risk assessments, contingency plans, and risk management: an exploratory study Supply Chain Manag An Int J 5:187–198. https://doi.org/10.1108/13598540010347307

E-Waste Management Strategies Across Recycling Industry of Northern India: An Empirical Investigation

Somvir Arya, Ajay Gupta, and Arvind Bhardwaj

Abstract E-waste management is a matter of concern for the industrial world for few years. This study is an attempt to assess the performance of e-waste management strategies. Percentage awareness and performance of e-waste management strategies have been calculated. Results indicated that the respondents have high knowledge about better channels available for disposing of e-waste rather than selling these to a normal rag picker. However, a null hypothesis has been assumed for correlation analysis and hypotheses have been framed for regression analysis. Results indicated that awareness highly contributes towards basic terminology of e-waste management followed by knowledge of benefits, knowledge of practices, and knowledge of barriers. The result of this test showed that there is a significant difference in the level of awareness about the hazardous effects of e-squander on the health of human beings and surroundings. Basic terminology of e-waste management contributes highly towards overall awareness.

Keywords E-waste · Performance · Correlation analysis and basic terminology

1 Introduction

Electronics waste (E-waste) is the fastest growing waste stream in the industrialized and urbanized world. Few decades back, the amount of waste generated was considered small enough to be diluted in the environment. With a massive growth in the electronics and hardware sector, the demand of the electronics products has been enhanced manifold. Faster change of features in the electronics devices and availability of improved products forcing the consumers to dispose of the electronics products rapidly. This has caused a generation of e-waste alarmingly. The major

S. Arya (✉) · A. Gupta · A. Bhardwaj
Department of Industrial & Production Engineering, Dr. B R Ambedkar National Institute of Technology, Jalandhar, Punjab, India
e-mail: somvirarya21@gmail.com

M. Jawaid and A. Khan (eds.), *Conversion of Electronic Waste in to Sustainable Products*, Sustainable Materials and Technology,
https://doi.org/10.1007/978-981-19-6541-8_9

source of e-waste is the disposal of the hardware and electronic items from government offices, public and private sectors, and academic and research institutes. Household consumers are also contributing a significant volume of end-of-life electronics products. Apart from domestic generation in India, the imported e-waste volume is also growing substantially, though, import is prohibited in India (Gao et al. 2004). Sources of e-waste are divided into those from the industrial sector as well as the household and institutional. The e-waste from the industrial sector includes electrical and electronic assemblies, while the household and institutions produce e-waste from the used and end-of-life electrical and electronic equipment (Mohan and Bhamawat 2008).

From past records, it seems certain that new problems of physical, biological, and social change, not now widely anticipated, will arise sooner than later. This is because our scientific knowledge of each of these systems is incomplete, the mass of human population and its demands are increasing relentlessly and the possible human adjustments and adaptations, including technology, are multiplying (White 1996). Only a few years ago, some of the environmental issues of concern included the trio: acid rain, stratospheric ozone layer depletion, and global warming. Today, waste electrical and electronic equipment (WEEE) or electronic waste (e-waste) generation, transboundary movement, and disposal are becoming issues of concern to solid waste management professionals, environmentalists, international agencies, and governments around the world (Musson et al. 2000). This study is an attempt to assess the performance of e-waste management strategies in the manufacturing industry of northern India.

2 Literature Review

Borthakur and Sinha (2013) performed a study on e-scrap management in India and highlighted the approaches used for the disposal, recycling, and storage which are causing great danger. The researchers tried to assess the existing state of e-trash management techniques in India. People lack in knowledge regarding the adverse impact of e-waste over environment, health of recyclers, and atmosphere. In the Indian context, the e-waste reutilization sector may be divided into formal and informal sectors.

Dwivedy and Mittal (2013) summarized the 3-R principle, which can be termed Reduce, Reuse, and Recycle. He also suggested two scenarios for India towards the assessment of the current and future e-waste outflows. The first framework represents the idealistic assumptions wherein the majority of the outdated items are reused and some are stored. The second scenario represents the pessimistic scenario, where a considerable amount of inventory remains in storage. Waste products are reused after dismantling and testing and, in the end some parts, if required, will be replaced to make the refurbished one.

Ryen et al. (2014) discussed that e-waste due to huge volume has become a challenge for domestic and global waste infrastructure. Material and flow analysis

(MFA) was conducted for discarded computers and other types of e-waste. Along with this, dynamic flow analñyses were conducted using the market supply method.

Wang and Xu (2014) consumption of electrical and electronic equipment has increased significantly in recent decades. The vital driving force involved in the recycling of discarded e-gadgets is the presence of valuable materials in them. In this paper, they focused on the methods and technologies to manage the e-scrap and especially on the recovery of non-metallic fraction. Pyrolysis technology, to recover the plastic, has the lowest energy consumption process but existence of brominated flame retardants makes the process problematic. Supercritical fluids (SCF) and gasification technology put a comparatively smaller impact on the environment than the pyrolysis process, but the energy utilization is on a higher side. Removal of lead metal from the glass parts of the discarded e-gadgets is the essential process before reusing the cathode ray tube (CRT) funnel glass and recycling of liquid crystal display (LCD) glass. However, it is necessary to assess the impact of recycling processes on environment before its industrialization.

Perkins et al. (2014) performed an investigation to document the scope of the issues linked with informal managing habits of e-scrap. The amount of e-waste being generated is increasing and is compounded by illegal exportation from developed to developing countries. They told that recycling of e-waste is necessary but recycling must be carried out in an environmental-friendly and safe method. They emphasized on refurbishing and reusing of the complete products instead of disposal. Disposal should be carried out only when refurbishing and reusing of the gadgets are not possible. Skilled labors and proper protection must be considered for e-scrap recycling in developed and developing nations. Approximately 25% of e-waste of the total is managed by appropriate recycling facilities with sufficient worker safety. There must not be any difference between developed and developing nations in acceptable threat thresholds for dangerous and secondary e-scrap.

Garlapati (2016) presented a review paper that elaborates global e-waste stats, health concerns of e-waste components along with the waste management, recycling, legislative policies, and recommendations related to e-waste. Existing and future initiatives of e-waste management have been addressed by explaining the developed countries' initiatives towards e-waste management. The key to success in terms of e-waste management such as Extended Producer Responsibility (EPR) and Producer Responsibility Organization (PRO) initiatives have been presented in a lucid manner. E-waste arena is a platform for business initiatives for energy production (hydrogen and electricity) and precise metal recovery (gold, silver and platinum) through biotechnological approaches.

3 Research Design

For this survey, questionnaire has been designed that consists of two different sections including respondent designation, type of industry, and measurement of other sections is done on the five-point Likert scale as shown in Tables 1, 2, 3, 4 and 5. The questionnaire is pretested for content validity, ambiguity, and clarity by experienced managers of the industry of Punjab. Descriptive statistics and percentage contribution for knowledge level of e-waste management strategies have been calculated.

Table 1 Knowledge about basic terminology of individuals on e-waste disposal system

Sr. no	Basic terminology	No idea {1} (%)	Little knowledge {2} (%)	Average knowledge {3} (%)	High knowledge {4} (%)	Highly updated knowledge {5} (%)	Mean	SD
1	Do you have any idea of the term "e-waste"?	10.1	40.1	28.6	14.3	7.0	2.68	1.1
2	Do you have idea about how a common man can contribute towards managing e-waste efficiently?	7.0	31.0	27.9	24.0	10.1	2.99	1.11
3	Are you aware of hazardous materials present in e-waste?	7.0	30.3	29.3	23.3	10.1	2.99	1.11
4	Do you consider life of the electronic product as an important factor at the time of buying a new product for reducing e-waste generation?	10.1	39.4	25.8	14.3	10.5	2.76	1.14
Mean total							**2.86**	

Table 2 Knowledge of practices of individuals about e-waste management

Sr. no	Knowledge of practices	No idea {1} (%)	Little knowledge {2} (%)	Average knowledge {3} (%)	High knowledge {4} (%)	Highly updated knowledge {5} (%)	Mean	SD
1	How do you rate official take back system used by some e-products manufacturing companies, as a good e-waste management practice	22.0	29.3	17.8	17.8	13.2	2.71	1.34
2	Do you know that there are better channels available for disposing of e-waste rather than selling these to a normal ragpicker?	27.2	28.2	13.9	15.7	15.0	2.63	1.41
3	In Indian perspective, do you think that easy access to informal collection and recycling channels is a barrier in adopting formal e-waste management practices?	25.1	22.0	19.9	20.9	12.2	2.73	1.36

(continued)

4 Results and Discussion

Since the confirmatory factor analysis established the validity of various constructs and the measurement questions associated with each construct, response of individuals can be used to measure the average level of awareness of individuals about e-waste. The following tables present the summary statistics for each measured item of all the constructs.

Table 2 (continued)

Sr. no	Knowledge of practices	No idea {1} (%)	Little knowledge {2} (%)	Average knowledge {3} (%)	High knowledge {4} (%)	Highly updated knowledge {5} (%)	Mean	SD
4	Do you think that inadequate financial incentive offered by official take-back channels of the companies is a barrier to their adoption in India?	19.9	28.2	20.9	19.2	11.8	2.75	1.3
Mean total							**2.71**	

Table 3 Knowledge of benefits of e-waste management

Sr. no	Knowledge of benefits	No idea {1} (%)	Little knowledge {2} (%)	Average knowledge {3} (%)	High knowledge {4} (%)	Highly updated knowledge {5} (%)	Mean	SD
1	Do you think proper removal of hazardous materials from e-waste is beneficial in minimizing negative effect on human health and environment?	21.3	43.6	15.7	17.4	2.1	2.36	1.1
2	To what extent, do you think, formal e-waste management will reduce public health hazards?	24.7	42.2	14.6	14.3	4.2	2.31	1.12
3	Do you think formal recycling is helpful in reducing the final volume of e-waste (Recovery of components/parts)?	34.5	23.7	18.8	17.8	5.2	2.36	1.26
Mean total							**2.34**	

Table 4 Knowledge of barriers to e-waste management

Sr. no	Knowledge of barrier	No idea {1} (%)	Little knowledge {2} (%)	Average knowledge {3} (%)	High knowledge {4} (%)	Highly updated knowledge {5} (%)	Mean	SD
1	Do you think, limited access to formal recycling facility is a barrier in formal disposal of e-waste?	39.0	40.8	7.3	5.2	7.7	2.02	1.17
2	Do you think, easy access to informal collection channels is a barrier in adoption of formal e-waste disposal methods by individuals?	41.8	35.2	8.7	7.0	7.3	2.03	1.20
3	Do you think lack of knowledge about the hazardous effects of e-waste act as a barrier in adopting formal e-waste management practices?	46.7	24.4	11.8	11.5	5.6	2.05	1.25
4	To what extent do you think that inadequate legislations are resulting in non-adoption of formal e-waste recycling approach?	41.1	28.6	15.0	8.0	7.3	2.12	1.24

(continued)

Table 4 (continued)

Sr. no	Knowledge of barrier	No idea {1} (%)	Little knowledge {2} (%)	Average knowledge {3} (%)	High knowledge {4} (%)	Highly updated knowledge {5} (%)	Mean	SD
5	Do you think that inadequate collection efforts by recyclers/Govt. agencies are resulting in piling up of e-waste at home?	50.2	29.3	13.2	4.9	2.4	1.80	1.01
Mean total							**2.0**	

5 Results Discussion of the Findings

Life of the electronic product as an important factor at the time of buying a new product for reducing e-waste generation is the most important basic terminology of e-waste management; there are better channels available for disposing of e-waste rather than selling these to a normal ragpicker is rated most important factor in terms of knowledge of practices; formal recycling is helpful in reducing the final volume of e-waste (recovery of components/parts) is rated most important in terms of knowledge of benefits; lack of knowledge about hazardous effects of e-waste act as a barrier in adopting formal e-waste management practices is the most significant barrier in removing e-waste; toxic/hazardous materials from discarded e-products require special treatment for environmentally sound disposal is rated most important in terms of awareness.

6 Correlation Analysis

The study examines the relationship between *basic terminology, knowledge of practices, knowledge of benefits, knowledge of barriers, and awareness level of e-waste management from the perspective of recyclers.* The Pearson correlation analysis is used to analyze the statistical relationship between different selected variables. The correlation analysis using the Pearson statistics assumes that the variables are continuous. The conditions of correlation test are satisfied. The null hypothesis of Pearson correlation test is mentioned in Table 6:

Null hypothesis**: **There is no significant correlation between basic terminology, knowledge of practices, knowledge of benefits, knowledge of barriers, and awareness about e-waste management in the recyclers selected for the study.

Table 5 Awareness of individuals on e-waste management

Sr. no	Awareness	No idea {1} (%)	Little awareness {2} (%)	Average awareness {3} (%)	High awareness {4} (%)	Highly updated awareness {5} (%)	Mean	SD
1	To what extent do you know how to dispose of e-waste safely for human beings as well as for the environment?	30.0	44.9	16.7	4.9	3.5	2.07	0.99
2	Are you aware of any electronic waste management policy currently implemented in India for safe disposal of electrical and electronic items?	53.7	16.7	17.8	6.6	5.2	1.93	1.20
3	What is your perspective towards health and environmental hazards associated with e-waste?	46.0	30.3	14.6	4.2	4.9	1.92	1.10
4	Do you think toxic/hazardous materials from discarded e-products require special treatment for environmentally sound disposal?	35.9	33.8	17.8	9.1	3.5	2.10	1.1
Mean total							**2.01**	

Since all the correlation co-efficient are statistically significant, the null hypothesis of no significant correlation between basic terminology, knowledge of practices, knowledge of benefits, knowledge of barriers, and awareness about e-waste management cannot be accepted. Thus, it can be concluded from the results that there exists a significant relationship between basic terminology, knowledge of practices, knowledge of benefits, knowledge of barriers, and awareness about e-waste management. The positive correlation is found between the following pairs:

Table 6 Correlation estimates

		Awareness	Knowledge of benefits	Knowledge of practices	Knowledge of barriers	Basic terminology
Awareness	Pearson correlation (*P* value)	1	0.853 (0.000)	0.815 (0.000)	0.777 (0.000)	0.881 (0.000)
Knowledge of benefits	Pearson correlation (*P* value)	0.853 (0.000)	1	0.682 (0.000)	0.692 (0.000)	0.811 (0.000)
Knowledge of practices	Pearson correlation (*P* value)	0.815 (0.000)	0.682 (0.000)	1	0.547 (0.000)	0.721 (0.000)
Knowledge of barriers	Pearson correlation (*P* value)	0.777 (0.000)	0.692 (0.000)	0.547 (0.000)	1	0.696 (0.000)
Basic terminology	Pearson correlation (*P* value)	0.881 (0.000)	0.811 (0.000)	0.721 (0.000)	0.696 (0.000)	1

- Awareness level of recyclers and knowledge of benefits about e-waste management.
- Awareness level of recyclers and knowledge of practices about e-waste management.
- Awareness level of recyclers and knowledge of barriers about e-waste management.
- Awareness level of recyclers and basic terminology about e-waste management.

By looking at the standardized Pearson correlation coefficients' values, we can assess that basic terminology, knowledge of benefits, knowledge of practices, and knowledge of barriers and all are significantly correlated to the overall awareness level of the recyclers. In our analysis, basic terminology has the strongest relationship with the overall awareness level of recyclers. And in order to increase the overall awareness of the recyclers, basic terminology must be the most significant variable on which government needs to focus to reduce the adverse effects of e-waste. That does not mean that the other variables like knowledge of practices, knowledge of benefits, and knowledge of barriers are not important but basic terminology is the base of all the other variables and government must also focus on all these independent variables simultaneously.

7 Regression Analysis

The awareness level of recyclers about e-waste management is found to have a significant relationship between basic terminology of e-waste management, knowledge of practices about e-waste management, knowledge of benefits about e-waste management, and knowledge of barriers about e-waste management. Hence, the multivariate regression analysis is applied to the variables in order to test the cause and effect relationship between the four exogenous variables and one endogenous variable, i.e., overall awareness.

When independent variables in regression analysis are highly correlated then it shows the problem of multicollinearity. It is the basic assumption in the regression analysis that independent variables should not be highly correlated. All the independent variables should be independent in nature and there should not be any relation between them. The high value of correlation depicts that the independent variables are having some relationship among them. High correlation among the independent variables also affects the overall fitness of the regression model (Frost 2020). The presence of high correlation (generally 0.9 or higher) is the indication of multicollinearity (Field 2013). All the correlation coefficients in our analysis have been found less than 0.9 as shown in Table 7, so we can say that our regression model is not having a problem of multicollinearity.

Other ways to verify the multicollinearity in regression analysis is to check the values of tolerance statistics and Variance Inflation Factor (VIF). To ensure the absence of multicollinearity, the value of tolerance statistics should be higher than 0.1 (Uyanik and Guler 2013) or 0.2 (Menard 1995) and the values of VIF statistics should be less than 10 (Mooi and Sarstedt, 2014; Myers 1990). As depicted from Table 7, all the values of tolerance statistics are found more than 0.1 and the values of VIF are also found less than 10. For the same, we can confirm that there is no problem of multicollinearity in our regression model.

The following hypothesis is used to be tested with the help of regression analysis:

Hypothesis 1: "*There is no impact of awareness about basic terminology of e-waste management on overall awareness level of the recyclers.*"

Table 7 Regression estimates

Model	Regression coefficients	T statistics (*P* value)	F statistics (*P* value)	R square	Collinearity statistics	
					Tolerance	VIF
Constant	0.162	2.014 (0.046)	64.224 (0.000)	0.9		
Knowledge of benefits	0.170	2.124 (0.043)			0.296	3.384
Knowledge of practices	0.265	3.471 (0.002)			0.452	2.211
Knowledge of barriers	0.187	2.688 (0.012)			0.468	2.137
Basic terminology	0.289	2.707 (0.011)			0.266	3.761

(Dependent Variable/Constant: Overall Awareness)

Hypothesis 2: "*There is no impact of knowledge of e-waste management practices on overall awareness level of the recyclers.*"

Hypothesis 3: "*There is no impact of knowledge about benefits arising out of proper e-waste management on overall awareness level of the recyclers.*"

Hypothesis 4: "*There is no impact of knowledge of barriers about e-waste management on overall awareness level of the recyclers.*"

The regression model can be expressed as

$$\begin{aligned}&\text{Overall Awareness} = 0.162 + 0.170 \times \text{Knowledge of Benefits}\\&+ 0.265 \times \text{Knowledge of Practices} + 0.187 \times \text{Knowledge of Barriers}\\&+ 0.289 \times \text{Basic Terminology} + e\end{aligned}$$

The result of the regression analysis is shown below in the table.

Hypothesis 1 There is no impact of awareness about basic terminology of e-waste management on overall awareness level of the recyclers.

Conclusion: Overall awareness significantly predicted basic terminology, $F = 64.224, p < 0.05$, which indicates that the knowledge about the basic terminology of e-waste management plays a significant role in order to enhance the overall awareness level of the recyclers (regression coefficient $= 0.289, t = 2.707, p < 0.05$). These results clearly direct that *our null hypothesis of no significant effects of awareness of basic terminology on overall awareness level of the recyclers* can be rejected with a 95% confidence level.

Hypothesis 2 There is no impact of knowledge of e-waste management practices on overall awareness level of the recyclers.

Conclusion: The table depicts that the regression coefficient between knowledge of practices and overall awareness is 0.265 and t-value is 3.471 at a 5% level of significance. Thus, the hypothesis that *there is no significant impact of knowledge of practices on overall awareness level of recyclers* can be rejected with a 95% confidence level. The knowledge of e-waste management practices has a significant positive impact on the overall awareness level of the recyclers about e-waste management.

Hypothesis 3 There is no impact of knowledge about benefits arising out of proper e-waste management on overall awareness level of the recyclers.

Conclusion: Un-standardized Regression coefficient between knowledge of benefits and overall awareness level of the recyclers is 0.170 and t-value is 2.214 at a 5% level of significance. Thus, the hypothesis that *there is no impact of knowledge of benefits on overall awareness level of the recyclers* can be rejected. In other words, we can also say that there is positive significance between knowledge of benefits and overall awareness level of recyclers.

Hypothesis 4 There is no impact of knowledge of barriers about e-waste management on overall awareness level of the recyclers.

Conclusion: The value of regression coefficient between knowledge of barriers in implementing proper e-waste management and overall awareness level of the recyclers is 0.187. The value of t-statistics is 2.688 at a 5% level of significance. The results depict that knowledge of barriers affects the overall awareness level of recyclers positively and significantly. Thus, our *null hypothesis of no significant impact of knowledge of barriers on overall level of awareness of recyclers* can be rejected.

The regression weights obtained are shown in Table 7 above. The value R^2 always lies between 0 and 1, where a higher R^2 indicates a better model fit (Mooi and Sarstedt 2014). As can be seen from the R square values, four constructs account for around 90% variability in the overall level of awareness.

8 Conclusions and Limitation

It is clear from the regression analysis that all the independent constructs affect the dependent construct, i.e., the overall awareness level of recyclers towards e-waste management positively and significantly. Thus, the government needs to pay attention to all the four constructs simultaneously in order to increase the awareness of the recyclers about the ill effects of e-waste on human health and environment. Basic terminology of e-waste management contributes maximum towards overall awareness. Knowledge of benefits significantly aimed at improving awareness followed by basic terminology, knowledge of barriers, and knowledge of benefits. Knowledge of benefits and knowledge of barriers significantly aimed at improving awareness. Basic terminology significantly aimed at improving awareness followed by knowledge of benefits, knowledge of practices, and knowledge of barriers. There is a possibility of method variance as there is only one respondent from each company.

References

Borthakur A, Sinha K (2013) Electronic waste management in India: a stakeholder's perspective. Electron Green J 1(36):01–21

Dwivedy M, Mittal RK (2013) Willingness of residents to participate in e-waste recycling in India. Environ Develop 6: 48–68

Field A (2013) Discovering statistics using SPSS, 4th edn. SAGE Publications Limited, London

Frost J (2020) Hypothesis regression analysis: an intuitive guide for using and interpreting linear models. Statistics By Jim Publishing

Gao Z, Li J, Zhang HC (2004) Electronics and the environment. In: IEEE international symposium, pp 234–241

Garlapati VK (2016) E-waste in India and developed countries: management, recycling, business and biotechnological initiatives. Renew Sustain Energy Rev 54:874–881

Menard S (1995) Applied logistic regression analysis. Sage university paper series on quantitative applications in the social sciences. Sage, Thousand Oaks, CA, pp 07–106

Mohan D, Bhamawat PMK (2008) E-wastemanagementglobal scenario: a review. J Environ Res Develop 2(4):817–823

Mooi E, Sarstedt M (2014) A concise guide to market research—the process, data, and methods using IBM SPSS statistics. Springer Texts in Business and Economics

Musson SE, Jang Y-C, Townsend TG, Chung I-H (2000) Characterization of lead leachability from cathode ray tubes using the toxicity characterization leaching procedure. Environ Sci Technol 34:4376–4381

Perkins D, Drisse M, Nxele T, Sly P (2014) E-waste: A global hazard. Ann Glob Health 80(4):286–295

Ryen E, Babbitt C, Tyler A, Babbitt G (2014) Community ecology perspectives on the structural and functional evolution of consumer electronics. J Ind Ecol 18(5):708–721

Uyanik GK, Guler N (2013) A study on multiple linear regression analysis. Procedia – Social and Behavioral Sciences, pp 234–240

Wang R, Xu Z (2014) Recycling of non-metallic fractions from waste electrical and electronic equipment (WEEE): a review. Waste Manage 34:1455–1469

White GF (1996) Emerging issues in global environmental policy. Ambio 25(1):58–60

Circular E-Waste Supply Chains' Critical Challenges: An Introduction and a Literature Review

Pedro Senna, Lino G. Marujo, Augusto da Cunha Reis, and Ana Carla de Souza Gomes dos Santos

Abstract Electronic waste (e-waste) is gaining the attention of scholars since its supply chain offers valuable materials that can be recovered, generating resilience to supply chains and the environment. To recover these materials, it is necessary to establish closed-loop supply chains, enabling a circular economy logic. In this sense, supply chain flows must be designed to retrieve these values and mitigate the risks. Furthermore, collection points must be strategically positioned to make this operation feasible, integrating the concept of smart cities. Therefore, this article proposes a conceptual analysis of the literature and, as its main result, presents an integrated framework considering five dimensions: (i) E-waste management, (ii) Supply Chain Resilience—SCRes, (iii) Circular Economy, (iv) Closed-loop supply chains, and (v) Smart cities.

Keywords E-Waste management · Supply chain resilience · Circular economy · Closed-loop supply chains · Smart cities

P. Senna (✉) · A. da Cunha Reis
Postgraduate Program in Production Engineering and Systems (PPPRO), Federal Centre for Technological Education Celso Suckow da Fonseca (CEFET/RJ), Rio de Janeiro/RJ, Brazil
e-mail: pedro.senna@cefet-rj.br

A. da Cunha Reis
e-mail: augusto.reis@cefet-rj.br

P. Senna · L. G. Marujo
Production Engineering Program, PEP/COPPE-UFRJ, Rio de Janeiro/RJ, Brazil
e-mail: lgmarujo@ufrj.br

A. C. de Souza Gomes dos Santos
Industrial Production Management Department and Graduate Program in Production Engineering and Systems (PPPRO), Federal Institute of Education, Science, and Technology of Rio de Janeiro (IFRJ) and Federal Centre for Technological Education Celso Suckow da Fonseca (CEFET/RJ), Rio de Janeiro/RJ, Brazil
e-mail: ana.carla@ifrj.edu.br

M. Jawaid and A. Khan (eds.), *Conversion of Electronic Waste in to Sustainable Products*, Sustainable Materials and Technology,
https://doi.org/10.1007/978-981-19-6541-8_10

1 Introduction

The current production models are responsible for many harmful effects that threaten natural ecosystems as well as human health (Lüdeke-Freund et al. 2019). Driven by the increase in global population and economic growth, the continuous enhancement of people's living conditions, innovation, and shorter products lifespans, the demand for newer manufactured products is increasing, generating all types of wastes and incorrect discarding (Peng et al. 2018; Cai and Choi 2021). In addition, this industrial growth creates additional challenges for environmental agencies due to the increased pollution (Rezayat et al. 2020).

These conditions demand a massive amount of resource extraction and consumption of the supply chains (Garg 2021). In this context, electronic waste (e-waste, also named WEEE) has become a global issue due to its fast-growing rates (Thi et al. 2019). Nevertheless, operations management researchers and practitioners still understand little about e-waste processing supply chains (Esenduran et al. 2020). In this sense, sustainable development can be supported by WEEE recycling and refurbishment, which are important practices capable of mitigating WEEE hazards (Sharifi and Shokouhyar 2021; Miao et al. 2020). Moreover, reverse logistics (RL) emerges as an important enabler of reuse, refurbish, recycle, and remanufacturing, and proper disposal (Garg 2020). Due to the lack of sustainable practices, 44 of the 118 elements in the periodic table are endangered to become extinct in their ore forms within the next 80 years (Tansel 2020). Since these elements are crucial to many electronic products, serious risks affect these supply chains. Therefore, recovery and recycling-related research and practices have been increased by the concern on environmental sustainability (Ozgur Polat and Gungor 2021).

Among the main challenges for supply chains is the lack of infrastructure to collect and recycle discarded consumer goods, since in recent years, distribution channels are designed to be more compact and efficient (Tansel 2020). WEEE problem pressures environmental agencies, government, and original equipment manufacturers to adopt, design, and innovate environment-friendly e-waste mitigation strategies (Garg 2021). Additionally, consumers are considered to be the primary source of e-waste generation and their behavior influences e-waste management (Chen and Gao 2021).

A new concept that provides tools and definitions that converge with e-waste risk mitigation is the Circular Economy (CE) concept. CE refers to production systems that are restorative and regenerative, allowing products and materials to be kept in the market at their highest utility and value (Webster 2015). In order to design a supply chain where materials can reenter the production cycles, scholars have coined the concept of a closed-loop supply chain (CLSC). Concerns about the increasing release of consumer products to the environment, especially defective electronic products, helped the CLSC concept emerge (Rezayat et al. 2020). Furthermore, collection points must be strategically positioned to make this operation viable, needing to integrate the concept of smart cities. Smart cities can be understood as a concept that encompasses the utilization of information and communication technologies to provide services such as energy and mobility (Jnr et al. 2020).

Although some papers are emerging and providing local solutions for e-waste supply chain problems, no papers specifically analyze WEEE supply chain risks. On the other hand, the concept of supply chain risk management is considered as essential enabler of supply chain resilience. Moreover, some research fields such as Healthcare (Senna et al. 2020), Industry 4.0 (Ivanov et al. 2019), and CE (Yazdani et al. 2019) are among the research streams that least apply supply chain resilience and supply chain risk management.

Both concepts offer a large set of tools and practices to assist managers in supply chain risk identification, assessment, mitigation, and monitoring. To mitigate WEEE, supply chain risk managers must implement an efficient reverse channel, which is considered a significant challenge. A supply chain must be resilient in either direct or reverse flows; therefore, supply chain risks must be efficiently managed throughout the supply chain. Additionally, e-waste supply chains have the potential to generate resilience through the recovery of precious materials (Ozgur Polat and Gungor 2021). Moreover, it has the potential of reducing the rate of raw materials extinction, generating environmental resilience. In this respect, refurbishment activities (e.g., testing, inspection, repair, and polishing) bring new opportunities to maximize the total benefit (Ozgur Polat and Gungor 2021).

In this regard, this article has the objective of proposing a framework based on a conceptual analysis of the following constructs: (i) E-waste management (C1), (ii) Supply Chain Resilience—SCRes (C2), (iii) Circular Economy (C3), (iv) Closed-loop supply chains (C4), and (v) Smart cities (C5). To the best of our knowledge, only a few articles about this subject are corroborated by Rezayat et al. (2020). Moreover, we did not find any study that discusses how SCRM contributes to the resilience of WEEE supply chains and how the concept of smart cities can be a means of risk mitigation. There are no studies that propose a framework that integrates the constructs C1, C2, C3, C4, and C5. In this sense, this study is unique and presents relevance since it builds a resilient circular WEEE supply chain.

In addition to this introductory section, Sect. 2 presents a literature review, Sect. 3 presents the methodology, Sect. 4 presents the framework elements, Sect. 5 presents the discussion, and Sect. 6 closes the paper with the conclusions, limitations, and further research opportunities.

2 Literature Review

2.1 E-Waste

The world is facing an unprecedented production of e-waste (John et al. 2018). Such increase is due to the technological innovation and increasingly short life cycles of Electrical and Electronic Equipment (EEE), huge demand (Agrawal et al. 2018; Chen et al. 2012; Singhal et al. 2019a, b), which creates a significant amount of risks to human health and environment (Shumon et al. 2016). Marinello and Gamberini

(2021) define EEE as the equipment that depends on electric currents or electromagnetic fields to work and equipment for the generation, transfer, and measurement of such currents and fields and designed for use with a voltage lesser or equal to 1000 V for alternating current and lesser or equal to 1500 V for direct current. Nowadays, more than 900 different EEE items can be found, which most of the time, have no good repair options (Marinello and Gamberini 2021). WEEE typically includes end-of-life computers, televisions, photocopiers, and mobile phones, also including non-electronic goods such as ovens and refrigerators, since all can be classed as discarded appliances that use electricity (Reddy et al. 2019).

In this sense, WEEE reverse supply chain management has gained more attention due to environmental issues (Isernia et al. 2019). Scholars agree that landfills or incineration of WEEE are not acceptable end-of-use management options because they are not sustainable solutions (Geyer and Blass 2010; Singhal et al. 2019a, b). Economic and social concerns also justify the importance of researching WEEE (Ozgur Polat and Gungor 2021). Nevertheless, most WEEE discarded household waste in landfills without any treatment or recycling processes (Reddy et al. 2019). WEEE usually contains hazardous and non-hazardous materials, such as plastic, glass, wood and plywood, concrete and ceramics, and other materials besides metal components (Ghalehkhondabi and Ardjmand 2020), which can generate devastating effects on the environment (Rezayat et al. 2020).

2.2 *Strategies for E-Waste Management*

A growing demand for resources and products' shorter life cycles leads supply chain managers to develop practices of waste handling and rebounding valuable materials (Corsini et al. 2017; Sadeghi et al. 2019). Among the stakeholders, there are countries' governments that formulate legislation and customers that are increasingly demanding greener products and worrying about the depletion of natural resources, which leads industrialists to implement recycling and remanufacturing practices (Sadeghi et al. 2019). Recycling attracts much attention for collaborating with pollution reduction and conserving natural resources (Guo et al. 2018). Among the most viable options for WEEE are direct reuse and resale or remanufacturing via refurbishment (Bryan et al. 2020). Remanufacturing is a strategy that can extend the life cycle of e-waste and offer advantages in terms of carbon emission reductions saving material and energy consumption (Singhal et al. 2019b).

Moreover, WEEE recycling and remanufacturing practices are linked to a worldwide tendency of the establishment and implementation of the Extended Producer Responsibility system (ERP), which is becoming increasingly popular (Gong et al. 2019). Remanufacturing can make a significant positive impact on urban waste governance. Shi et al. (2019) and Ghalehkhondabi and Ardjmand (2020) affirm that a WEEE management system includes e-waste generation, collection centers, dismantling, separation, transportation, treatment, storage, distribution, recovery, and disposal processes. Therefore, WEEE management consists of a strategic supply chain problem (Ghalehkhondabi and Ardjmand 2020). In summary, product recycling has become an emerging field of supply chain management research (Gong et al. 2019).

2.3 E-Waste Supply Chains

Original Equipment Manufacturers (OEMs) have implemented Extended Producer Responsibility (EPR) programs for the safe disposal of end-of-life (EOL) products (Reddy et al. 2019). EPR has been an important innovation in terms of product life cycle management. However, real-life implementation has multiple impacts and supply chain risks that consist of a challenging management effort (Corsini et al. 2017b). Creating greener supply chain channels is imperative to reduce environmental pollution and incentivize a green product design (Wang et al. 2019b).

Governments and industries must work together and look for sustainable production to mitigate environmental concerns and safety issues, and preserve resources (Singhal et al. 2019b). Strategies such as reuse, recycle, and refurbishment need a well-established reverse supply chain management (RSCM) (Kianpour et al. 2017). In this sense, EEE supply chains have been the focus of attention for policymakers, academics, and manufacturers (Kumar 2019). Effective WEEE management systems must be linked with WEEE processing plants. Nevertheless, it is imperative to prioritize the supply chain network's primary constructs to be able to address the main challenges and risks (Baidya et al. 2020).

In terms of techniques to deal with WEEE management, Safdar et al. (2020) developed a multi-objective, reverse logistics network model for e-waste management, including the triple bottom line concept in their study. Their model considered new goods customers, collection centers, distribution centers, recycled goods customers, and reprocessing centers (consisting of return evaluation centers, recycling centers, and refurbishing centers) as variables in their scenarios.

In summary, the reverse WEEE supply chain flows must deal with two different issues: (i) how to regulate the recycling and disposal of these electronic wastes avoiding environmental and health hazards (Zhu and Li 2020); and (ii) Design supply chain processes that can recover materials, including critical minerals and precious metals with limited global reserves (Shevchenko et al. 2021).

3 Methodology

Our entire research methodology is presented in the workflow seen in Fig. 1.

3.1 Formulate Research Question

The research question must be clear and highlight an objective research stream; in this sense, the research question investigated in this paper is: RQ What are the main building blocks of a circular e-waste ecosystem?

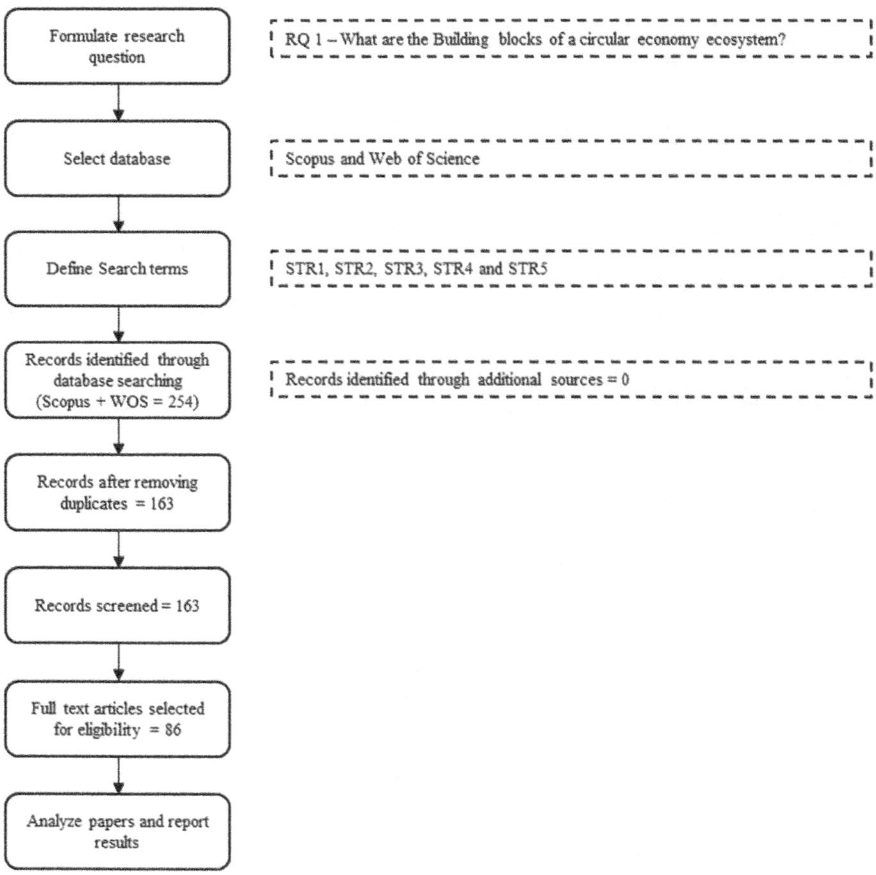

Fig. 1 Research methodology

3.2 Database Selection

In this study, we chose both Scopus and Web of Science (WoS) databases. Scopus database is the most extensive searchable citation and abstract source, while WoS was the only citation database and publication that covers all domains of science for many years (Chadegani et al. 2013). Among the most relevant scientific databases in the world are Scopus and WoS (Wang et al. 2019a). About 2/3 of the studies can be found in both databases and 1/3 only in one database, in addition, both databases provide ".bib" files providing full search data (such as references, article authors, etc.) that can be further analyzed in bibliometric packages (Vieira and Gomes 2009). In addition, Scopus and WoS as search engines allow locating papers of Science Direct, Taylor and Francis, Emerald, Springer, and other important bases.

3.3 Search Terms Definition

Our search strategy consisted in verifying how e-waste and supply chain subjects are related in the literature. Previous research revealed that e-waste supply chains are prone to significant supply chain risks. In this sense, our first string (STR1) was *(e-waste AND "supply chain risk management") OR ("electronic waste" AND "supply chain risk management")*. STR1 returned zero occurrences in both bases, showing that academia still did not produce studies that formally consider supply chain risk management in e-waste supply chains. STR2 and STR3 encompass supply chain risk management and supply chain risk search terms combined with "refurbishment," whereas STR3 returned one paper. Moreover, we broadened the search by combining e-waste with supply chains, resulting in STR4 and STR5. Table 1 summarizes the investigation.

Table 1 Search terms

STR	Strings	Scopus	WOS
STR1	(E-waste AND "supply chain risk management") OR ("eletronic waste" AND "supply chain risk management")	0	0
STR2	Refurbishment AND "supply chain risk management"	0	0
STR3	Refurbishment AND "supply chain risk"	1	0
STR4	(E-waste AND "supply chain") OR ("eletronic waste" AND "supply chain")	97	102
STR5	Refurbishment AND "supply chain"	29	25

3.4 Screening Process

In the Search engines, we considered articles or reviewed final papers in the English language from 2010 to now. Then we read the article titles and abstracts and selected 86 papers. The reading of the 86 documents revealed that there is not a single paper that proposes an integrated framework that identifies the main building blocks of a circular economy ecosystem. In this sense, this paper proposes a conceptual analysis where we place these building blocks based on the literature.

4 The Constructs of a Circular WEEE Economy

The reading of the 86 papers revealed that a robust Circular e-waste supply chain (CEWSC) is composed of different agents and stakeholders. This paper identified the main building blocks of a CE "ecosystem": (i) E-waste management, (ii) Closed-loop supply chains, (iii) Supply chain resilience, and (iv) Smart cities.

The relationship between these blocks can be characterized as follows. CEWSC is considered the universe, it is a broad concept formed by several building blocks. For an economy to be circular, the various supply chains that feed this economy need to design reverse logistics processes. These recovered wastes can reenter as raw materials in the beginning of the chain, thus, closing some loops in the supply chains. However, closing supply chain loops are not sufficient to build a CEWSC. There is a need to create resilience for these supply chains, both in reverse and direct flows, which is achieved by managing the risks of these supply chains.

Government is also a key stakeholder in this equation, especially about the concept of smart cities and urban planning. The last mile has been considered an important challenge, increasingly granular deliveries in crowded urban centers where trucks have very specific time windows to be able to unload their products (often in shopping centers and homes without proper docks) generate both challenges for direct and reverse flows. In this sense, which channels can be used to engage and motivate the customer to be actively involved with this reverse logistics process? In this sense, this section characterizes each of these building blocks and their main variables, closing with a conceptual figure where the fit between these blocks is made. Moreover, the integration of a circular economy with supply chain processes is beneficial from a sustainability as well as a business point of view (Dhonde and Patel 2020).

4.1 Circular Economy

The usual model of value creation is normally based on a one-directional flow of primary activities (Mishra et al. 2018). Scholars are increasingly studying the concept of circular economy since it presents means to change the current consumption paradigm, which is only focused on continuous growth (Ghisellini et al. 2016). CE concept is rooted in very diverse theoretical backgrounds: ecological economics, environmental economics, and industrial ecology (Ghisellini et al. 2016) and is being considered a solution for balancing ambitions for economic growth and environmental protection (Lieder and Rashid 2016). The CE economic system is defined by Fonseca et al. (2018) as a way to conciliate economic and environmental performance while showing innovation in the way of relating business and environment. Geissdoerfer et al. (2017) define CE as a regenerative system in which resource input and waste, emission, and energy leakage are minimized by slowing, closing, and narrowing material and energy loops. Bag et al. (2018) consider the CE as a system that applies manufacturing principles and sustainable manufacturing practices. For some authors, CE is associated with the development of waste and resource management, increased resource productivity, and new business models (Howard et al. 2018).

The CE perspective focuses on restoring the value of used resources (Jabbour et al. 2018). CE differs from the linear economy because it disassociates economic growth from resource extraction and environmental losses (Elia et al. 2017). Moreover, CE can help organizations to obtain sustainable development (McDowall et al. 2017) and can be achieved through product design that aims to extend EOL, maintenance, repair, reuse, remanufacturing, refurbishing, and recycling (Geissdoerfer et al. 2017; Garrido-Hidalgo et al. 2020). The circular economy (CE) requires companies to rethink their supply chains and business models (Lüdeke-Freund et al. 2019). CE provides society with a new economic methodology that reintroduces waste as raw materials, transforming production systems into circular chains (Nascimento et al. 2019).

CE is supported by a complex network of direct and reverse flows, composing closed-loop supply chains (CLSCs), which are also called circular supply chains and involve inter-firm relationships considering the utilization of waste as inputs to further production processes (Batista et al. 2018). The main aim of the CE framework is to create a regenerative system able to ensure optimal reuse, renovation, remanufacturing, and recycling of products, materials, and waste by handling them in closed loops (Isernia et al. 2019). From the CE perspective, the reverse supply chain and the reverse logistics can be considered necessary approaches to "close the loops" of end-of-life (EOL) products (Isernia et al. 2019). In this sense, reverse logistics starts from end users where the products are collected from customers (return products) and then attempts to manage EOL products through different decisions such as remanufacturing, repairing, and finally, disposing of some used parts (Govindan et al. 2015).

4.2 E-Waste Management

E-waste management consists of a very challenging task for scholars and managers. WEEE is a very specific type of waste composed of ferrous metals, non-ferrous metals, glass, plastics, and other materials (Marinello and Gamberini 2021). Consumers are considered to be the primary source of e-waste generation, in this sense, their recycling behavior has an important influence on e-waste management. (Chen and Gao 2021). If consumers can be appropriately persuaded to use remanufactured products, managers will minimize resource consumption and negative environmental impacts (Nascimento et al. 2019).

E-waste management faces some very sensible questions, such as transport and the proper remanufacturing infrastructure (Shevchenko et al. 2021). The commitment of top management concerning returns management is considered a crucial strategy in terms of e-waste management (Garg 2021). Garg (2021) affirms that to obtain effective e-waste management, managers should implement strategies such as extended producer responsibility (EPR).

A meaningful way to address WEEE is the recycling platforms, which involve three constructs: consumers, manufacturers, and platforms (Zhu and Li 2020). Any time a consumer needs to recycle e-waste, they can upload information to specialized websites responsible for finding the adequate manufacturer (Zhu and Li 2020). In general, OEMs can create significant value from product recovery. Nevertheless, managers must design an effective reverse logistics network incorporating remanufacturing facilities (Reddy et al. 2019). Moreover, 3D printing and industry 4.0 are seen as essential enablers of e-waste management, automating processes, therefore, allowing the stakeholders to focus on the technical parts of recycling and innovation (Nascimento et al. 2019).

4.3 Closed-Loop Supply Chains

The idea of slowing and particularly closing resource loops has led to the concept of closed-loop supply chains (CLSCs), which are supply chains, that, in addition to typical forward flows, present reverse flows of used products back to manufacturers (Souza 2013) and generate revenue by taking back products from customers and recovering the remaining added value (Atasu et al. 2008). Closed-loop supply chain management (CLSCM) addresses the negative implications of business operations leading to excessive waste generation and resource depletion (Lüdeke-Freund et al. 2019).

Guide and Van Wassenhove (2009) define three pillars for CLSC: (i) product return management, (ii) remanufacturing operational issues, and (iii) remanufactured products market development. In CLSCs, there is a significant acquisition of EOL and end-of-use (EOU) products, which enter the forward supply chain to make both production and consumption sustainable (Singhal et al. 2019a).

Besides minimizing resource consumption and reducing production costs and negative environmental impacts, the circular model also helps keeping hazardous materials out of landfills and oceans (Nascimento et al. 2019). Moreover, remanufacturing of WEEE reduces production costs and mitigates the harmful environmental effects of e-waste (Wang et al. 2019c). In this context, the Extended Producer Responsibility (EPR) offers various practices that make the producers responsible for the treatment or disposal of post-consumer products (Corsini et al. 2017a). Nowadays, recycling is the primary waste treatment strategy (Garrido-Hidalgo et al. 2020).

Online to Offline (O2O) is a new recycling mode that is capable of integrating upstream and downstream resources on the network platform, creating a recycling and processing mode for the industry chain of waste sorting and developing a circular development mode featuring "resources-products-waste-renewable resources" to realize the recycling of resources (Miao et al. 2020). Such platforms connect (i) Consumers that voluntarily recycle WEEE and (ii) manufacturers who buy e-waste for remanufacturing (Zhu and Li 2020). In addition to recycling, there are methodologies to extend the useful life of products and recover values from WEEE (Garrido-Hidalgo et al. 2020). For example, indium, a critical metal that is scarce in primary sources but abundant in e-waste, represents both a supply chain risk and mitigated opportunity (Akcil et al. 2019). Nowadays, product refurbishment is one of the most profitable and environmental benefit processes, drawing more and more attention from both product manufacturers and customers (Chen et al. 2018).

4.4 Supply Chain Resilience

Many specialists are understanding the concept of supply chain resilience (SCRes) understand the concept of supply chain resilience (SCRes) as the ability of a supply chain network to return to the previous state (or even a better one) after a disruption while avoiding failure modes (Bhamra et al. 2011; Senna et al. 2020; Carvalho and Cruz Machado 2007). Moreover, supply chains must apply all efforts to be resilient in order to achieve competitiveness (Barroso et al. 2011). Among the most significant barriers to obtain supply chain resilience is the lack of management of supply chain risks.

Companies and researchers are increasingly paying more attention to SCRM, which is motivated by the frequency and intensity of catastrophes, disasters, and crises increasing on a global scale (Fan et al. 2011). Supply chain risk management is known as a roadmap to obtain SCRes (Senna et al. 2020). Juttner et al. (2003), Juttner (2005), and Thun and Hoenig (2011) agree that the first stage concerning SCRM analysis is to visualize a supply chain as a set of cross-functional processes, therefore, avoiding local solutions that do not result in supply chain optimum.

In addition, responding to these risk drivers, imply that supply chains should develop strategic response capabilities to assess and mitigate disruptions (Singh and Singh 2019). Managing supply chain risks often obligates managers to assess the probability of undesirable events and the magnitude of such events to estimate potential losses (Doan et al. 2021).

The increased social attention towards ecological conservation, resource preservation, and waste reduction has directed to integrate RL operations with the traditional supply chain (Garg 2020). In this sense, e-waste supply chains experience more difficulties in terms of uncertainties, for example, OEMs cannot manufacture products if their facility is bounded by fluctuating inventory (Chen et al. 2012). Sustainability allows companies opportunities to save costs, increase efficiency, gain new customers, and incorporate the potential to gain a competitive advantage (Kumar et al. 2012). Nevertheless, managing risks that comprise the whole supply chain can become an exceedingly difficult task. To make this hard task viable, Senna et al. (2020) propose the following steps to manage supply chain risks: (i) Identify risks, (ii) assess risks, (iii) mitigate risks, and (iv) monitor risks.

4.5 Smart Cities

Smart cities are increasingly being considered a viable solution to address problems related to last-mile logistics and reverse logistics. Moreover, sustainable logistics plays a major role in the development of smart cities (Lan et al. 2020). Smart cities can be understood as a concept that encompasses the utilization of information and communication technologies to provide services such as energy and mobility (Jnr et al. 2020). The citizens play a very important role in smart cities and must be encouraged to perform sustainable practices (Ma et al. 2016).

Waste accumulation points are a relevant problem in modern smart cities (Toutouh et al. 2020). Big data associated with computation methods can diminish environmental impacts by providing inputs to optimize waste collection points (Toutouh et al. 2020).

The collection of large datasets is supported by devices such as surveillance equipment, smart meters, smartphones, and many other devices. These datasets provide significant information that helps designing reverse channels and collection points. However, these devices are vulnerable to security and privacy attacks (Makhdoom et al. 2020). In order to launch and technically maintain the smart e-waste reverse system, it is advisable to create a local e-waste reverse operator at the specialized processing enterprise (Shevchenko et al. 2021).

Vehicles of delivery companies can function as mobile collection points, collecting and delivering WEEE to processing points (Shevchenko et al. 2021). Nevertheless, literature still lacks studies that relate smart city practices to WEEE.

Smart cities are considered crucial to solve waste management problems providing solutions that include e-waste collection and transport, which are supported by technologies such as the Internet of Things or IoT, that provide data for machine learning algorithms (Shevchenko et al. 2021).

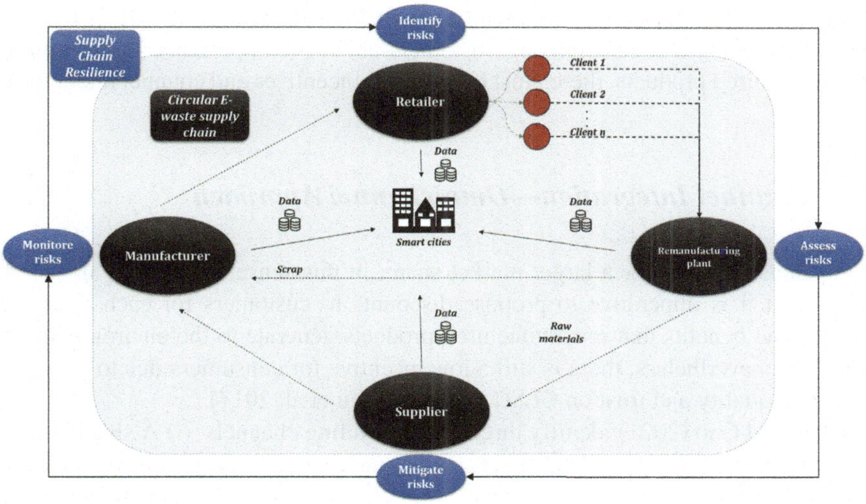

Fig. 2 Integrated framework

4.6 *Integrated Framework*

Our integrated framework is presented in Fig. 2.

5 Discussion

The proposed framework is discussed in this section under two main dimensions, (i) Culture—Consumers must contribute and (ii) Channel integration—Omnichannel approach.

5.1 *Culture—Customers Must Contribute to the Process*

All the stakeholders must be imbued in creating a culture of recycling behavior, which is considered key for saving the value of materials in the economy as long as possible (Shevchenko et al. 2021). In this sense, there is an imperative need to understand consumers' recycling behavior to improve e-waste management initiatives, which have been ignored in the reverse supply chain management literature, especially in the cross-cultural context (Kumar 2019). Lack of customer's education, motivation, and responsibility of the producers can be considered important drivers of a failed e-waste management policy (Condemi et al. 2019; Kianpour et al. 2017). In addition to educating and sensibilizing customers about the importance of consuming remanufactured products, there must be financial incentives and channel integration.

5.2 *Channel Integration—Omnichannel Approach*

Companies compete for a larger market share; in this sense, Sadeghi et al. (2019) affirm that it is imperative to propose discounts to customers for each returning device. The benefits that remanufactured products generate to the environment are notorious, nevertheless, there is still a low incentive for consumers due to concerns regarding quality and trust on CLSC processes (Xu et al. 2017).

Chen and Gao (2021) identify three main recycling channels: (i) A single offline recycling channel, (ii) a single online recycling channel, and (iii) a hybrid dual-recycling channel. In this sense, managers should create an omnichannel focused on recycling. Companies should provide a diverse set of channels to facilitate customer's inclusion in this process. There should be online platforms available in apps and websites that motivate clients to contribute and inform the available recycling points.

CEOs must have viable startups that can aggregate existing services to the reverse channels. Car service apps can motivate the passengers to take small detours to deliver EOL products to recycling plants or leave with the driver that consolidates the products and delivers to the plants by the end of his shift. The apps may create point reward systems and long-term benefits for clients that consistently return EOL products. In this sense, the government has a great power of motivation, creating laws and tax benefits.

6 Conclusions

This article presented the following RQ: What are the main building blocks of a circular e-waste ecosystem? In this regard, we propose a framework based on a conceptual analysis of 86 papers from 2010 to 2021 aiming to identify the main building blocks of a circular e-waste economy. In this sense, this paper offers a unique contribution proposing a framework based on the constructs: (i) E-waste management (C1), (ii) Supply Chain Resilience—SCRes (C2), (iii) Circular Economy (C3), (iv) Closed-loop supply chains (C4), and (v) Smart cities (C5). To the best of our knowledge, there is no similar framework in the literature. As the main limitation, we highlight that this paper is a product of our method of choice and other methods in other supply chains could obtain different results. Our review considered only published papers in English, excluding doctoral theses, congress papers, and documental analyzes. As future work, we recommend interviewing the stakeholders of these supply chains for empirical validation of the framework.

References

Agrawal S, Singh RK, Murtaza Q (2018) Reverse supply chain issues in Indian electronics industry: a case study. J Remanufacturing 8:115–129. https://doi.org/10.1007/s13243-018-0049-7

Akcil A, Agcasulu I, Swain B (2019) Valorization of waste LCD and recovery of critical raw material for circular economy: a review. Resour Conserv Recycl 149:622–637. https://doi.org/10.1016/j.resconrec.2019.06.031

Atasu A, Sarvary M, Van Wassenhove LN (2008) Remanufacturing as a marketing strategy. Manag Sci 54:1731–1746

Bag S, Gupta S, Foropon C (2018) Examining the role of dynamic remanufacturing capability on supply chain resilience in circular economy. Manag Decis

Baidya R, Debnath B, Ghosh SK, Rhee SW (2020) Supply chain analysis of e-waste processing plants in developing countries. Waste Manag Res 38:173–183. https://doi.org/10.1177/0734242X19886633

Barroso AP, Machado VH, Machado VC (2011) Supply chain resilience using the mapping approach. InTech 161:161–184

Batista L, Bourlakis M, Smart P, Maull R (2018) In search of a circular supply chain archetype—a content-analysis-based literature review. Prod Plan Control 29:438–451

Bhamra R, Dani S, Burnard K (2011) Resilience: the concept, a literature review and future directions. Int J Prod Res 49:5375–5393

Bryan CG, Williamson BJ, Całus-Moszko J et al (2020) CEReS—co-processing of coal mine & electronic wastes: novel resources for a sustainable future. Hydrometallurgy 197. https://doi.org/10.1016/j.hydromet.2020.105444

Cai YJ, Choi TM (2021) Extended producer responsibility: a systematic review and innovative proposals for improving sustainability. IEEE Trans Eng Manag 68:272–288. https://doi.org/10.1109/TEM.2019.2914341

Carvalho H, Cruz Machado V (2007) Designing principles to create resilient supply chains. In: Proc 2007 Ind Eng Res Conf Nashv

Chadegani AA, Salehi H, Yunus M et al (2013) A comparison between two main academic literature collections: web of science and scopus databases. Asian Soc Sci 9:18–26

Chen L, Gao M (2021) Optimizing strategies for e-waste supply chains under four operation scenarios. Waste Manag 124:325–338. https://doi.org/10.1016/j.wasman.2021.02.012

Chen YJ, Sheu JB, Lirn TC (2012) Fault tolerance modeling for an e-waste recycling supply chain. Transp Res Part E Logist Transp Rev 48:897–906. https://doi.org/10.1016/j.tre.2012.02.005

Chen YT, Chan FTS, Chung SH, Park WY (2018) Optimization of product refurbishment in closed-loop supply chain using multi-period model integrated with fuzzy controller under uncertainties. Robot Comput Integr Manuf 50:1339–1351. https://doi.org/10.1016/j.rcim.2017.05.005

Condemi A, Cucchiella F, Schettini D (2019) Circular economy and e-waste: AN opportunity from RFID TAGs. Appl Sci 9. https://doi.org/10.3390/app9163422

Corsini F, Rizzi F, Frey M (2017) Extended producer responsibility: the impact of organizational dimensions on WEEE collection from households. Waste Manag 59:23–29. https://doi.org/10.1016/j.wasman.2016.10.046

Dhonde B, Patel CR (2020) Implementing circular economy concepts for sustainable urban freight transport: case of textile manufacturing supply chain. Acta Logist 7:131–143. https://doi.org/10.22306/al.v7i2.172

Doan LTT, Amer Y, Lee SH et al (2021) Optimizing a reverse supply chain network for electronic waste under risk and uncertain factors. Appl Sci 11:2–21. https://doi.org/10.3390/app11041946

Elia V, Gnoni MG, Tornese F (2017) Measuring circular economy strategies through index methods: a critical analysis. J Clean Prod 142:2741–2751

Esenduran G, Lin YT, Xiao W, Jin M (2020) Choice of electronic waste recycling standard under recovery channel competition. Manuf Serv Oper Manag 22:495–512. https://doi.org/10.1287/msom.2018.0750

Fan W, Gu J, Tang H, Gao X (2011) Risk management in end-to-end global supply chain. In: Proc 11th Int Conf Chinese Transp Prof (ICCTP), Nanjing

Fonseca LM, Domingues JP, Pereira MT, Martins FF, Zimon MD (2018) Assessment of circular economy within Portuguese organizations. Sustainability 10

Garg CP (2021) Modeling the e-waste mitigation strategies using grey-theory and DEMATEL framework. J Clean Prod 281:124035. https://doi.org/10.1016/j.jclepro.2020.124035

Garg CP (2020) A robust hybrid decision model to evaluate critical factors of reverse logistics implementation using Grey-DEMATEL framework. Springer India

Garrido-Hidalgo C, Ramirez FJ, Olivares T, Roda-Sanchez L (2020) The adoption of internet of things in a circular supply chain framework for the recovery of WEEE: the case of lithium-ion electric vehicle battery packs. Waste Manag 103:32–44. https://doi.org/10.1016/j.wasman.2019.09.045

Geissdoerfer M, Savaget P, Bocken NM, Hultink EJ (2017) The circular economy—a new sustainability paradigm? J Clean Prod 143:757–768

Geyer R, Blass VD (2010) The economics of cell phone reuse and recycling. Int J Adv Manuf Technol 47:515–525. https://doi.org/10.1007/s00170-009-2228-z

Ghalehkhondabi I, Ardjmand E (2020) Sustainable E-waste supply chain management with price/sustainability-sensitive demand and government intervention. J Mater Cycles Waste Manag 22:556–577. https://doi.org/10.1007/s10163-019-00952-z

Ghisellini P, Cialani C, Ulgiati S (2016) A review on circular economy: the expected transition to a balanced interplay of environmental and economic systems. J Clean Prod 114:11–32

Gong Y, Chen M, Zhuang Y (2019) Decision-making and performance analysis of closed-loop supply chain under different recycling modes and channel power structures. Sustainability 11. https://doi.org/10.3390/su11226413

Govindan K, Rajendran S, Sarkis J, Murugesan P (2015) Multi criteria decision making approaches for green supplier evaluation and selection. J Clean Prod 98:66–83

Guide VDR, Van Wassenhove LN (2009) The evolution of closed-loop supply chain research. Oper Res 57:10–18

Guo L, Qu Y, Tseng ML et al (2018) Two-echelon reverse supply chain in collecting waste electrical and electronic equipment: a game theory model. Comput Ind Eng 126:187–195. https://doi.org/10.1016/j.cie.2018.09.036

Howard M, Hopkinson P, Miemczyk J (2018) The regenerative supply chain: a framework for developing circular economy indicators. Int J Prod Res 57:7300–7318

Isernia R, Passaro R, Quinto I, Thomas A (2019) The reverse supply chain of the e-waste management processes in a circular economy framework: evidence from Italy. Sustainability 11. https://doi.org/10.3390/su11082430

Ivanov D, Dolgui A, Sokolov B (2019) The impact of digital technology and Industry 4.0 on the ripple effect and supply chain risk analytics. Int J Prod Res 57:829–846. https://doi.org/10.1080/00207543.2018.1488086

Jabbour AB, Jabbour CJ, Godinho Filho M, Roubaud D (2018) Industry 4.0 and the circular economy: a proposed research agenda and original roadmap for sustainable operations. Ann Oper Res 270

Jnr BA, Petersen SA, Ahlers D, Krogstie J (2020) API deployment for big data management towards sustainable energy prosumption in smart cities-a layered architecture perspective. Int J Sustain Energy 39:263–289

John ST, Sridharan R, Ram Kumar PN (2018) Reverse logistics network design: a case of mobile phones and digital cameras. Int J Adv Manuf Technol 94:615–631. https://doi.org/10.1007/s00170-017-0864-2

Juttner U (2005) Supply chain risk management: understanding the business requirements from a practitioner perspective. Int J Logist Manag 16:120–141

Juttner U, Peck H, Christopher M (2003) Supply chain risk management: outlining an agenda for future research. Int J Logist Res Appl 6:197–210

Kianpour K, Jusoh A, Mardani A et al (2017) Factors influencing consumers' intention to return the end of life electronic products through reverse supply chain management for reuse, repair and recycling. Sustainability 9.https://doi.org/10.3390/su9091657

Kumar A (2019) Exploring young adults' e-waste recycling behaviour using an extended theory of planned behaviour model: a cross-cultural study. Resour Conserv Recycl 141:378–389. https://doi.org/10.1016/j.resconrec.2018.10.013

Kumar S, Teichman S, Timpernagel T (2012) A green supply chain is a requirement for profitability. Int J Prod Res 50:1278–1296

Lan S, Tseng M-L, Yang C, Huisingh D (2020) Trends in sustainable logistics in major cities in China. Sci Total Environ 712

Lieder M, Rashid A (2016) Towards circular economy implementation: a comprehensive review in context of manufacturing industry. J Clean Prod 113:36–51

Lüdeke-Freund F, Gold S, Bocken NMP (2019) A review and typology of circular economy business model patterns. J Ind Ecol 23:36–61. https://doi.org/10.1111/jiec.12763

Ma L, Liu N, Zhang J et al (2016) Energy management for joint operation of CHP and PV prosumers inside a grid-connected microgrid: a game theoretic approach. IEEE Trans Ind Inform 12:1930–1942

Makhdoom I, Zhou I, Abolhasan M, et al (2020) PrivySharing: a blockchain-based framework for privacy-preserving and secure data sharing in smart cities. Comput Secur 88

Marinello S, Gamberini R (2021) Multi-criteria decision making approaches applied to waste electrical and electronic equipment (WEEE): a comprehensive literature review. Toxics 9:1–20. https://doi.org/10.3390/toxics9010013

McDowall W, Geng Y, Huang B et al (2017) Circular economy policies in China and Europe. J Ind Ecol 21:651–661

Miao S, Liu D, Ma J (2020) O2O Recycling closed-loop supply chain modeling based on classification process considering environmental index. Discret Dyn Nat Soc 2020.https://doi.org/10.1155/2020/5737514

Mishra JL, Hopkinson PG, Tidridge G (2018) Value creation from circular economy led closed loop supply chains: a case study of fast moving consumer goods. Prod Plan Control 29:509–521

Nascimento DLM, Alencastro V, Quelhas OLG et al (2019) Exploring Industry 4.0 technologies to enable circular economy practices in a manufacturing context: a business model proposal. J Manuf Technol Manag 30:607–627. https://doi.org/10.1108/JMTM-03-2018-0071

Ozgur Polat L, Gungor A (2021) WEEE closed-loop supply chain network management considering the damage levels of returned products. Environ Sci Pollut Res 28:7786–7804. https://doi.org/10.1007/s11356-020-10249-6

Peng B, Tu Y, Wei G (2018) Governance of electronic waste recycling based on social capital embeddedness theory. J Clean Prod 187:29–36. https://doi.org/10.1016/j.jclepro.2018.02.265

Reddy KN, Kumar A, Ballantyne EEF (2019) A three-phase heuristic approach for reverse logistics network design incorporating carbon footprint. Int J Prod Res 57:6090–6114. https://doi.org/10.1080/00207543.2018.1526422

Rezayat MR, Yaghoubi S, Fander A (2020) A hierarchical revenue-sharing contract in electronic waste closed-loop supply chain. Waste Manag 115:121–135. https://doi.org/10.1016/j.wasman.2020.07.019

Sadeghi R, Taleizadeh AA, Chan FTS, Heydari J (2019) Coordinating and pricing decisions in two competitive reverse supply chains with different channel structures. Int J Prod Res 57:2601–2625. https://doi.org/10.1080/00207543.2018.1551637

Safdar N, Khalid R, Ahmed W, Imran M (2020) Reverse logistics network design of e-waste management under the triple bottom line approach. J Clean Prod 272:122662. https://doi.org/10.1016/j.jclepro.2020.122662

Senna P, Reis A, Santos IL et al (2020) A systematic literature review on supply chain risk management: is healthcare management a forsaken research field? Benchmarking 28:926–956. https://doi.org/10.1108/BIJ-05-2020-0266

Sharifi Z, Shokouhyar S (2021) Promoting consumer's attitude toward refurbished mobile phones: a social media analytics approach. Resour Conserv Recycl 167

Shevchenko T, Saidani M, Danko Y et al (2021) Towards a smart E-waste system utilizing supply chain participants and interactive online maps. Recycling 6:1–14. https://doi.org/10.3390/recycling6010008

Shi J, Zhou J, Zhu Q (2019) Barriers of a closed-loop cartridge remanufacturing supply chain for urban waste recovery governance in China. J Clean Prod 212:1544–1553. https://doi.org/10.1016/j.jclepro.2018.12.114

Shumon MRH, Ahmed S, Ahmed S (2016) Fuzzy analytical hierarchy process extent analysis for selection of end of life electronic products collection system in a reverse supply chain. Proc Inst Mech Eng Part B J Eng Manuf 230:157–168. https://doi.org/10.1177/0954405414553980

Singh NP, Singh S (2019) Building supply chain risk resilience role of big data analytics in supply chain disruption mitigation. Benchmarking Int J 26:2318–2342

Singhal D, Jena SK, Tripathy S (2019a) Factors influencing the purchase intention of consumers towards remanufactured products: a systematic review and meta-analysis. Int J Prod Res 57:7289–7299. https://doi.org/10.1080/00207543.2019.1598590

Singhal D, Tripathy S, Jena SK (2019b) Sustainability through remanufacturing of e-waste: examination of critical factors in the Indian context. Sustain Prod Consum 20:128–139. https://doi.org/10.1016/j.spc.2019.06.001

Souza GC (2013) Closed-loop supply chains: a critical review, and future research. Decis Sci 44:7–38

Tansel B (2020) Increasing gaps between materials demand and materials recycling rates: a historical perspective for evolution of consumer products and waste quantities. J Environ Manag 276:111196. https://doi.org/10.1016/j.jenvman.2020.111196

Thi L, Doan T, Amer Y, et al (2019) E-Waste reverse supply chain: a review and future perspectives. Appl Sci

Thun JH, Hoenig D (2011) An empirical analysis of supply chain risk management in the German automotive industry. Int J Prod Econ 131:242–249

Toutouh J, Rossit D, Nesmachnow S (2020) Soft computing methods for multiobjective location of garbage accumulation points in smart cities. Ann Math Artif Intell 88:105–131

Vieira ES, Gomes JANF (2009) A comparison of Scopus and Web of Science for a typical university. Scientometrics 81:587–600

Wang M, Liu P, Gu Z, et al (2019a) A scientometric review of resource recycling industry. Int J Environ Res Public Health 16. https://doi.org/10.3390/ijerph16234654

Wang Y, Xin B, Wang Z, Li B (2019b) Managing supplier-manufacturer closed-loop supply chain considering product design and take-back legislation. Int J Environ Res Public Health 16:1–26. https://doi.org/10.3390/ijerph16040623

Wang Z, Huo J, Duan Y (2019c) Impact of government subsidies on pricing strategies in reverse supply chains of waste electrical and electronic equipment. Waste Manag 95:440–449. https://doi.org/10.1016/j.wasman.2019.06.006

Webster K (2015) A wealth of flows, 2nd edn. Ellen MacArthur Foundation Publishing

Xu X, Zeng S, He Y (2017) The influence of e-services on customer online purchasing behavior toward remanufactured products. Int J Prod Econ 187:113–125. https://doi.org/10.1016/j.ijpe.2017.02.019

Yazdani M, Gonzalez EDRS, Chatterjee P (2019) A multi-criteria decision-making framework for agriculture supply chain risk management under a circular economy context. Manag Decis. https://doi.org/10.1108/MD-10-2018-1088

Zhu X, Li W (2020) Research on the pricing strategy of "Internet+" recycling platforms in a two-sided network environment. Sustainability 12.https://doi.org/10.3390/su12031001

Sustainable Use of Plastic E-Waste with Added Value

Yamila V. Vazquez, Marcos Volpin, and Silvia E. Barbosa

Abstract The suitability of using plastic from Waste Electrical and Electronic Equipment (WEEE) for the manufacturing of new products, closing the loop of circular economy will be analyzed in this chapter. In this way, two business models were identified for market opportunities, gross margin-low-turnover and low-margin high-turnover products. High Impact Polystyrene (HIPS) WEEE, Acrylonitrile–Butadiene–Styrene (ABS) WEEE, and an equitable blend of these materials (H50/A50 blend) were considered for the study. From the initial characterization of HIPS WEEE and ABS WEEE, it was found the presence of different kinds of mineral fillers and additives that give them UV resistance, flame retardance, and specific mechanical properties to each material, that favored certain characteristics depending on the final application. From the study, it is possible to claim that plastics from WEEE are suitable for the manufacturing of different kinds of products, since they can be easily processed, achieving a good overall performance, including UV and flame retardance. These results are very promising for the recycling of this complex plastic waste stream with profit, promoting sustainable methodologies and, consequently, closing the loop of circular economy.

1 Introduction

Recycling of plastic from Waste Electrical and Electronic Equipment (WEEE) has become a challenge during the last years. It is estimated that 13 million tons of plastic WEEE were generated worldwide in 2018 and it is expected to increase up to 14 million tons by 2021 (Baldé et al. 2017). This particular plastic waste stream

Y. V. Vazquez (✉) · M. Volpin · S. E. Barbosa
Planta Piloto de Ingeniería Química, PLAPIQUI (UNS-CONICET), Camino "La Carrindanga" Km 7, CP: 8000 Bahía Blanca, Argentina
e-mail: yvazquez@plapiqui.edu.ar

Departamento de Ingeniería Química, Universidad Nacional del Sur, Av. Alem 1253, CP: 8000 Bahía Blanca, Argentina

M. Jawaid and A. Khan (eds.), *Conversion of Electronic Waste in to Sustainable Products*, Sustainable Materials and Technology,
https://doi.org/10.1007/978-981-19-6541-8_11

is composed of several kinds of thermoplastic resins like Acrylonitrile–Butadiene–Styrene (ABS), High Impact Polystyrene (HIPS), and Polycarbonate (PC), among others (Tansel 2017). The composition of plastics WEEE depends on the source of origin (television, cellphones, computers, fridges, air conditioners, etc.) although in general ABS and HIPS represent the major fractions. Moreover, these two plastics are styrenic copolymers usually black colored (Lepawsky 2020; Turner 2018). These facts became a problem during the sorting step by using different characterization methods and techniques because their precision strongly depends on the composition. Most popular technique is Near-Infrared Spectroscopy (NIR) and also density separation and impact milling are used too (Öztürk 2015; WRAP 2009). Classification by NIR is not precise since materials are mainly dark-colored which makes detection difficult by the equipment. Additionally, as it was aforementioned, they are mainly styrenics resins and, consequently, their chemical similarities also contribute to complicate the detection (Da Silva and Wiebeck 2020; Arends et al. 2015). In this way, sorting is commonly performed manually which also is not accurate and increases labor costs. Moreover, this kind of methodology implies unsafe and unhealthy conditions for workers since plastics can contain contaminants and hazardous substances (Wannomai et al. 2020; Sutawon et al. 2020). In this regard, a more sustainable and safer alternative is recycling the major fractions of plastic WEEE (ABS and HIPS), and also more complicated to separate, together as a blend. Several studies of ABS and HIPS blends have been carried out indicating that blends with similar properties to single ABS or HIPS can be obtained (Vazquez and Barbosa 2018, 2016; Tostar 2016).

Plastics WEEE can be considered as an innocuous waste stream. However, since they contain hazardous additives like brominated substances as flame retardant, they also can be treated as special waste. In this way, they must be analyzed in order to determine the concentration of hazardous components. Then, if it is necessary, they should be treated to reduce or eliminate those additives in order to be able for recycling (Haarman and Gasser 2016; Peeters et al. 2014; European Union 2011). Besides these particular additives, it is well known that plastics from WEEE contain several kinds of mineral fillers such as calcium carbonate, talc, silica, aluminum silicates, titanium dioxide, and carbon black, among others (Rothon DeArmitt 2017; Wagner et al. 2019). In the case of ABS and HIPS, it is known that the total amount of fillers is around 10 wt% and 5 wt%, respectively (Vazquez and Barbosa 2016; Tostar 2016). This fact indicates that they actually are composite materials with a complex polymer matrix. It is important to note that these fillers are aggregated onto the polymer matrix in order to achieve specific properties for different requirements (e.g., impact resistance, tensile and flexural strength, ductility, etc.). For example, besides the use of carbon black to give dark colors, it could also be incorporated in plastic products or devices parts for outdoors uses (e.g., housing for electrical and electronic devices, automobile components, piping, etc.), providing UV and thermal stability (Turner 2018). Moreover, in the case of white plastic, the color is given by the addition of titanium dioxide. However, this additive also provides UV resistance since it absorbs UV radiation protecting outdoor plastic devices/products from these rays (Buxbaum 2008).

From the aforementioned characteristics, it is possible to appreciate that these materials could be suitable for the manufacture of new different kinds of products. In this way, two business models are identified for market opportunities, high-margin low-turnover and low-margin high-turnover products (Sahajwalla and Gaikvad 2018; Besley and Brigham 2014). The first one, high-margin low-turnover products, involves a business model in which products are sold for higher costs than those associated with the acquisition, and the expenses related to the product usually are covered with low sales volumes. Regarding the type of products involved in this category, one clear example are the design objects such as trays or boxes for cosmetics. On the other hand, low-margin high-turnover products represent the opposite scenario. Within this business model, products are sold at prices close to the production ones and the benefit is obtained by high volume sales in a short period of time with low costs. Taking into account the source of the materials analyzed in this chapter, an example of this product are housings for new electrical and electronic equipment (EEE), like outdoor luminaries. These alternatives indicate that, if recycled materials could be introduced into the market like new EEE or other types of products, the loop of the circular economy would be closed. In this way, the aim of this chapter is to analyze the suitability of using plastic WEEE in the fabrication of different kinds of products. Two case studies will be considered as examples: housings for public luminary as low-margin high-turnover products, and design objects for high-margin low-turnover products. Since they constitute the major fraction of plastic WEEE, HIPS and ABS are the materials considered for the study. Moreover, taking into account that, within plastic WEEE, the amounts of ABS and HIPS are similar (29% and 26%, respectively) and their separation by type is difficult (Cardamone et al. 2021), equitable ABS WEEE/HIPS WEEE blends are included in the study. This blend is considered in order to analyze the feasibility of avoiding sorting these plastics and recycling them together.

Plastic from WEEE—Sustainable Alternatives of Use

Plastic materials considered in this work came from WEEE and not only from a single source. According to the supplier, plastic came mainly from computer screens and keyboards but also, may come from refrigerators, air conditioners, and televisions, among others. Consequently, each plastic WEEE material is actually a mix of different types of ABS or HIPS, respectively. It is important to mention that these materials were manually sorted by type and ground into flakes. In this way, and in order to obtain a representative sample of each plastic WEEE and analyze their homogeneity, materials were arranged separately in rafts to perform a quartering. For each material, eight samples of 5 g were taken from different sectors of the draft. They were processed and injected one by one in a mini mixer, at 10 rpm and 180 °C for 5 min. Then, in order to analyze the relative copolymer content of ABS WEEE and HIPS WEEE, modulated high-resolution thermogravimetric analysis (HiRes™-MTGA) was carried out for each obtained specimen of both materials. Tests were performed at a heating rate of 5 °C/min from room temperature to 650 °C, with a sensitivity of 1.00 and a resolution of 6.00. Regarding the modulation parameters, an amplitude of 5 °C with a period of 200 s was used. This technique allows to separate different degradation events which occur within a degradation process (TA Instruments 1992). As a result, a semiquantitative composition of each material was obtained. In this way, approximate percentages of each block within ABS and HIPS are presented in Table 1 meanwhile the corresponding thermograms can be observed in Fig. 1. It is important to mention that only one curve of each material is presented in a representative way. From this information, it can be seen that both plastic WEEE can be considered as homogeneous materials, since the obtained errors are less than 5%. In the case of ABS, it contains approximately 20 wt% acrylonitrile (AN), 56 wt% of styrene (St), and approximately 13 wt% of butadiene (Bu). Moreover, a 9 wt% of mineral fillers was observed as was expected. On the other hand, in the case of HIPS, the composition was 52 wt% of St and 34 wt% of Bu, approximately. Also, 5 wt% of mineral fillers were detected. Additionally, the presence of 6 wt% of AN was found in HIPS. This result evidenced that separation by type is not accurate, as was expected. The presence of mineral fillers in both materials corroborates that WEEE plastics are actually composite materials with a very complex polymer matrix. These results agree with the literature regarding other case studies (Tostar 2016; Vazquez and Barbosa 2016).

As it was aforementioned, the presence of considerable amounts of fillers was determined by HiRes™-MTGA. In this regard, it is important to know which kind of minerals they are in order to analyze their influence on the final properties of the processed materials. Figure 2 shows Wide-Angle X-ray Scattering (WAXS) spectra for ABS WEEE and HIPS WEEE, with the corresponding diffraction peaks of mineral fillers, and the carbon black halo. From this study, it was possible to detect the presence of talc, antimony trioxide (Sb_2O_3), titanium dioxide (TiO_2), calcium carbonate

Table 1 Relative mass content of volatiles, acrylonitrile, styrene, butadiene, and fillers in ABS and HIPS from WEEE determined by HiRes™—MTGA

Material	Volatiles (wt%)	AN (wt%)	St (wt%)	Bu (wt%)	Filler (wt%)
ABS WEEE	0,7 ± 0,1	20.3 ± 1.7	56.4 ± 3.1	13.7 ± 0.5	8.9 ± 0.5
HIPS WEEE	1,4 ± 0,2	6.7 ± 1.3	52.9 ± 3.2	34.4 ± 3.2	4.6 ± 0.4

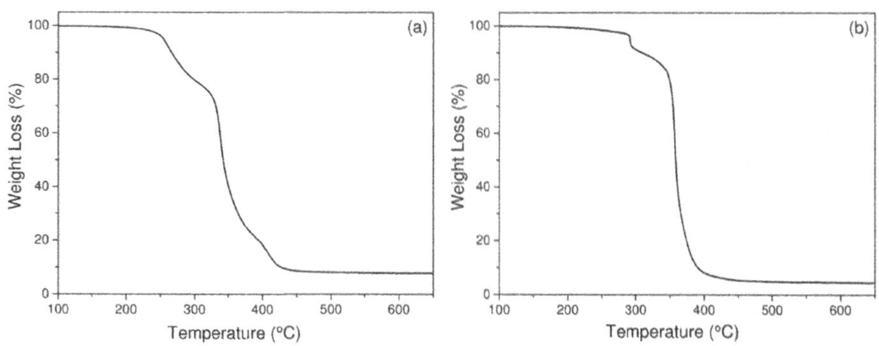

Fig. 1 HiRes TGA thermograms for **a** ABS and **b** HIPS

($CaCO_3$), and carbon black. These types of mineral fillers correspond to the typical additives used in this kind of material (Al-Malaika et al. 2017; Maris et al. 2015). In thermoplastics, different fillers are incorporated in order to achieve specific characteristics, such as enhanced processing, improved mechanical properties, flame retardancy, and UV resistance, among the most important. TiO_2 is often used as a colorant in light-colored plastics, but it also acts as a thermal and UV stabilizer (Turner 2018). Moreover, the use of carbon black has similar features to TiO_2, since it is used as black colorant and additionally confers UV resistance properties (Buxbaum 2008). Regarding talc and $CaCO_3$, they are the most common fillers used to improve impact resistance, hardness, and stiffness of thermoplastics like ABS and HIPS (Rothon and Dearmitt 2017). On the other hand, the detection of Sb_2O_3 also indicates the presence of brominated flame retardants since antimony is widely used as a synergistic additive for this kind of flame retardant additive (Arduin et al. 2020; Hahladakis et al. 2018). It is important to note that the presence of antimony and consequently brominated compounds can be used as an opportunity considering the final application of recycled plastic from WEEE, as long as they are within the allowed limits. In this way, X-ray Fluorescence (XRF) analysis was performed in order to corroborate the presence of bromine and its amount. Results indicate that ABS WEEE bromine content is 1750 mg/kg, while HIPS WEEE contains less than 1000 mg/kg, both under the maximum amount allowed of 2000 mg/kg (Hennebert and Fillela 2018; CLC/TS 50,625-3-1 2015).

From the characterization results, it can be concluded that plastic WEEE considered in the study are actually composite materials with several types of mineral fillers

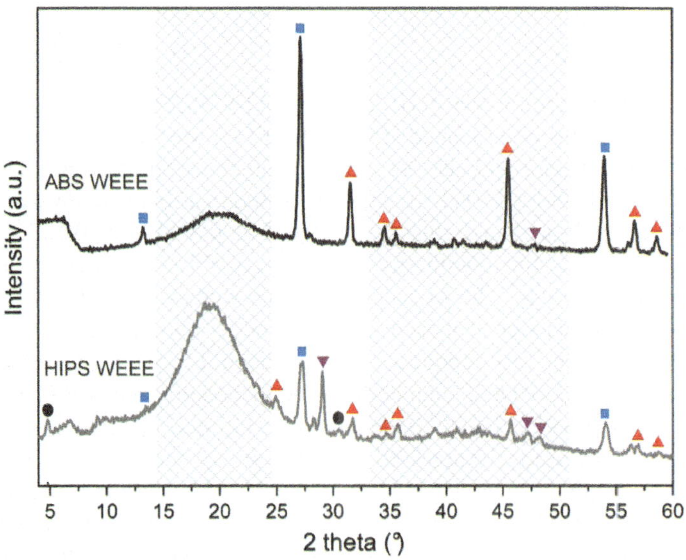

Fig. 2 WAXS spectra for ABS and HIPS with the corresponding identified peaks and the carbon black halo. Ref: Talc (●), Sb_2O_3 (■), TiO_2 (▲), $CaCO_3$ (▼), and carbon black (crosshatch pattern)

and additives. Amount and types of these components indicate that recycled plastics would have certain properties that, depending on the final application, can be taken advantage of. In this way, the proposal of this chapter is to analyze the use of recycled ABS and HIPS from WEEE in the development of housings for outdoor luminaires and design objects, as opportunities for low- and high-margin products, respectively. In this way, considering the final application of each product, recycled plastic will need different characteristics. In the following sections, they will be analyzed according to the requirements of each product.

2 Case Study of Low-Margin High-Turnover Products: Outdoor Luminaries

Main requirements of housings for outdoors luminaries include flame retardant properties, UV and thermal resistance, and good mechanical properties, since they are electrical articles exposed to environmental factors and mechanical and thermal conditions from the electrical parts of the equipment (Wang et al. 2020; Turner 2018). In this way, as it was aforementioned, the presence of additives to achieve flame retardant, UV and thermal resistance could be useful and it is expected that would not be necessary to add new additives, which is also an economic advantage. In order to analyze the suitability of using HIPS WEEE and ABS WEEE in the manufacturing of housings for outdoor luminaries, tensile mechanical performance will be analyzed along with material morphology and environmental conditions' influence will be assessed.

Mechanical Behavior

Tensile mechanical performance of HIPS WEEE, ABS WEEE, and the corresponding 50/50 blend (from this point named H50/A50 WEEE) was first performed, in order to analyze their behavior and also, equitable blend compatibilization effectiveness. Initial HIPS WEEE and ABS WEEE were processed in a single-screw extruder with a flat nozzle at 180ºC and an extrusion rate of 1.5 kg/h. H50/A50 WEEE blend was also processed at the same conditions in order to perform an accurate comparison with the initial materials. The extruded materials were ground into flakes. Specimens for tensile tests were cut from plates prepared by compression molding at 180ºC with the flakes. Test conditions and specimen dimensions were determined according to the ASTM D638 standard for thermoplastics and were performed at room temperature (ASTM D638 2010). Modulus, ultimate strength, and elongation at break were comparatively assessed from the stress–strain curves. In this way, mechanical properties which depend on phase adhesion would indicate if compatibilization was achieved. Young Modulus (E), a low strain property, is related with the internal structure of the species and relative concentration of the components in blends. However, variations in high strain properties like ultimate strength (σ_u)

Table 2 Tensile mechanical properties (E, ε_b, and σ_u) of HIPS WEEE, H50/A50 WEEE blends, and ABS WEEE

Sample	E (Mpa)	ε_b (%)	σ_u (Mpa)
HIPS WEEE	1256 ± 34	11.0 ± 2.1	17.3 ± 0.4
H50/A50 WEEE	1430 ± 31	2.5 ± 0.3	23.1 ± 0.6
ABS WEEE	1390 ± 44	2.6 ± 0.3	26.0 ± 2.0

and elongation at break (ε_b) are the best evidence of compatibilization effectiveness. Then, blend compatibility is assessed through the analysis of changes in σu and εb of H50/A50 WEEE blend with respect to ABS WEEE and HIPS WEEE. Tensile mechanical properties (E, σ_u, and ε_b) for HIPS WEEE, H50/A50 WEEE, and ABS WEEE are shown in Table 2, while the representative stress–strain curves are presented in Fig. 3. It is evident that ABS WEEE is more rigid and resistant than HIPS WEEE since its E and σ_u are greater, which is expected due to the presence of more rigid chains from the acrylonitrile phase in ABS WEEE. Moreover, in both cases, it is possible to observe that ABS WEEE has lower ε_b than HIPS WEEE which indicates a less ductile behavior. The difference in ductility between ABS WEEE and HIPS WEEE is related to the presence of higher amounts of mineral fillers in ABS WEEE. This fact is also associated with higher content of butadiene in HIPS WEEE (twice the corresponding in ABS), the rubber component in these materials, which gives more ductility.

Regarding the equitable blend, H50/A50 WEEE, it is possible to note that its mechanical behavior is similar to the ABS WEEE one (Table 2 and Fig. 3). Values of ε_b and E are close to the corresponding ABS WEEE meanwhile, mechanical resistance is lower than the ABS WEEE. However, considering standard deviations

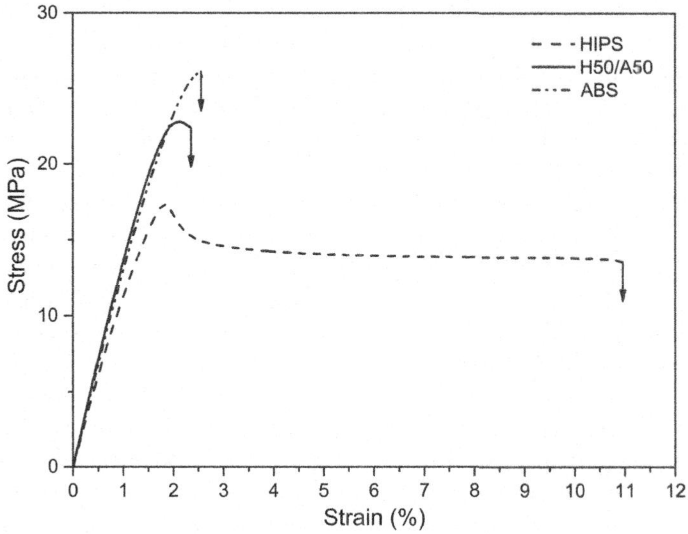

Fig. 3 Tensile stress–strain curves for HIPS WEEE, H50/A50 WEEE blends, and ABS WEEE

this change is less than 10%. On the other hand, H50/A50 WEEE σ_u is higher than the HIPS WEEE one. This indicates that mechanical resistance is between those values of both initial materials, which is expected to be seen in compatibilized blends (Subramanian 2017; Utracki and Wilckie 2014). It is important to note that equitable blends generally present phase segregation and very complex morphology which results in poor mechanical properties. However, from this mechanical analysis, it is possible to say that H50/A50 WEEE blend was compatibilized indicating that sorting of these plastics, ABS WEEE and HIPS WEEE, can be avoided in order to reduce the cost of workforce and improve workers' labor conditions. Sorting is not accurate, as it was demonstrated previously with the presence of AN in HIPS WEEE. Even using specific techniques and equipment these difficulties persist since ABS and HIPS are very similar resins from the compositional point of view and they are generally dark-colored, which difficult their differentiation (Da Silva and Wiebeck 2020; Arends et al. 2015). Because of these reasons, obtained results are very promising for the recycling industry.

Environmental Degradation Analysis

As it was previously explained, recycled materials used in the manufacturing of housings for outdoor luminaries will be exposed to environmental conditions which could affect their performance. In this way, accelerated aging tests were performed on HIPS WEEE, H50/A50 WEEE blends, and ABS WEEE. Experiments consisted of alternating cycles of 4 h of light at 60°C with 1.23 W/m^2 of irradiance and 4 h without light at 50°C with 100% relative humidity, according to the ASTM G53-96 standard (ASTM G53 1996). The total duration of the test was 25 days (600 h), rotating the samples every 4 days to ensure that all of them received the same irradiation and humidity. These tests allow the analysis of materials deterioration when they are exposed to the environment for a period of approximately 11 years (Perez et al. 2010). One of the best evidence of material degradation is mechanical performance deterioration. In this way, after aging all samples were subjected to mechanical tensile tests in order to analyze changes in their mechanical properties through a comparison with no-aged specimens. Moreover, mechanical behavior after and before aging was comparatively analyzed with morphology.

Mechanical properties of aged HIPS WEEE, H50/A50 WEEE, and ABS WEEE compared with no aged tensile properties (from Table 2) are presented in Fig. 4 for HIPS WEEE, H50/A50 WEEE, and ABS WEEE. It is possible to appreciate that aged HIPS WEEE became more rigid and resistant, increasing approximately 13% of its elastic modulus (Fig. 4a) and mechanical resistance (Fig. 4b). Also, its ductility (Fig. 4c) was reduced by around 80%. Regarding ABS WEEE, it can be observed an increment of approximately 10% in mechanical resistance value and a decrement of 5% in the elongation at break, evidencing that material is more resistant but less ductile after aging. In the case of the H50/A50 WEEE blend, mechanical resistance and rigidity did not suffer notable changes. However, this sample presents a reduction of 25% in ductility. All these results agreed with polymer crosslinking, resulting in harder and less ductile materials, as a consequence of UV exposure (Lagern 2020; Signoret et al. 2020).

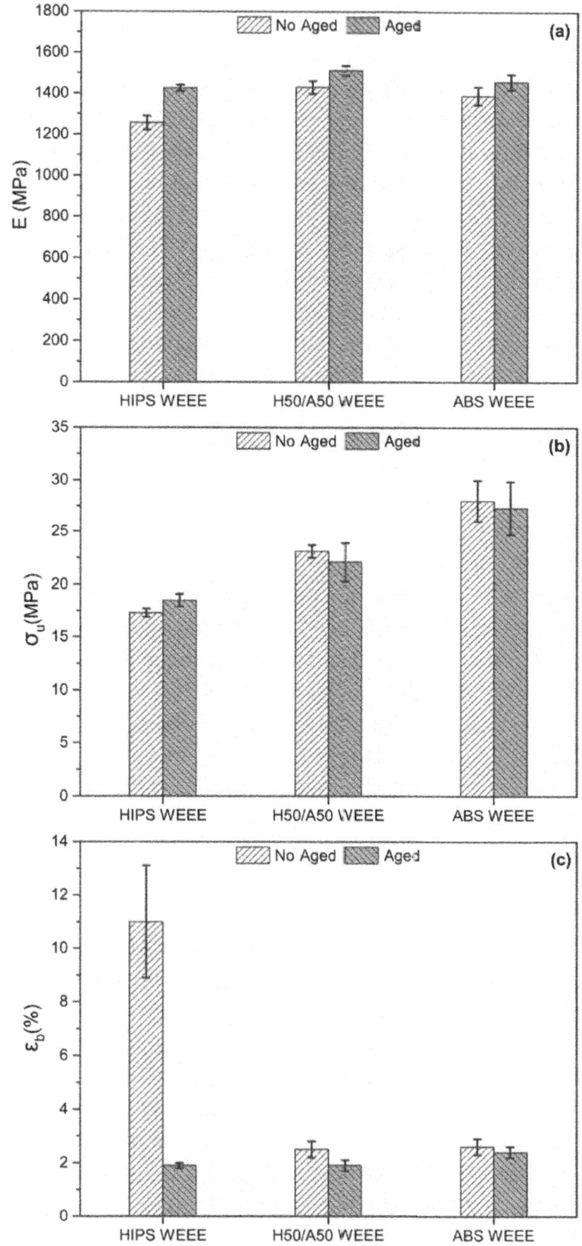

Fig. 4 Mechanical properties comparison between No-Aged and Aged samples: **a** Elastic Modulus (E), **b** Mechanical Resistance (σ_u), and **c** Ductility (ε_b)

Mechanical performance after and before aging, as well as structural characterization of materials, is complemented through materials morphology analysis. In this way, morphology analyses were performed by Scanning Electron Microscopy (SEM) in an electron microscope, operated at 10 kV. Same type of specimens used for mechanical tests were prepared for this study. These samples were cryo-fractured by immersion in liquid nitrogen, mounted on bronze stubs, and then coated with a gold layer (30 Å), using an argon plasma metallizer. In Fig. 5, SEM micrographs (1000x) of cryo-fracture surfaces for HIPS WEEE, H50/A50 WEEE, and ABS WEEE after and before aging are presented. For no-aged samples, it can be appreciated that HIPS presents bigger and less homogeneous domains than ABS; meanwhile, ABS evidence has sharper edges than HIPS. This fact corroborates that HIPS is a more ductile and less resistant material than ABS. Moreover, H50/A50 blend present sharpened fracture edges and also there are unmasked fillers. These facts corroborate mechanical performance, where it was observed a similar behavior to the ABS one but slightly less ductile and resistance.

On the other hand, if no-aged micrographs are now compared with aged ones, for HIPS WEEE it can be appreciated that soft fracture edges of no-aged HIPS WEEE became sharp after aging, which corroborates the reduction in elongation at break and the increment of mechanical resistance of this material. Regarding ABS WEEE, sharpen edges in the aged sample are more evident than in the corresponding no-aged one, agreeing mechanical resistance increment as a consequence of aging. For H50/A50 WEEE samples, slight differences in edges and the presence of mineral fillers along the fracture surface can be appreciated, consistent with the reduction in ductility after aging. In all cases, morphology agrees with mechanical behavior and also verifies the presence of mineral fillers. Additionally, it is evident that ABS WEEE and H50/A50 WEEE present similarities which is a notable result, since it is possible to claim that separation can be avoided.

Flame Retardance

Since housings for outdoor luminaries will be exposed to heat from electric/electronic components, it is necessary to analyze material performance under combustion. In this way, burning time of the materials involved was assessed. The criteria proposed by Underwriters Laboratories according to the UL 94 Standard (UL 94 1996) and the ASTM D 635 98 Standard (ASTM D 635 1998) were considered for this analysis. Specimens were cut from plates obtained by compression molding at 180 °C. According to this criteria, two marks were included in HIPS WEEE, H50/A50 WEEE, and ABS WEEE specimens. Flame resistance will be classified depending on flame behavior during test and if ignition reaches or not the second mark. Results showed that in all cases ignition did not reach the second mark of the specimen. Regarding ABS WEEE, it did not hold the flame when the ignition source was removed. In the case of HIPS WEEE, it maintained the flame alive when the source was removed, but the combustion did not spread to the second mark. The H50/A50 WEEE blend showed a similar behavior to ABS WEEE but it maintained a slight flame when the ignition source was taken out. Based on these results, both standards indicate that all materials can be classified as Horizontal Burning (HB). This means that they

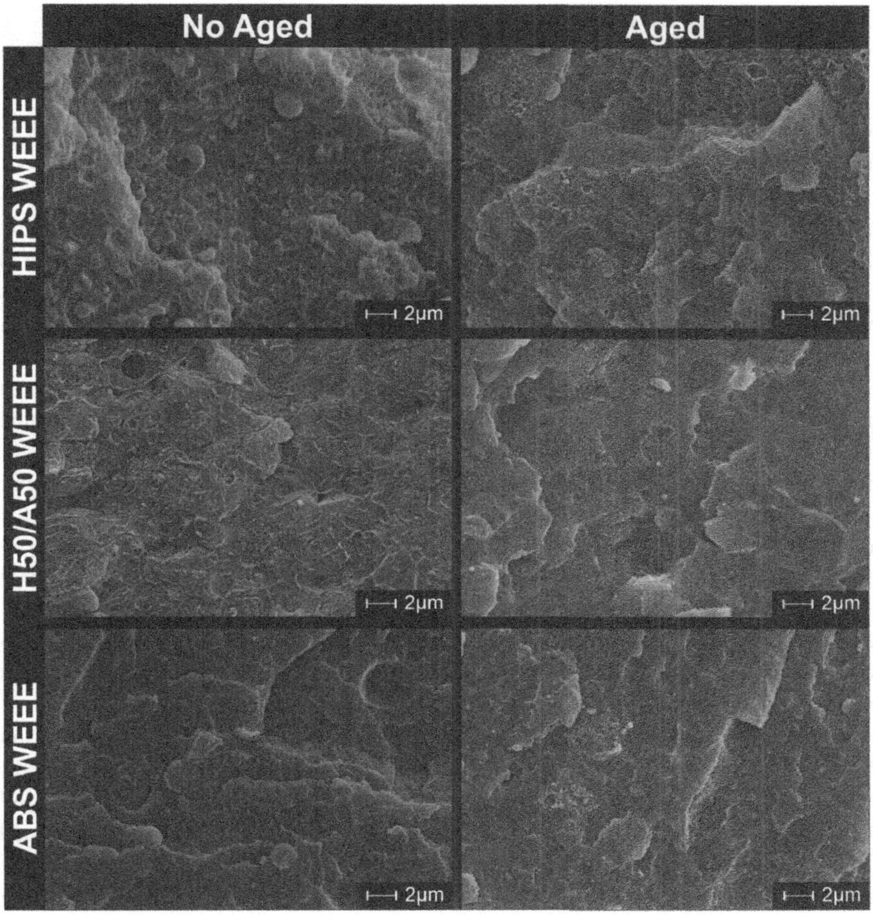

Fig. 5 Cryo-fracture surface SEM micrographs for HIPS WEEE, H50/A50 WEEE, and ABS WEEE, before and after aging

present a slow flame spread. This result was expected considering the presence of Br and Sb (from flame retardant additives) that WEEE plastics contain presenting low flammability and slow combustion (Mao 2020; Buezas Sierra 2010). Also, it is possible to claim that low quantities of these substances can confer flame retardancy without the addition of extra additives.

3 Case Study of High-Margin Low-Turnover Products: Design Objects

Requirements for this final application are different in comparison with the public luminaries' ones. In this case, considering that objects will not be exposed to environmental conditions and also, will not be subjected to high temperatures, UV resistance, and flame retardance are not priority requirements. Design and image are the most important characteristics of these products, along with a good mechanical performance. In this way, design objects, such as trays to put cosmetics or toiletries products, guest trays, holders, and external containers, among others, are considered.

Taking advantage of the gray color scale of HIPS WEEE and ABS WEEE flakes, samples were prepared by compression molding and injection molding, without previous mixing. Compression molding was carried out at 180 °C for 3 min and injection molding was performed on a mini-injector at 180 °C for 5 min for stabilization. Photographs of square plates obtained by compression molding, and rectangular sticks from injection molding, for HIPS WEEE, H50/A50 WEEE, and ABS WEEE are shown in Fig. 6. As it is possible to observe, this kind of processing without mixing gives the final piece a texture and a distribution of colors similar to marble (injection molding) or granite (compression molding). Moreover, it can be appreciated that HIPS WEEE contains more black-colored flakes and, on the contrary, ABS WEEE more white-colored ones. These differences allow the obtention of different color combinations to get lighter or darker textures.

On the other hand, despite the fact that appearance is very important in this kind of product, appropriate mechanical performance needs to be achieved. These objects will not be subjected to great mechanical stresses, however, they must resist falls and bending stresses produced by their handling. In this way, drop tests were performed from an average height of 1 m. This value was chosen taking into account the average height of people and the position of hands during handling the piece at the center of the upper part of the body. All samples, compression molding and injection molding, were dropped 20 times each and no damage was observed. These results indicate a good impact on mechanical performance for both kinds of processing.

Moreover, manual bending experiments were carried out. In this way, female and male volunteers between 30 and 34 years old, manually bend plates and sticks in order to analyze their resistance to this kind of manipulation. Results indicate that both types of samples are resistant to manual bending since no one was broken with the application of an average effort. When the applied force is considerably increased, according to volunteers, samples could be intentionally broken. In this way, from this study, it is possible to claim that obtained pieces resist an average force of 37 kgf, considering estimated force values for females and males between 30 and 34 years old (Wang et al. 2018).

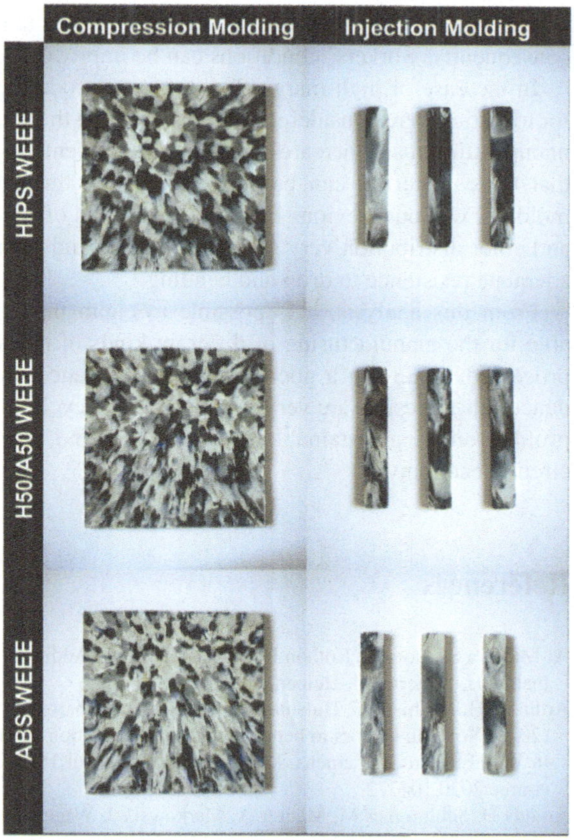

Fig. 6 Samples of HIPS WEEE, H50/A50 WEEE, and ABS WEEE obtained by compression

4 Concluding Remarks

The aim of this work was to analyze the suitability of using plastic from WEEE for the manufacturing of new products. Two business models were identified for market opportunities, high-margin low-turnover and low-margin high-turnover products. HIPS WEEE, ABS WEEE, and an equitable blend of these materials (H50/A50 WEEE blend) were considered for the study. From the initial characterization of HIPS WEEE and ABS WEEE, it was found the presence of different kinds of mineral fillers and additives that give them UV resistance, flame retardance, and specific mechanical properties to each material. Regarding low-margin high-turnover products, housings for outdoor luminaries were taken into account as an example Main materials requirements for this product can be fulfilled using recycled plastic from WEEE. All considered materials, HIPS WEEE, H50/A50 blend, and ABS WEEE, showed good processability, mechanical performance, and stability under UV and flame exposure. Additionally, it was demonstrated that H50/A50 WEEE blend presents similar behavior to ABS WEEE, which indicates that sorting can be avoided and, consequently, workers' conditions can be improved.

In the case of high-margin low-turnover products, design objects such as trays for toiletries were considered as an example. In this case, image, texture, and certain manipulation resistance are the main requirements to achieve. It was demonstrated that these materials can be easily processed through compression and injection molding, without previous blending. This kind of process allows to obtain textures and color distribution very similar to marble and granite. Also, samples present an adequate resistance to drop and bending.

From this analysis, it is possible to claim that plastics from WEEE are suitable for the manufacturing of different kinds of products, since they can be easily processed, achieving a good overall performance, including UV and flame retardance. These results are very promising for recycling this plastic waste stream with profit, promoting sustainable methodologies and, consequently, closing the loop of circular economy.

References

Al-Malaika S, Axtell F, Rothon R, Gilbert M (2017) Additives for plastics. In: Brydson's plastics materials. Butterworth-Heinemann, pp 127–168

Arduin RH, Mathieux F, Huisman J, Blengini GA, Charbuillet C, Wagner M, Baldé CP, Perry N (2020) Novel indicators to better monitor the collection and recovery of (critical) raw materials in WEEE: focus on screens. Resour Conserv Recycl 157:104772. https://doi.org/10.1016/j.resconrec.2020.104772

Arends D, Schlummer M, Mäurer A, Markowski J, Wagenknecht U (2015) Characterisation and materials flow management for waste electrical and electronic equipment plastics from German dismantling centres. Waste Manage Res 33(9):775–784. https://doi.org/10.1177/0734242X15588585

ASTM D 635 (1998) Standard test method for rate of burning and/or extent and time of burning of plastics in a horizontal position

ASTM D638 (2010) Standard test method for tensile properties of plastics

ASTM G53 (1996) Practice for operating light- and water-exposure apparatus (Fluorescent UV-Condensation Type) for exposure of nonmetallic materials (Withdrawn 2000)

Baldé CP, Forti V, Gray V, Kuehr R, Stegmann P (2017) The global e-waste monitor–2017, United Nations University (UNU), International Telecommunication Union (ITU) & International Solid Waste Association (ISWA), Bonn/Geneva/Vienna. Electronic Version, pp 978–992

Besley S, Brigham EF (2014) Principles of finance. Cengage Learning

Buezas Sierra N (2010) Guía: plásticos y fuego. Programa de fomento de la innovación en Institutos Tecnológicos. AIMPLAS, Instituto Tecnológico del Plástico - Generalitat Valenciana - UE Fondo Europeo de Desarrollo Regional

Buxbaum G (2008) Industrial inorganic pigments. John Wiley & Sons

Cardamone GF, Ardolino F, Arena U (2021) About the environmental sustainability of the European management of WEEE plastics. Waste Manage 126:119–132. https://doi.org/10.1016/j.wasman.2021.02.040

CLC/TS 50625-3-1 (2015) Requirements for the collection, logistics and treatment of WEEE—part 3-1: Specification relating to depollution—general. CENELEC, Brussels, Belgium da Silva DJ, Wiebeck H (2020) Current options for characterizing, sorting, and recycling polymeric waste. Prog Rubber, Plast Recycl Technol 36(4):284–303. https://doi.org/10.1177/1477760620918603

European Union (2011) Directive 2011/65/EC of the European Parliament and of the Council of 8 June 2011 on the restriction of the use of certain hazardous substances in electrical and electronic equipment (recast), RoHS 2. Off J Eur Union 174:88–110

Haarman A, Gasser M (2016) Managing hazardous additives in WEEE plastic from the Indian informal sector. A study on applicable identification & separation methods. Sustain Recycl Ind Swiss Fed Labor Mater Sci Technol

Hahladakis JN, Velis CA, Weber R, Iacovidou E, Purnell P (2018) An overview of chemical additives present in plastics: Migration, release, fate and environmental impact during their use, disposal and recycling. J Hazard Mater 344:179–199. https://doi.org/10.1016/j.jhazmat.2017.10.014

Hennebert P, Filella M (2018) WEEE plastic sorting for bromine essential to enforce EU regulation. Waste Manage 71:390–399. https://doi.org/10.1016/j.wasman.2017.09.031

Lagern V (2020) Ultraviolet stability of mixed waste plastic composites. Master's thesis, Lappeenranta-Lahti University of Technology LUT. School of Energy Systems and Mechanical Engineering

Lepawsky J (2020) Sources and streams of electronic waste. One Earth 3(1):13–15. https://doi.org/10.1016/j.oneear.2020.07.001

Mao S, Gu W, Bai J, Dong B, Huang Q, Zhao J, Zhuang X, Zhang C, Yuan W, Wang J (2020) Migration characteristics of heavy metals during simulated use of secondary products made from recycled e-waste plastic. J Environ Manage 266:110577. https://doi.org/10.1016/j.jenvman.2020.110577

Maris E, Botané P, Wavrer P, Froelich D (2015) Characterizing plastics originating from WEEE: a case study in France. Miner Eng 76:28–37. https://doi.org/10.1016/j.mineng.2014.12.034

Öztürk T (2015) Generation and management of electrical–electronic waste (e-waste) in Turkey. J Mater Cycles Waste Manage 17(3):411–421. https://doi.org/10.1007/s10163-014-0258-6

Peeters JR, Vanegas P, Tange L, Van Houwelingen J, Duflou JR (2014) Closed loop recycling of plastics containing Flame Retardants. Resour Conserv Recycl 84:35–43. https://doi.org/10.1016/j.resconrec.2013.12.006

Pérez JM, Vilas JL, Laza JM, Arnáiz S, Mijangos F, Bilbao E, León LM (2010) Effect of reprocessing and accelerated weathering on ABS properties. J Polym Environ 18(1):71–78. https://doi.org/10.1007/s10924-009-0154-7

Rothon R, DeArmitt C (2017) Fillers (including fiber reinforcements). In: Brydson's plastics materials. Butterworth-Heinemann, pp 169–204

Sahajwalla V, Gaikwad V (2018) The present and future of e-waste plastics recycling. Curr Opin Green Sustain Chem 13:102–107. https://doi.org/10.1016/j.cogsc.2018.06.006

Signoret C, Edo M, Lafon D, Caro-Bretelle AS, Lopez-Cuesta JM, Ienny P, Perrin D (2020) Degradation of styrenic plastics during recycling: Impact of reprocessing photodegraded material on aspect and mechanical properties. J Polym Environ 28:2055–2077. https://doi.org/10.1007/s10924-020-01741-8

Subramanian MN (2017) Polymer blends and composites: chemistry and technology. John Wiley & Sons

Sutawong P, Wattanachaiyingcharoen D, Itsubo N, Suvannarat K (2020) The risks to human health and environmental impacts from community e-waste separation. Geographia Technica 15. https://doi.org/10.21163/GT_2020.151.33

TA Instruments (1992) High resolution thermogravemetric analysis: a new technique for obtaining superior analytical results. Therm Anal Rheol, TA-023 (nd)

Tansel B (2017) From electronic consumer products to e-wastes: Global outlook, waste quantities, recycling challenges. Environ Int 98:35–45. https://doi.org/10.1016/j.envint.2016.10.002

Tostar S (2016) Mechanical and thermal properties of recycled WEEE plastic blends. Chalmers University of Technology

Turner A (2018) Black plastics: Linear and circular economies, hazardous additives and marine pollution. Environ Int 117:308–318. https://doi.org/10.1016/j.envint.2018.04.036

UL 94 (1996) Tests for flammability of plastic materials for parts in devices and appliances. Underwriters Laboratories Inc.

Utracki LA, Wilkie CA (eds) (2014) Polymer blends handbook. Springer, Dordrecht, The Netherlands

Vazquez YV, Barbosa SE (2018) Compatibilization of HIPS/ABS blends from WEEE by using Styrene-Butadiene Rubber (SBR). J Environ Manage 217:381–390. https://doi.org/10.1016/j.jenvman.2018.03.124

Vazquez YV, Barbosa SE (2016) Process window for direct recycling of acrylonitrile-butadiene-styrene and high-impact polystyrene from electrical and electronic equipment waste. Waste Manage 59:403–408. https://doi.org/10.1016/j.wasman.2016.10.021

Wagner F, Peeters JR, De Keyzer J, Janssens K, Duflou JR, Dewulf W (2019) Towards a more circular economy for WEEE plastics–part a: development of innovative recycling strategies. Waste Manage 100:269–277. https://doi.org/10.1016/j.wasman.2019.09.026

Wang Y-C, Bohannon RW, Li X, Sindhu B, Kapellusch J (2018) Hand grip strength: normative reference values and equations for 18- to 85-year-olds residing in the United States. J Orthop Sports Phys Ther 1–34. https://doi.org/10.2519/jospt.2018.7851

Wang S, Su D, Wu Y, Chai Z (2020) Application of life-cycle assessment to the eco-design of LED lighting products. Euro-Mediterr J Environ Integr 5(2):1–10. https://doi.org/10.1007/s41207-020-00180-0

Wannomai T, Matsukami H, Uchida N, Takahashi F, Viet PH, Takahashi S, Kunisue T, Suzuki G (2020) Bioaccessibility and exposure assessment of flame retardants via dust ingestion for workers in e-waste processing workshops in northern Vietnam. Chemosphere 251:126632. https://doi.org/10.1016/j.chemosphere.2020.126632

WRAP (2009) Separation of mixed WEEE plastics final report (WRAP Project MDD018 and MDD023). Report prepared by Axion Consulting

The manufacturer's authorised representative in the EU is Springer Nature Customer Service Centre GmbH, Europaplatz 3, 69115 Heidelberg, Germany. If you have any concerns regarding our products, please contact ProductSafety@springernature.com

Printed and bound by CPI Group (UK) Ltd, Croydon, CR0 4YY
30/09/2026
02225242-0004